The College "Y"

The College "Y"

Student Religion in the Era of Secularization

David P. Setran

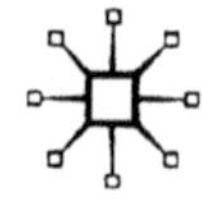

THE COLLEGE "Y"

First published in 2007 by
PALGRAVE MACMILLAN™
175 Fifth Avenue, New York, N.Y. 10010 and
Houndmills, Basingstoke, Hampshire, England RG21 6XS
Companies and representatives throughout the world.

PALGRAVE MACMILLAN is the global academic imprint of the Palgrave Macmillan division of St. Martin's Press, LLC and of Palgrave Macmillan Ltd. Macmillan® is a registered trademark in the United States, United Kingdom and other countries. Palgrave is a registered trademark in the European Union and other countries.

ISBN 978–1–4039–6125–9

Library of Congress Cataloging-in-Publication Data is available from the Library of Congress.

A catalogue record for this book is available from the British Library.

Design by Newgen Imaging Systems (P) Ltd., Chennai, India.

First edition: January 2007

To my parents,
Raymond and Evelyn Setran

Contents

Acknowledgments

In a project of this scope, there are numerous people to thank for their contributions along the way. By allowing me to serve as his research assistant for a book on D.L. Moody, Lyle Dorsett opened my eyes to the spiritual potency of the college YMCA movement. As mentor and friend, Lyle fueled my interest in religious history and in the history of college ministry. I am grateful to Wheaton College for the provision of an Aldeen grant that financially supported this research. I am deeply indebted to my doctoral professors and fellow students at Indiana University who read, critiqued, and discussed issues relevant to this work. I owe a special debt of gratitude to my major professor and dissertation advisor, B. Edward McClellan, who awakened my interest in the history of character education in the United States and also convinced me that there was much terrain to be explored in this area. I am also grateful to Dr. Andrea Walton who read the very earliest versions of my work and invited me to discuss my preliminary findings in an interdisciplinary seminar at Indiana University's Center on Philanthropy.

Countless librarians and archivists from various colleges and universities have also assisted in the research dimensions of this project. In particular, I wish to thank Dagmar Getz and her staff at the Kautz Family YMCA Archives in Minneapolis and Martha Smalley and her staff at the Yale University Divinity School. Both helped enormously in the process of tracking down materials, answering questions, and recommending unique source materials. Corey Thomas served brilliantly as my research assistant, and Kathleen Cruse helped enormously in footnote formatting. Don Warren provided support, encouragement, and invaluable advice in the sometimes painful process of editing. The editorial staff at Palgrave provided numerous helpful suggestions that improved the quality of this book.

This book also bears the mark of a tremendously supportive family. I am grateful to my parents for their unending support as I was immersed in this process. I also want to thank my four children, Parker, Anna Joy, Owen, and Emily for providing countless "study breaks" and reminders of the really important things in life. Finally, I thank my wife Holly, my partner and my best friend, whose love and encouragement have kept me afloat during this project. Through many late nights and blank stares at the computer screen, she has been there to support, challenge, laugh, and provide perspective. The words from Proverbs that began our journey are more accurate now than ever: "Many women do noble things, but you surpass them all" (Prov. 31:29).

Abbreviations

AYSD	Archives of the YMCA — Student Division, Record Group No. 58, Special Collections, Yale Divinity School Library
AWSCF	Archives of the World Student Christian Federation: General Records, Record Group No. 46/1, Special Collections, Yale Divinity School Library
CPSP	Clarence Prouty Shedd Papers, Record Group No. 35, Special Collections, Yale Divinity School Library
JRMP	John R. Mott Papers, Record Group No. 45, Special Collections, Yale Divinity School Library
APJRM	*Addresses and Papers of John R. Mott*, vol. 3 (New York: Association Press, 1947)
Kautz Archives	Kautz Family YMCA Archives, University of Minnesota Libraries
SCARP	Student Christian Association Records, Department of Rare Books and Special Collections, Princeton University Library.
CB	*College Bulletin*
Int	*The Intercollegian*
NAS	*The North American Student*

INTRODUCTION

> *Mr. Mott has said that this movement has done more to bring the colleges into sympathy with each other and into comradeship with each other than any other movement ever started. Of course it has done so.*
>
> —Woodrow Wilson, 1902[1]

At the 1902 ceremonies commemorating the twenty-fifth anniversary of the intercollegiate YMCA, former Princeton University President Francis L. Patton addressed the role of this flourishing organization in the evolving religious culture of higher education. An ardent supporter of the college YMCA and a frequent contributor to its work both nationally and on the Princeton campus, Patton said that the movement was the "inevitable" result of the changing religious patterns of college life. Mentioning a host of forces, such as the growing appeal of voluntary over compulsory religion, the disdain for sectarian religiosity, and the loss of faculty moral and religious oversight, Patton stated with great excitement that "at the most opportune time in the history of university development there has grown up an agency, in obedience to that great law of specialization of function, that must hold this responsibility of the religious life of the organization of the university." Stating that the YMCA had "consolidated" and "vitalized" the religious life of the university while establishing "a shelter and refuge and a home for those who come from Christian homes," Patton claimed that the preservation of a Christian witness in higher education was absolutely dependent upon these clubs. The fact that this "agency" had been birthed at Princeton in 1877 led Patton to wistfully proclaim, "I think I never was prouder of Princeton than I am at this moment when I think of this magnificent movement, the extent of it, the work it has already accomplished, and the boundless possibilities ahead of it."[2]

Historians have frequently rehearsed the variables attending this transformation. Traditional accounts of the late nineteenth and early twentieth centuries suggest that institutions of higher learning, sparked by the diffusion of German intellectual models, were dramatically altered to reflect the scientific proclivities of an influential coterie of educational reformers. Driven by a radical new agenda that highlighted scholarly research and professional preparation, required core classes were gradually replaced with free elective courses chosen by undergraduate students. New courses in the natural and social sciences, history, and modern language emerged alongside standbys in Greek, Latin, mathematics, geography, English grammar and composition, and rhetoric. Inquiry-oriented lectures, labs, and seminars displaced recitations as the dominant methodological paradigm, reflective of the prioritization of knowledge production over knowledge acquisition. Faculty members, previously cast as moral guardians of the campus community, were redefined as academic experts in highly specialized intellectual domains. Whether enacted gradually in existing colleges or more rapidly in newly established universities, such changes were portrayed as the logical concomitants of a culture more in need of trained technical experts than cultured gentlemen.[3]

As might be expected, such historical interpretations left little room for a continued robust religious presence in higher educational institutions. Many of the early narratives of the growth of the university paradigm in the United States argued in straightforward fashion that emphases on evolution, utility, and scientific inquiry overwhelmed goals of religious piety and nurture that had dominated the antebellum college.[4] This rapid religious disestablishment was accompanied by the removal of theological themes and emphases in the curriculum, the replacement of pious academic generalists with secular specialists on university faculty, the decline in clergy presidents and board members, and the elimination of university-sponsored religious activities such as mandatory chapel services and church attendance. For such authors, the growth of the university, marked by a shift from revelatory to empirical ways of knowing, was attended by an overarching "war" between science and religion. As one would imagine, historians saw religion as the big loser in the battle between premodern and modern *Zeitgeists*. When religious belief was defined as a block to unfettered and objective inquiry, they surmised, the marginalization of faith—often defined as institutional secularization—was assured.

More recently, revisionist historians have suggested that the notion of a dramatic revolution has obscured as much as it has revealed. Scholars such as George Marsden, Julie Reuben, Jon Roberts, and James Turner, for example, have begun to demonstrate the complexity of the secularization process by focusing on the role of religious and intellectual shifts in these institutions.[5] While still acknowledging the substantive changes that took place in this era, they also suggest that new leaders did not seek a radical dismissal of religious ideals. Rejecting the facile oppositional dualism implied by the rhetoric of a "war" between science and religion, these historians have contended that secularization was often the gradual by-product of religious and scientific change rather than a unilateral denial of religious perspectives. All acknowledged that a number of attempts were made to reconcile religion and science within the curriculum. The very means of doing so, however, often cemented the marginalization of religious beliefs within the intellectual core of the institutions. Largely through the venue of an assimilationist liberal Protestantism, these scholars argue, leaders and faculty members adopted scientific and democratic values that blunted the distinctiveness of the faith, replacing doctrinal orthodoxy with a vague ethical naturalism. These authors maintain that liberal religion, despite its "relevance," laid the foundation for its own demise.[6]

While recent historical accounts have persuasively documented the role of intellectual, curricular, and administrative trends in this larger process of secularization, there has been no concentrated attempt to demonstrate the positioning and influence of students in this religious transformation. There were certainly a number of internal and intellectual struggles that defined the path to religious marginalization, but collegians were themselves important players in this larger narrative. Individual schools could not neglect the needs of religious students, and many parents still desired a college education that linked academic excellence with moral and religious virtue. More importantly, students in this era crafted a significant peer culture, often removed from the supervisory presence of faculty and administrators, which had a powerful influence on the nature of "lived religion" in campus settings. Patton's comments, in fact, were indicative of the fact that the "location" of religious activity was indeed shifting into the student realm. As the curriculum was increasingly defined as the proper setting for factual scientific inquiry, the extracurriculum was often labeled as the best location for value-laden personal religious growth. In fact, Marsden himself has maintained that "during the decades around the turn of the century, when formal disestablishment was taking place at the fastest rate, voluntary Christianity on campuses was probably at its most vigorous ever."[7] It is here, as well as in prescribed courses and activities, that the historian must search for the changing nature of religion in American higher education.

While the religious dimensions of student culture are in some ways more difficult to interpret than formalized programs and administrative actions, this book is designed to initiate such an analysis by looking specifically at one organization that served as the primary clearinghouse for student religiosity in the era of college and university secularization. During the last four decades of the nineteenth century, the YMCA emerged as the focal point of student religious life in American higher education. Beginning both gradually and haphazardly throughout the 1860s and 1870s, the growth and consolidation of college chapters throughout the 1880s and 1890s generated a legitimate intercollegiate movement of national scope. Bolstered by the formal backing of the parent YMCA in the 1870s and by a warm reception within the colleges themselves, these extracurricular clubs appeared on nearly every major college and university campus in this era. While in 1877 only forty chapters existed, by 1900 there were 559 local chapters and 31,901 college and university students claiming campus YMCA membership. By 1920, YMCA chapters could be found in a staggering 764 institutions of higher education, boasting a total student membership of 80,649. These statistics are perhaps even more impressive when one considers YMCA membership from the perspective of the relatively small total student population of the early twentieth century. Between 1900 and 1920, the organization regularly enrolled between 25 and 30 percent of all male students, thus playing a significant role in shaping the male extracurricular experience. Either replacing preexisting religious societies or initiating attempts to facilitate student religious work, these associations grew so rapidly that it was no hyperbole for Patton to proclaim at the end of the nineteenth century that the YMCA had achieved a "near monopoly" on student religious life in America.[8]

Despite this enormous influence, the college YMCA movement has received scant scholarly attention. In both grand historical narratives and in monographs devoted to student culture, the YMCA is often either completely ignored or mentioned only in passing as an abortive effort to withstand the ceaseless flow of secular ideals on the college campus.[9] The pieces that have been written, such as Clarence Shedd's *Two Centuries of Student Christian Movements* (1934) and William Morgan's *Student Religion during Fifty Years* (1935), display common limitations in providing a more substantive analysis of college student religious life at this time.[10] First, their breadth—marked by a sweeping focus on the YMCA, the Young Women's Christian Association (YWCA), the missionary Student Volunteer Movement (SVM), and the global World's Student Christian Federation (WSCF)—does not allow for a focused evaluation of the American, male campus movement. In addition, the reliance upon national source materials, specifically the pronouncements of organizational leaders, renders these works incapable of demonstrating the lived realities of campus chapters. Finally, and most importantly, these authors do very little to place the intercollegiate YMCA within the larger context of the history of higher education. While details of organizational meetings and executive decisions are critical, there is little provided to demonstrate the broader reasons for changes in the midst of a secularizing university context. Drawing upon the excellent archival holdings of the Yale University Divinity School, the Kautz national YMCA archives, and local materials acquired from 112 local campuses, this study attempts to address these deficiencies while building upon the important work initiated by these scholars.

In exploring the history of the college YMCA movement between its inauguration in 1858 and its steep decline during the 1920s and early

1930s, this book attempts to trace the organization's growth, influence, and changing orientation in the context of the broader changes in higher education. The narrative presented here depicts an organization characterized by three distinct eras. The first, beginning with the inaugural formation of college YMCA chapters in 1858, was marked by a strong evangelistic consciousness. Anchored by a primary—in some ways singular—commitment to student evangelism, the movement in its early years aimed squarely at saving collegiate souls and training students to share the Christian faith with others on the campus, in the local communities, and across the globe. Spurred by Protestant revivalist fervor in these years, leaders of the college YMCA desired to foster spiritual awakening and preserve the spiritual health and vitality of the nation's elite, most far from home for the first time. Hoping to help students see the need for practical Christian soul work even in these largely academic settings, the YMCA gradually began a process of eclipsing earlier campus religious societies and establishing a hold as a national intercollegiate organization. Under the decade-long leadership of Luther D. Wishard (1877–1888), the YMCA maintained a solidly evangelical posture even as it spread to the majority of the nation's institutions of higher learning.

A second era, spanning the period from Wishard's departure until about 1915, witnessed a number of significant deviations from this earlier paradigm. As the movement grew both numerically and in influence under the leadership of John R. Mott, there was an emerging desire to pursue, not primarily student salvation, but rather the development of character and service. Progressive-era staples in higher education, these emphases reflected changes in the collegiate landscape and in the broader understanding of religion on the campuses of this period. Promulgating a "muscular Christianity" that was increasingly common among Protestants, the faith championed by college YMCA leaders was rooted self-consciously in action rather than theology, and behavior rather than belief. Locating the core of Christianity in ethical living rather than doctrinal reflection or soul-focused introspection, the YMCA found itself quite at home within a university culture that also wished to distinguish between inappropriate theologizing and more palatable emphases on service and religious living. In fact, YMCA chapters in this era provided valuable student housing, recreational, and employment services for local campuses, thus meeting significant institutional needs and facilitating an enhanced reputation among students and university administrators.

The juxtaposition of Mott's departure and the onset of World War I set the stage for another significant transition between 1915 and 1934. Under the leadership of David R. Porter, the college association made substantial revisions of policy and practice to coordinate with altered institutional realities, changing student ideals, and new postwar religious and educational philosophies. Fostered by a new affiliation with the forces of liberal Protestantism and progressive education, the organization substituted attempts to secure Christian character and service with aspirations toward the construction of a new Christianized world. Dropping much of the traditional program and adopting instead a Christ-inspired devotion to a democratic social order, students were directed toward a thoughtful engagement with vexing social issues of war, race, and economic justice. Devoted to intellectual work bathed in the social sciences, the organization set forth an expansive agenda that sought to raise a "prophetic minority" of socially radical students poised to establish the Kingdom of God on earth. While the earlier mission had attracted universal support, this revised vision alienated church leaders, university leaders, and conservative religious students. Combined with the new competition from denominational groups and from the universities themselves (in the provision of social and religious student services), these changes sponsored a rapid numerical downturn and hastened the decline of the organization nationwide. The YMCA's monopoly, and in many ways its position as a central arbiter of student religious life, was over.

In the end, therefore, I contend that the YMCA passed through a series of successive stages with regard to core themes, stressing first evangelism and soul saving (1858–1888), then character and service (1888–1915), and finally Christian social reconstruction (1915–1934). The notion of three distinct eras, marked by changes in national leadership, obviously creates artificial divisions between these periods. Significant changes in movement policy and procedure were never quite so precise, and such delineations mask the often gradual and halting nature of organizational transformation. At the same time, those eras and their leaders did represent unique perspectives on the nature of college student religious training. Certainly, each leader brought to his position a novel vision for the best means of achieving Christian growth in these settings. In addition, however, these periods were also marked by substantial changes in higher education institutions, student demographics, American religious and educational thought, and the larger social order. Thus, despite the artificial nature of such segmentation, these eras do constitute a useful organizing framework for analyzing the changes in the movement over time.

Looking at campus religion through the lens of a student organization highlights a variety of critical themes in the relationship between faith and higher education in the "era of secularization." First of all, the story of the college YMCA reveals the inextricable links between student religion and campus life and culture in these years. While secularizing trends were facilitating a loss of formal religiosity within the compulsory machinery of the institutions, Protestant Christianity still shaped much of the general character of student life in the late nineteenth and early twentieth centuries. While much attention has been given to the rise of the Greek fraternities and sororities and to the emergence of intercollegiate athletics in this era, the position and popularity of the YMCA at the turn of the century in many ways matched these better-known attractions. YMCA-sponsored student Bible study groups expanded greatly in these years, even within the fraternities. Missionary interest also reached a peak, spawning a host of mission study groups on the nation's campuses. Scores of students flocked to YMCA summer conferences and to the quadrennial assemblies of the SVM. In addition, in the absence of extensive collegiate facilities and personnel, the YMCA often served as the de facto student services office of the universities, providing undergraduates with assistance in housing, part-time employment, career guidance, recreation, counseling, and many other necessities. YMCA buildings on local campuses emerged as university social centers, facilities centralizing student entertainment and services in one locale. In many cases, campus administrators added YMCA secretaries to the institutional payrolls to serve in capacities that would later be associated with the position of "dean." Church-state separation notwithstanding, this religious movement, particularly prior to World War I, was indeed a primary locus of college student life.

An examination of the YMCA in these years also provides a window into the often tense battles to define the best locations to train students in religious faith. The YMCA, local community churches, expansive denominational agencies, and the institutions themselves often vied for control over the collegiate religious terrain. While the YMCA held a position of unqualified dominance and leadership in campus religious work up until World War I, the story after this date reveals increasing competition over the proper organizational channels for student moral and religious growth. These skirmishes represented more than simple "turf wars" on local campuses. At their core, the contests over venue became central locations for dialogues on larger issues of church and state, sectarian versus nonsectarian religious expression, and the theological grounding of religious teaching. The answers given to such boundary-defining questions have had long-term implications for our present understanding of the place of religion in higher education.

Finally, and perhaps most importantly, the history of the intercollegiate YMCA between 1858 and 1934 provides a new vantage point to examine the secularization of the academy during this era. Marsden has argued that secularization was often facilitated in these institutions through a dual process that was both methodological (the suspension of religious belief in academic work for the purpose of retaining scientific objectivity) and ideological (the active dismissal of creedal religion).[11] Interestingly, the intercollegiate YMCA accelerated both forms of university secularization. By providing a setting for vigorous extracurricular religion on the nation's campuses, the YMCA made it easier for professors and administrators to sanction an institutional separation of facts and values in these settings. When concerned parents raised questions relating to the apparent loss of formal religion in these institutions, administrators were always poised to articulate the popularity of the YMCA. Professors, similarly, could pursue their scholarship with scientific objectivity, knowing that students' moral and religious welfare was being cared for by the association. Devout faculty members could also join the YMCA chapters in this extracurricular work, serving as speakers, Bible study leaders, and advisory board members.

As students flocked to join this popular movement, in fact, the separation of curricular facts and extracurricular values in higher education appeared to be a brilliant application of the law of specialization of function, providing for both enhanced scholarship and enhanced religious development in the nation's colleges and universities. Despite the decline in religious perspectives in the curriculum and the gradual elimination of mandatory religious activities, administrators were thus able to convince critics and concerned parents that the campuses were actually more religious than they had ever been.

Interestingly, once the college YMCA had assumed this important position, it also served to accentuate the ideological secularization taking place in these institutions. Between 1888 and 1915, influenced by a movement for muscular Christianity, the college chapters redefined religion in terms of character and service, viewing faith as practical rather than creedal. Such a perspective prompted a gradual decline in the belief of the revelatory nature of the Bible, elevated service over doctrine, and judged Christianity on the basis of pragmatic results rather than divine origin. After World War I, influenced by the liberal Protestant social gospel movement and progressive educators, religion was conceptualized more in terms of the creation of a democratic kingdom of God on earth, marked by a reconstruction of the social order in light of the radical social teachings of Jesus. In either case, religion no longer resided within an adherence to substantive beliefs derived from Scripture or Protestant doctrinal statements. By divorcing Christian practice from its groundedness in a particular faith tradition, the YMCA fostered, many times self-consciously, a loss of concrete belief among its members. Marked by the growing religious pluralism of the campuses and by a consequentialist ethic that placed little value in the source of one's religious actions, the organization progressively rejected the importance of Christian faith as a necessary component of Christian living. Without question, the intercollegiate YMCA movement was a result of the secularization of higher education, filling a void created by the elimination of formal religious expression on the campuses of the United States in this era. At the same time, the YMCA was also a source of secularization, both for the institutions and for the individual students who joined its rolls.

Patton, like other college leaders of his time, increasingly viewed a voluntary, nonsectarian student agency like the YMCA as the best, and perhaps the only, means of securing student religious culture on the campuses. While such a change was by no means rapid or clear-cut, the gradual emergence of this idea did represent a revolution in institutional thought. Whether or not this shift was inevitable, it did lay the groundwork for our current understanding of the place of religion within the larger structure of American higher education.

Section 1
Evangelizing the Campus, 1858–1888

Chapter One
Organizing a College Revival

Collegiate life in antebellum America was designed, at least rhetorically, to conform to a familial pattern of living, defined by secluded residential proximity, common meals, shared religious experiences, and a collective submission to the paternal discipline of college administrators and faculty. This model of the "collegiate way," though rarely observed fully in practice, was an indication of the degree to which colleges sought to model themselves after the in loco parentis patterns of English institutions such as Oxford and Cambridge. Prescribing intimate oversight, not only of student academic life, but also of recreational, dietary, moral, and religious practices, leaders aspired to a kind of domestic caretaking role that would ensure both behavioral compliance and a propriety appropriate to their social status. The institutions existing in these years—typically small-scale enterprises enrolling fewer than a hundred students and employing fewer than ten instructors—were crafted to develop not only scholars but also "gentlemen" in the fullest sense of that term.[1]

Such oversight was in part maintained by faculty responsibility for student conduct, a reality that inextricably merged academic and disciplinary tasks for college professors. Attempting to protect students from moral blight while promoting righteous living on the campus, teachers were called upon and empowered to regulate, not only classroom decorum, but also choices of clothing, attendance at religious meetings, and punishment for moral and social affronts.[2] Such purported paternalism, of course, implied neither perfect execution by faculty nor perfect compliance by students. As recent historians have suggested, the decline of the intensive community of the collegiate way, at least in New England, had already begun before mid-century. The presence of more "mature" (enrolling at age twenty-five or later) collegians made disciplinary appeals less credible. In addition, many poor students were forced by economic necessity to take jobs and secure housing in local towns, removing them from comprehensive campus influence. Over time, poor students took up residence in the increasingly dilapidated campus dorms (or with charitable townspeople) while wealthier students found more expensive rooms in town. Such dispersion, a combination of disparate needs and tastes, thus facilitated a fragmentation of lived social life and a decline in supervisory oversight.[3] Furthermore, when such oversight was deemed intrusive, students (particularly wealthy students) often demonstrated active resistance to the heavy-handedness of their superiors. In the years between 1800 and mid-century, many colleges witnessed intense struggles among collegians, faculty, and administrators, some taking the form of large-scale riots.[4]

However tempered by living conditions and active student resistance, the larger goals of the collegiate way certainly influenced antebellum student religious life. Collegiate religiosity in this era was expressed in a rich, if somewhat formal and circumscribed, institutional ethos. Leading up to the 1860s, American colleges were quite explicitly Christian in orientation. Many schools were affiliated with particular Protestant denominations, approximately two-fifths Presbyterian and Congregationalist with a growing number allied with Methodists and Baptists newly emboldened by gains in the Second Great Awakening.[5] At the same time, and opposed to some traditional interpretations, the colleges were rarely marked by deep denominational loyalty. While religious affiliation was common, market pressures and local boosterism often blunted the sharp edges of doctrinal boundaries and allowed for a significant blending of Protestant groups in these institutions. Denominational rivalries continued to be fierce, to be sure, but the realities of enrollment pressures for financially strapped institutions made restrictive creedal codes untenable. In addition, with limited transportation options and with colleges committed to local pride, students were more likely to be regionally than denominationally selected.[6] Thus, while Cornell President White's description of the antebellum era as one defined by "the regime of petty sectarian colleges" was a helpful apologetic for his new vision of the research university, it was likely an inappropriate descriptor of early- to mid-nineteenth-century college life. Needing to attract a broader constituency and attempting to serve a local public of mixed denominational heritage, colleges were far more likely to highlight the general truths of Protestantism than the more detailed theological distinctives of specific denominations.[7]

Intellectual changes accompanying the shift from Calvinist orthodoxy to a more "Enlightened Christianity" derived from Scottish Common Sense philosophy also eroded strict creedal sensitivities prior to the Civil War. As American educators took part in what historian George Marsden has called a "Whig republican consciousness," many increasingly sought to promulgate a Christian faith that was essentially a nonsectarian public piety defined by moral (rather than doctrinal) affirmations.[8] Bathed in Enlightenment notions of science and natural moral law, Common Sense theorists and those influenced by them proposed that moral ideals could be derived from universally self-evident principles of nature. While all subjects were to be studied from a Christian vantage point, the crux of intellectual integration was reserved for courses in natural theology and moral philosophy. Natural theology courses were designed to probe nature as a veritable text for empirically demonstrating the wisdom and beneficence of the Creator, linking general revelation with the special revelation of God found in the Bible.[9] Moral philosophy, typically taken during a student's senior year and often taught by the president of the college, examined the classical disciplines studied over the previous years from the perspective of a Protestant worldview, providing normative guidelines for ethical action in the world. Such courses assumed a unity of truth that viewed Christian teachings and other disciplines within a common epistemological universe, anchored by natural law.[10] At the same time, they were not explicitly theological. While such principles were always said to be consistent with biblical revelation and therefore affirming of divine truth, it is clear that such an epistemology was a step outside of the dependence upon revelation and theological a priori argumentation marking earlier commitments.[11]

While this generic Protestantism lacked a stiff sectarian consciousness, the atmosphere on the campus was nonetheless self-consciously bathed in religious ideals and practices. Many of the leaders of these institutions were formally trained in theology. While religious tests of orthodoxy for faculty and students were growing less common, most of the colleges maintained long-standing traditions of formal religiosity. Required chapel was a reality in most collegiate settings, and many institutions mandated Sunday morning and evening church attendance. On some campuses, Monday classes were prohibited so that students would not be tempted to profane the Sabbath by studying. Annual "Day of Prayer" gatherings, typically held in late January or early February and marked by the cancellation of classes, were broadly observed for the purpose of pleading for students' salvation and victory over temptation. Spontaneous revivals were also common in the antebellum era, occurring with escalating frequency between 1810 and 1860. Closely associated with the Second Great Awakening, these periods of fervor were characterized by prayer, confession, and the relinquishing of items—playing cards, brandy bottles, and hazing canes in one instance—known to promote moral dissolution.[12] Not all, of course, responded favorably to these formalized appeals to evangelical piety, and stories of student irreverence during chapel and morning prayers demonstrated that rebellion against paternalism could also extend to religious domains. Yet while accounts of apathy and resistance were commonplace, there was little doubt that student religious life leading into the 1860s was inextricably intertwined with the colleges' compulsory machinery.[13]

Even within this hothouse of required religiosity, students in the antebellum years created voluntary religious societies designed to enhance Christian growth. As Clarence Shedd has documented, colleges between the American Revolution and the Civil War proved to be fertile seed beds for manifold religious agencies to meet the spiritual needs and proclivities of students. Some, fashioned in a manner similar to college literary societies, assumed an intellectual bent, hosting frequent disputations on religious topics. Harvard's Adelphoi Theologia, for example, served as a secret society for Harvard students who were interested in, among other things, providing "critical dissertations" on the Christian religion.[14] In other cases, student groups emphasized devotional topics and aimed primarily at fostering the spiritual growth of their charges. In Yale's Moral Society, members promised to resist profanity, gambling, and drinking while also informing college officials of moral breaches witnessed on campus. Despite its title, the Theological Society of Dartmouth also saw such guidance as its primary role, specifying its purposes in terms of moral "watchfulness" and punishing infringements ranging from drunkenness to "nocturnal cow hunting."[15] Adding to this diversity, student missionary societies also emerged as key channels of religious expression in the colleges. Tracing their origins to the famous Haystack Revival at Williams College, such groups analyzed the spiritual condition of foreign lands while also strategizing new methods for missionary labor. By the time of the Civil War, about a hundred "Societies of Missionary Inquiry" had been established nationwide, making them the most prominent form of religious club in American higher education.[16]

Prior to the advent of the college YMCA, therefore, American colleges supported a wide variety of voluntary theological, devotional, and missionary groups, sometimes housing many such agencies in single institutions. Clarence Shedd, in fact, has estimated that, between 1810 and 1858, there were approximately a hundred such societies within a group of seventy New England colleges, reflecting a multiplicity of such clubs and a competitive marketplace for student interest and attention. There were certainly common features characterizing groups in all regions of the nation. Even when schools used similar titles to mark their agencies, however, these clubs were decidedly local, reflecting institutional character and working in complete independence from similar groups elsewhere. With the possible exception of correspondence between some of the missionary agencies of different schools, therefore, voluntary student religious life up until the Civil War remained a self-contained campus affair.[17]

Reviving the Colleges

In the late 1850s, however, religious impulse spilled beyond the boundaries of the campuses and linked the colleges to larger national spiritual currents. On the heels of the "bank panic" of 1857, a wave of revivalistic prayer meetings spread across the nation in both rural and urban contexts. By 1858, while most journalistic attention was given to the various prayer meetings among businessmen in New York, Philadelphia, and other metropolitan centers, it was clear that this ferment had extended rather indiscriminately to all regions of the country. The revival cut across all of the major Protestant denominations, sparking awakenings and "protracted meetings" in Baptist, Methodist, Presbyterian, Dutch Reformed, Episcopalian, and Congregationalist churches and setting the stage for what historian Kathryn Long has called a "public, transdenominational religious identity among evangelicals." In her estimation, the 1857–1858 awakening was "perhaps the closest thing to a truly national revival in American history."[18]

Without question, the chief para-church organization within which the spirit of the revival was institutionalized was the YMCA. This Protestant voluntary society had its origins in the work of dry-goods merchant George Williams who, in the 1840s, established the first association in London to serve as an agency of both evangelistic revival and moral improvement. Transplanted to the American context in Boston in 1851, the movement spread quickly among middle-class men in urban centers and by 1860, 205 such associations with over 25,000 members existed nationwide. Designed to foster "piety, character, and spiritual resolve in young men," these organizations were allied closely with the Protestant evangelical churches and yet interdenominational from the outset, highlighting points of Christian unity and resolving "never to admit any intermeddling with those matters of faith and polity on which the churches differ." While some were hesitant to endorse an organization working toward religious aims outside of the Church, in general ministers and laypeople across the denominational spectrum assisted the work of the association and embraced its participatory and voluntary impulses as appropriate to the urban setting. Like the revival itself, the YMCA became an important agent of an integrated lay Protestant evangelical identity.[19]

As Long suggests, the YMCA "came into its own" during the 1857–1858 revival.[20] The early associations were envisioned as agencies bent upon stimulating, nurturing, and protecting the moral and religious ideals of young middle-class men located in the morally turbulent cities. For those already in the faith, the YMCA could serve as a safeguard against urban vice and provide outlets for evangelism to lost peers. For those outside of the safety net of Christianity, the organization desired to create an inviting space to bring such individuals under the influence of Christian principles. The program of these early associations therefore included such staples as evangelistic work, biblical instruction, lectures, reading rooms, and, as the constitution of the chapter in Richmond, Virginia put it, "any other agencies in accordance with the Scriptures."[21] Active membership was reserved for members of orthodox evangelical congregations, but most chapters also made provision for "any young man of good moral character" to be received as an associate member without voting or leadership privileges. As the idea spread within the cities during the 1850s, YMCA chapters also began to spring up among more specialized groups, most notably African Americans (1853) and German immigrants (1856).[22]

Less predictable, however, was the concomitant appearance of YMCA chapters on select college campuses in this decade. While little attention has been given to the place of higher education in the 1857–1858 revival, it is clear that the colleges were deeply affected by the larger national religious upsurge. Widespread revivals were reported in institutions throughout the country. Though the settings were quite diverse, representing both small and large colleges in rural and urban contexts, the patterns of awakening were quite similar, spawning intense periods of prayer, repentance, and conversions to Christian faith.[23] At Yale, nearly one half of the student body "professed faith" and 120 students, more than one quarter of the campus population, came into membership at the college church. At Amherst, President Augustus Stearns noted that the wave of religious renewal on his campus deepened until "the entire collegiate community was brought under its influence." In addition to conversions (it was noted that only three or four seniors remained unconverted), Stearns noted that the "reformation in the character and manners" of the student body was no less dramatic than the "renewal of hearts." College discipline, he noted "seemed to lose its office," while "sacred psalms took the place of questionable songs, and the social revelries gave way to heavenly friendships." After years of student rebellion, this was revival indeed.[24] While higher education historian Fredrick Rudolph may have been shortsighted in claiming 1858 as the "last great revival year," there is little doubt that this represented the greatest concentration of such fervor in the nineteenth century.[25]

It was the 1858 revivals at the University of Virginia and the University of Michigan, however, that merged these impulses on the campuses with the lay movement of the YMCA. At Virginia, revival struck in the wake of a typhoid epidemic on campus, elevating the tone of religious intensity at the institution. While a local Student's Prayer Meeting and a Society for Missionary Inquiry did exist, students desired to coordinate the Christian forces on the campus for more unified and effective religious work. Throughout 1858, many blueprints for potential organizations were proposed, each failing due to the difficulty of forming a single society that would embrace all denominational perspectives. Because they agreed that there could be "no ecclesiastical organization for so many minds in a State University," they were excited to discover the platform of the new YMCA, which required only a "vital experience of faith in one Lord and Savior" as well as a desire for Christian fellowship and holiness.[26] With such a generic basis for union that could be endorsed by members of all Protestant groups, chaplain Dabney Harrison organized students into a YMCA club in October of 1858, the initial constitution stating that they would work for "the improvement of the spiritual condition of the students and the extension of religious advantages to the destitute points in the neighborhood of the University."[27]

Under the leadership of President Arthur Tappan, the University of Michigan also witnessed the effects of revival on the Ann Arbor campus during the 1857–1858 school year. *The Evangelist*, a New School Presbyterian religious periodical reported "a deep and solemn thoughtfulness among students of the State University," prompted by Tappan's teaching and by the "superadded influence of the Holy Spirit." Among the students most deeply affected was Adam K. Spence. Later to assume roles as a professor at Michigan and president of Fisk University, Spence hungered for the revivalist currents on the campus to be channeled into voluntary student religious enthusiasm on an ongoing basis. While a Society of Missionary Inquiry did exist, Spence and many of his more zealous friends (including his brother E.A. Spence) were disturbed that reports on missionary subjects had become little more than "papers on geographical, historical, ethnological, and similar subjects." The absence of religious passion, practical zeal, and prayerful intensity was deeply troubling, to be sure. Even more troubling, however, was the fact that those giving the reports, often "irreligious" students, were doing so in a "flippant" manner, harming the advance of the gospel and injuring the consciences of the devout.[28] Such conduct pressed Spence to conduct "earnest conferences" with President Tappan, a Dr. Ford, and professor Erastus Haven in order to initiate a new religious organization in place of the Society. After much deliberation, interested collegians formed a "Students' Christian Association" in the winter of 1857–1858, eventually bringing this group into affiliation with the YMCA.[29]

The desire for a unified, practical faith demonstrated at both Virginia and Michigan served as the groundwork for subsequent efforts. Because revivals on local campuses emphasized religious conversion, moral transformation, and interdenominational fellowship, students and leaders in select colleges and universities viewed the YMCA as a fitting conduit for channeling this renewed enthusiasm. Throughout the 1860s, eight other campuses witnessed the formation of independent YMCA chapters, most citing the need for more explicit attention to the experiential and practical aspects of the faith. During 1864–1865, students at the University of Rochester, "feeling that among students the tendency to be satisfied with the theoretical in religion in place of the experimental is always strong," transformed their Judson Society of Missionary Inquiry into a YMCA. Recognizing that students possessed a tendency "to give undue prominence to religious convictions apart from religious actions and emotions," their stated goal was to "form in its members regular habits of Christian work, and to keep their hearts warm with Christian love by mutual encouragement and sympathy."[30] The development in 1869 of the first YMCA chapter at a black institution, Howard University, was similarly anchored in the need to provide students a location to promote evangelism, practical Christian work, and personal holiness outside of the formal curriculum. Students at Grinnell College, Olivet College, the University of Mississippi, the College of the City of New York, Washington College (later Washington and Lee), and Cornell University also formed nascent chapters in the 1860s, most fresh off revivals and citing a desire to conserve the fruit of these awakenings on their respective campuses.[31]

Though the growth of isolated campus chapters was clearly haphazard, some ardent proponents saw within this grassroots development the potential for an expanded work. Attempting to capitalize on the momentum fostered by the new campus chapters, Adam K. Spence, charter member of the 1858 University of Michigan chapter and now a professor at that institution, made an appeal to the delegates of the 1868 International YMCA Convention in Detroit to consider the collegiate field as a viable extension of the organization's urban focus. Desiring a decisive endorsement of a movement along these lines, Spence provided the Convention with a ready-made resolution for discussion and vote:

> Whereas, during a portion of each year, and at a period of life most critical to moral and religious character, thousands of our young men are away from their homes at our universities, colleges, and other schools of learning; and whereas, Christian Associations have been formed and successfully maintained in some of these institutions; therefore, be it resolved that, as a Convention, and individuals, we will seek to plant a Christian Association in each and all of our universities, colleges, and seminaries of learning, to the end that we may bring to Christ our Savior as many as we can of the previous youth now in process of education, and soon to wield a vast influence in our land.[32]

The task of convincing YMCA leaders of the necessity of such work, however, was by no means straightforward. Some leaders, fearful of a diluted focus, were reticent to support the development of local clubs in settings that appeared wholly incommensurate with the aims of the larger movement. Many claimed that the secluded and cultured confines of academe represented a spirit antithetical to the practical and cosmopolitan character of the YMCA. Some contended additionally that, unlike independent and isolated urban youth, college men were well provided for as recipients of the parental oversight of college faculty and administrators and therefore less in need of external assistance. While the parallels between freshmen and new city dwellers made logical sense—both were often recently separated from home and therefore subject to new temptations—collegians seemed far less threatened than their urban counterparts. With many doubtful of this need, the resolution failed to receive the necessary votes for approval.[33]

Spence, however, would not relent. He suggested that the practical and cosmopolitan nature of YMCA religion was precisely what collegians needed in order to resist overintellectualizing the gospel. He also appealed to the influential nature of the college years in transforming student character and faith. Students, he suggested, were "at the critical formative period of life, the period of impulse and passion, and of skepticism too, when a traditional faith received from others must be replaced by a faith of one's own."[34] Since higher educational institutions provided ample environmental opportunities for both advance and decline with freedom to elect a personal destiny, Spence portrayed collegians as perched on the edge of a personal precipice, ready to promote righteousness or dissolution in the broader culture. Finally, Spence suggested that, while these years were critical for all young men, college students demanded particular attention because of their future leadership potential in church and state. With the necessary caveat that all men were important in the eyes of God, this "vast army of youth," he contended, was the "very flower of the land" and the "hope of the nation."[35] If the association desired to influence the culture as a whole, therefore, this was the group of potential-laden men upon which to focus directed energy.

While such arguments were surely persuasive for men who were interested in both leadership and moral influence, far more attention was actually given to the changing scope and nature of higher education in these years. Between 1860 and 1875, the student collegiate population doubled and the number of existing colleges increased by 75 percent. While 44 colleges had been chartered in the first three decades of the nineteenth century, a full 175 were chartered between 1860 and 1870 alone.[36] Sparked by the confluence of geographical expansion, denominational growth, and the more widespread availability of rail travel in this era, this tremendous increase was also accompanied by an expanding diversity of institutional types. Many of the schools started in this era were founded as what historian

Roger Geiger has called "multipurpose colleges," institutions offering an eclectic blend of classical and modern subjects, but women's colleges and scientific schools also emerged at this time. The growth of universities added to this complex array of available options, staking a more directed commitment to research and professional preparation, especially at the graduate level.[37] Many states utilized money from the Morill Land Grant Act of 1862 either to form "agricultural and mechanical" colleges and new universities or to help existing institutions add components to encourage such interests. While these trends would not come to full fruition until the turn of the century when economic need bolstered reformist zeal, institutions began to attract, not only those pursuing careers in law, medicine, and ministry, but also those in such fields as engineering, education, journalism, and agriculture. Although it is likely that only 2 percent of eligible men actually attended college in this era, such institutional expansion provided a growing market and a new clientele more in keeping with the practical proclivities of YMCA leaders.[38]

YMCA leaders also looked to the collapsing infrastructure of the in loco parentis college ideal. Expanding enrollments, combined with limited budgetary resources for dormitory construction, rendered schools less able than their English exemplars to fashion common residential space. In addition, the willingness of faculty to enforce holistic oversight was clearly on the wane. Yale's Lyman Bagg noted in 1871 that "Where a man is, or what he is doing, outside the hours when his presence is required at recitation, lecture, or chapel, the faculty make no effort to enquire. The 'paternal' theory of government is not much insisted upon by them."[39] In fact, many schools during this era—including Harvard in 1869—began separating behavioral from academic evaluations in student records, determining student ranking purely on the basis of classroom performance rather than moral excellence. As professors began retreating from such responsibilities and as increasing wealth opened up opportunities for social independence, college students were given more latitude to create their own social worlds, free from the oversight of their elders and defined by a more pervasive self-regulation.[40] Student literary societies were declining by the 1870s and 1880s, but they were replaced by a number of alternative activities emphasizing social over academic pursuits. Fraternities, present on campuses since the 1830s and 1840s, were just beginning to emerge as pivotal extracurricular forces in campus life. Similarly, athletic events, barely existent prior to the Civil War, were becoming prominent features of the collegiate social calendar. The YMCA, as a student-led religious organization, thus merged well with these new ideals. In the absence of strong faculty religious guidance, many felt that this domain must also be regulated through voluntary efforts.[41]

Finally, one of the most critical rationales for a growing YMCA presence related to the appeal of nonsectarian forms of campus religious faith. President of the University of Michigan Erastus Otis Haven, who had earlier assisted Spence in his attempts to fashion a new voluntary religious organization on that campus, addressed the 1868 International YMCA convention and described the ways in which declining tests of creedal orthodoxy had rendered a robust Christian identity increasingly problematic on his campus. At the same time, he evinced a supreme optimism regarding the Christian character of his institution, giving credit chiefly to the new college YMCA for upholding righteousness and safeguarding Christian morality in this setting. "But the question may be asked," he posited, "Is this a Christian university? Can the great system of schools be a Christian system? If it can be, it must be by the efforts of the Christian Association. No religious denomination should have a predominance in a state institution, and no one should be a professor who cannot sink his sectarianism. We have a flourishing Christian Association in the University and more than a hundred have been brought to Christ during the past year."[42] University of Rochester President M.B. Anderson similarly suggested to the gathering that the provision of pastors and "college churches" was open to "serious and obvious objections" in public institutions devoid of strict denominational control. "The most natural and efficient means of attaining the moral ends of a Christian school and avoiding the impropriety of trenching upon the conscience of either the pupils or their parents," he noted, "is through the organization of the Christian Association."[43]

The arguments regarding the need for a YMCA offered by Haven and Anderson reflected a common belief that student religious development would come, not through church and denominational influence but rather through a voluntary nonsectarian gathering of faithful Christians representing diverse denominational backgrounds. It was this belief that framed the formal arguments for a YMCA presence on the campus. Spence, writing for the *Association Monthly* just before the 1870 convention, argued that the YMCA possessed a "catholic character" that would allow students to transcend differences of creed and promote collective effort on a greater scale. While denominational distinctives had been a defining element for "the pious fathers of a past generation," Spence argued, the "children of today" would "grasp hands over the walls of sectarian separation." Allowing students to pursue the church of their own denomination, the YMCA would unite them together for joint encouragement, teaching, and practical work, thus "accomplishing that which seems to be God's law in the universe—*Unity in Variety*." Noting that the movement "liberalizes in the right way," downplaying "minor distinctions" and focusing instead on essentials, Spence suggested that the association would serve as the means of "begetting union in the hearts of the young" on college campuses.[44]

For an organization devoted to such interdenominational "liberalization," this appeal was indeed persuasive. Spence's literary campaign bore fruit at the 1870 convention in Indianapolis, where delegates, without discussion, adopted an updated resolution, "That this convention hails with joy the organization, in some of our academies and colleges, of YMCAs, and commends this feature of our work in behalf of the young men of America, and hopes that Christian Associations may be planted wherever practicable in our Academies, Colleges and Universities, and that we urge especially such Societies already existing that they seek to extend their work in this important field."[45] Such a pronouncement, while it did not specify a commitment of financial resources or personnel, did at least demonstrate a recognition that the YMCA might bring the fruits of an interdenominational Protestant revivalism to the campuses.

The Origins of a Movement

Much of the actual enthusiasm for this work came from Robert Weidensall, hired in 1868 as the first paid field secretary of the YMCA International Committee. Claiming that "One good college graduate has as much influence in his life work . . . as five hundred ordinary uneducated men," Weidensall, a graduate of Gettysburg College, took the Indianapolis resolution as an invitation to aggressive work among these potential-laden men. Traveling extensively across the Midwest and South, Weidensall met with college presidents, spoke to students, invited delegates to YMCA conventions, and attempted to organize new chapters. In some cases, he worked to form religious organizations where no such group had previously existed. In other cases, extant societies were restructured along more practical and evangelical lines and renamed as YMCAs, thereby enacting affiliation with the larger organization. In settings where YMCAs had already been formed, Weidensall worked to strengthen local programs, link them to state officers, and ensure compliance with the aims of the broader movement. In some cases, such as at the University of Virginia, Weidensall spent time dissuading local chapters from leaving the YMCA by convincing them of the growth of the movement nationwide.[46] Finding particular success in state institutions, which Weidensall called "the peculiar province of our associations," by 1872 twenty-six new collegiate chapters had been formed.[47]

The establishment of YMCA state secretaries in the 1870s furnished an important means of expanding the nascent college organizations. Early state secretaries gave great impetus to college chapters, and many organized associations in schools that lacked such groups. Indiana State Secretary L.W. Munhall, for example, established organizations in eleven colleges in 1872 alone. In Michigan, student YMCA leaders from the University of

Michigan and Olivet College traveled together with the state secretary as a "deputation team" establishing chapters at the colleges in Ypsilanti, Kalamazoo, Adrian, Albion, and Hillsdale. In addition to forming new chapters in their states, thus allowing Weidensall to delegate such responsibilities, state secretaries also provided critical ongoing support for these groups through campus visits. They served as a permanent link between these college groups and the parent YMCA, securing compliance with association policies and preventing isolation from the larger aims of the movement. These state secretaries constituted, in the early years, a kind of organizational umbilical cord for these embryonic collegiate chapters.[48]

While Weidensall was pleased that thirty-five college chapters had been formed by 1877, he also recognized the potential fragility of this experiment.[49] He was concerned, for example, that the movement would lose momentum as the larger YMCA scrambled to keep pace both with burgeoning urban chapters and with newer opportunities among railroad workers. State secretaries were unable to focus direct attention on the colleges, and even his own significant efforts were scattered among multiple fields. The result, he predicted, would be a fragmented movement marked by diverse and haphazard local clubs receiving little guidance on issues of policy and program. Later descriptions largely confirmed this opinion. Future general secretary of the movement John Mott, speaking in 1895 of these early associations, noted that they were "not bound together by any tie whatever; in fact, they did not know of the existence of each other. They had, as a rule, narrow and widely divergent purposes and methods of work." For Weidensall, such "criminal" neglect would continue unless the association hierarchy demonstrated a willingness to commit resources and "employ a special agent to give all his time to this much-needed work."[50]

One individual who affirmed Weidensall's position was Princeton undergraduate Luther Deloraine Wishard. Born in 1854 in Danville, Indiana, Wishard suffered from near blindness during his childhood and was able to read and recognize faces with great difficulty throughout the remainder of his life. Despite this struggle, however, he attended Indiana University in the fall of 1870 and accomplished himself as a baritone soloist and member of the most prestigious university literary society. Following a winter of employment at a district school, Wishard transferred to the Presbyterian Hanover College and forged an immediate bond with members of the college YMCA, founded just months earlier by Weidensall. He threw himself into Bible study and service activities in the surrounding neighborhoods and was chosen, as a freshman, to represent the school at the 1872 International YMCA convention in Lowell, Massachusetts. It was here that the young Wishard met Weidensall, who recounted to the gathering the success of his first two years planting YMCAs in the nation's colleges.

Inspired by this report and by a growing personal relationship with Indiana YMCA State Secretary Munhall, this zealous student continued his vigorous YMCA work at this institution, helping to build the fledgling organization into a strong campus chapter. After his sophomore year, however, he sensed that his academic preparation at Hanover was inadequate for his burgeoning intellectual aspirations. After speaking with a friend about the combined piety and brilliance of Princeton President James McCosh, Wishard sensed that the "pillar of cloud" was moving and he enrolled for his final two years at Princeton.[51]

Wishard's tenure at Princeton proved to be a time of intense religious zeal. The Christian fervor of Scottish Common Sense philosophy that had flourished under John Witherspoon's administration largely continued with McCosh at the helm.[52] This often polemical leader preserved at Princeton an ardent belief in the unity of truth and continued traditions of mandatory chapel and Sunday services, preaching regularly in these settings. Coursework in ethics and natural theology under such luminaries as Francis Patton and Charles W. Shields provided steady doses of evangelical Christian teaching. Furthermore, in the winter of 1876, just shortly after Wishard's arrival, a series of meetings led by D.L. Moody sparked a major campus revival. Reporting excitedly to the Board of Trustees, McCosh detailed that, in addition to 100 conversions, 70 backsliders had been "aroused" and 100 additional students exposed to "some heat from the burning fire."[53] This event also sparked a notable evangelistic fervor as students, including Wishard, began spreading the gospel message to nearby towns and campuses. Princeton was not immune to the changes influencing other schools—enrollment pressures, for example, had forced many students to off-campus housing—but it was more resistant than most to the forces eroding the religious character of antebellum colleges.[54]

While he thrived in this hothouse of religious zeal, Wishard was dismayed that the college had no YMCA. Outside of the formal curriculum, many students participated in the Philadelphian Society, founded in 1825 and by this era serving as what historian P. C. Kemeny called the "de facto campus ministry" of the institution.[55] While well integrated into campus life and culture, however, Wishard lamented its local character and failure to connect with practical Christians through the national YMCA and its inspirational conventions. Because of this reality, Wishard noted that Princeton's religious tenor betrayed a "gowneshness fostered by the exclusiveness of college life." His own experience with the broader YMCA while at Hanover had revealed the benefits of outside contact with those engaged in wider practical ministry, and he began persuading some of his closest friends, leaders of the Philadelphian Society, to consider affiliating with the national YMCA. While such a move was vigorously contested due to the long history of the Philadelphian Society, Princeton did approve YMCA affiliation in 1875.[56]

Since one of the primary reasons for joining the YMCA was the development of cooperative relationships, many in the organization desired to find ways to connect with other campus chapters. Proposals regarding the character of such relationships, however, were by no means straightforward. Some, including Wishard's own college roommate Robert Mateer, favored an exclusively student organization, an "Intercollegiate Christian Union" dedicated to student religious development in the nation's colleges. Wishard, however, vigorously opposed such a plan, stating that it would deprive the collegians of relationships with practical Christians within the YMCA network. By restricting the movement to higher education, Wishard predicted a tendency to remain introspective, theoretical, and prone to the same narrowness of vision that characterized many of the older religious societies. In addition, an organization relegated to college students alone would prevent city YMCAs from receiving the college-trained leadership that might rise through the ranks of those participating in the campus movements. For the mutual benefit of both college and city, Wishard was convinced that intercollegiate contact should be centered in an external and intergenerational agency such as the YMCA.[57]

At the recommendation of philanthropist William Earl Dodge, the Philadelphian Society committee composed a circular letter inviting delegates from 200 institutions of higher learning to attend the 1877 International YMCA convention in Louisville. The circular revealed much of the rationale for connection, lamenting that, while intercollegiate activity had energized student life nationwide through literary societies, athletic competition, and fraternities, student spiritual life was still relegated to haphazard and isolated expression. The committee offered a bleak picture of "cold intellectuality," minimal zeal for evangelism, and declining numbers of ministerial candidates, placing the blame squarely on the lack of coordinated spiritual efforts among students. Wishard implored the twenty-five delegates (representing twenty-one colleges) who attended the conference to begin cross-institutional correspondence, leaders in each college informing others about the work going on in their local settings. He urged them to subscribe to *The Watchman*, the general YMCA news source. Finally, he shared his own vision regarding the need for a "corresponding secretary," a worker who would extend the movement to other colleges and serve as a medium of communication between local campuses and the larger association. While intergroup contact was not completely novel in the history of student religious organizations, the appeal for a centralizing agent was a notable departure. Never had there been an individual whose responsibility it was to oversee the voluntary religious development of the nation's collegians. Despite previous plans to pursue state YMCA or foreign missionary work, Wishard agreed to fill this position, stating that "I could no longer doubt that the internal call which had come to me was the call of God."[58]

Becoming a YMCA

As a part-time "corresponding secretary," Wishard believed that movements would be assisted and new colleges brought into the fold through a simple campaign of letter writing and the dissemination of YMCA publications. His contract for the first year, in fact, provided largely for weekend correspondence, to be completed concomitantly with his graduate divinity degree program at Union Theological Seminary. Wishard took up this work with fervor, writing 700 letters to local campuses in 1877 and 1,500 in 1879. In addition to sending *The Watchman* to every college, he also began publishing the *College Bulletin*, the first intercollegiate religious publication in the United States.[59] While several colleges did indeed begin corresponding with Wishard, however, personal contact generated far more interest and enthusiasm than a long-distance missive. While shared letters were commonplace, he noted, "the coming and appeal of a man who came because he was sent by his college, yes, by a combination of colleges, this was something new; this arrested attention; this compelled reflection; this started the student body into discussion and this resulted in action." Wishard traveled 7,000 miles in 1878 and 12,000 miles in 1879, visiting fifty-four colleges and a host of student gatherings at state YMCA conventions. It was indeed no exaggeration for Wishard to call himself an "itinerant preacher" for the students of the nation.[60]

In most cases, the formation of YMCA chapters on local campuses represented not a de novo creation of new organizations but rather a replacement or a consolidation of the preexisting religious groups that dotted the nineteenth-century collegiate landscape. At Bowdoin College, for example, the YMCA that was formed in 1882 grew out of the "praying circle" that had existed since 1815. At Wooster University in Ohio, the YMCA grew out of a preexisting "Brainerd Society," which had been devoted to the cultivation of the missionary spirit through monthly meetings. Yet while new YMCA chapters frequently replaced existing single organizations, equally common was the consolidation of theological, devotional, and missionary clubs into a campus YMCA. At Rutgers, for example, a YMCA was formed in 1876, combining, and therefore eclipsing, a Bible Society, temperance society, and missions band. At the University of Virginia, contemporaries spoke of the fact that, while the campus temperance club, prayer meeting, and missionary society were all helpful, "the spirit of the hour . . . began to grope about for some form of organization that would give it more powerful and effective expression," thus resulting in affiliation with the YMCA. Writing about the formation of the YMCA at Yale, three alumni contested that, because of the manifold religious options on campus prior to this time, student efforts were "scattered." The "great lack in these years," they noted, "was some one co-ordinating and energizing agency for the forces of good. . . . It was each man for himself."[61] In most cases, Wishard sought to demonstrate to students that joining the YMCA would allow them to continue established practices while also adding new programs and links to Christian students on other campuses.[62]

Even with all of these compelling arguments regarding expansion and consolidation, the move to YMCA affiliation was rarely uncontested. Perhaps the most common source of resistance emerged from emotional attachments to preexisting religious societies. For many, the YMCA appeared as an unwelcome intruder, an external threat to the rich traditions of extant student religious organizations. As would probably be expected, such resistance was strongest in schools with either long-standing campus traditions or more particularistic denominational ties. Harvard students finally established a YMCA in 1887, but only after years of discussion over whether to give up their traditional "Society of Christian Brethren." At Yale, where class prayer meetings and student deacons had been in place for some time and where a Christian Social Union had already been formed in 1879, there was significant resistance to the emergence of a campus YMCA.[63] In institutions with commitments to distinct religious traditions, on the other hand, the YMCA seemed overly noncommittal, its generalized Protestantism viewed as a "lowest common denominator" faith. At Wesleyan University, for example, where faculty-initiated Methodist class meetings were offered, the YMCA had difficulty taking hold.[64]

This hesitancy reflected, not a fear of innovation, per se, but rather a fear that YMCA affiliation would diminish local tradition, local control, local campus flavor, denominational ties, and the intimate bonds associated with college "classes." In place of the local societies that, despite similar titles, reflected campus uniqueness and "personality," the YMCA offered a standardized agency external to the institution. What Wishard heralded as expansion, consolidation, and heightened efficiency, in other words, could just as readily be interpreted as heavy-handed regimentation imposed by an outside organization. For his part, Wishard was not persuaded by such resistance, and in fact saw the desire for continued local tradition as one of the primary barriers to religious progress in American student life, decimating opportunities for external, broadening perspectives.[65]

In addition to this localist resistance, another source of tension in the initial years of the fledgling movement related to the specified criteria for student membership. Leaders in the general YMCA mandated that associated campus chapters assume membership regulations commensurate with the parent movement. Specifically, this meant the formal adoption and enforcement of evangelical church membership as a condition for joining local chapters as active members. While such a requirement had served as the informal policy of many local YMCA chapters beginning in the 1850s, it was formally embraced in 1868 at the Detroit International Convention. A resolution was passed at this time specifying that active voting members would be taken from those who "profess to love and publicly avow their faith in Jesus, the Redeemer, as divine, and who testify their faith by becoming and remaining members of churches held to be Evangelical."[66] There were no specific guidelines as to the exact nature of an "Evangelical church" until the following year, when at the International Convention in Portland organizational leaders developed a complete theological rubric qualifying congregations that, "maintaining the Holy Scriptures to be the only infallible rule of faith and practice, do believe in the Lord Jesus Christ (the only begotten of the Father, King of Kings and Lord of Lords, in whom dwelleth the fullness of the Godhead bodily, and who was made sin for us, though knowing no sin, bearing our sins in his own body on the tree) as the only name under heaven given among men whereby we must be saved from everlasting punishment."[67] By drawing clear boundary lines of orthodoxy linked to the church, movement leaders were able both to secure doctrinal homogeneity and to demonstrate a commitment to local congregations, particularly important since many in the organization feared that students would view chapters as church replacements. While "any young man who was moral in character and in sympathy with the aims of the organization" could join as an associate member devoid of voting or leadership privileges (subject to a two-thirds vote), this "Portland test" became the critical membership criteria of the movement.[68]

While in the 1870s such a test often reflected and confirmed a Protestant consensus, this issue did prove to be a source of contention in the process of forming chapters in some institutions. Events at the University of Wisconsin revealed these tensions in dramatic fashion. While the organization had been formed on an evangelical basis, the secretary lost the original constitution and constructed a new document eliminating such a requirement. By 1877, the same year the intercollegiate movement was initiated, student leaders of the organization wrote to their campus peers that the YMCA would not become "a band of holy brethren desirous of shutting ourselves away from all 'wicked sinners.' " Instead, the association would admit to full membership, not only professing Christians, but "all who desire to see a healthful moral sentiment propagated in our midst." The chapter changed its name to the Student Christian Association (SCA) and still maintained its affiliation with the YMCA.[69]

This more liberal membership policy opened the chapter to nearly all students and also created a context that eventually erupted in controversy. In order to link the group more closely to the larger campus, student leaders in the new SCA asked university President John Bascom to provide Saturday evening "expository lectures" on the Bible to organization members. Bascom assumed the presidency in 1874, and while his New England heritage placed him squarely within the mainstream of presidents in this era, his religious perspective represented an evolutionary and progressive optimism more characteristic of liberal Protestantism. Frustrated by what they saw as a slide from orthodoxy, several conservative students and professors met with Weidensall and resolved to form a new organization that would hold the doctrinal line. Led by professor W.H. Williams of the classical Greek and Hebrew departments, this small group established a new YMCA in 1881, embracing church membership requirements and a clear statement of belief in biblical infallibility and the deity of Christ. Reflecting back on this need, Professor Williams suggested that "The fact that this liberalistic tendency prevailed in the University Christian Association . . . to such an extent as to wholly check the progress of evangelical truth, and the fact that the President of the University with prominent members of the faculty sanctioned this tendency, made necessary the organization of our Y.M.C.A."[70]

While both organizations coexisted on the campus and within the YMCA state structure for a time, the continued theological differences between the groups soon became more overtly hostile. It was widely claimed that Bascom discriminated against the YMCA by not allowing it to use desirable university facilities. In 1881, the YMCA asked a Professor Freeman to lecture before its membership at the same hour on Saturday evenings that Bascom was appearing before the Christian Association. Not recognizing the overlap, Freeman agreed, thus setting up competing "Christian" lectures each weekend. When Bascom complained, members of the Christian Association proposed alternating lectures, with Bascom and Freeman appearing before both organizations on alternate months. Leaders of the YMCA, however, flatly rejected this idea, noting that "The views of our Associations and our lecturer differ at the very center of Christianity from the views of the President. His Christ is not our Christ. His Christ is an inspired human soul. Our Christ is the infinite God himself." Matters grew even more dire when the YMCA accused Bascom of his "dictatorial power" in suppressing YMCA lectures, positing that he must be fearful of the competition from professor Freeman. While Freeman himself took Bascom's side and noted that the president was "not afraid of anyone in Madison or in the whole state," the damage had already been done.[71]

This impasse was only surmounted through the unbending state committee regulations regarding evangelical membership. Before 1883, state YMCA leaders allowed the SCA to maintain an active affiliation with the association because organizations formed prior to 1877 were given an exemption from the Portland test. By 1883, however, the state committee began requiring the evangelical church membership criteria of all affiliated chapters regardless of founding date. Thus the two organizations merged on an evangelical basis so as to maintain recognition within the national movement.[72] The story of this era at Wisconsin documents the potential tensions implicit in the strict evangelical membership requirements of the movement. Those on the campus opposed to these doctrinal limitations found themselves having to decide between a desired theological breadth and the preservation of their position within the growing YMCA movement. On some campuses, such as the University of Minnesota, enough opposition existed to preclude affiliation, at least for a time.[73] In most cases, however, students simply compromised and adopted the Portland test even if they disagreed with some of its restrictions. Without question, membership emerged as one of the most pervasive challenges both to the formation of new chapters and to the ongoing relationship between the campus chapters and the larger YMCA.

Making a Men's Movement

While the evangelical requirement proved to be the most vexing membership issue in forming new associations, another concern that proved equally controversial was that of gender. Wishard's crusade to expand the YMCA movement in colleges and universities coincided with the growing acceptance of coeducation in the nation's institutions of higher learning. In addition to the formation of elite women's colleges and coordinate colleges in this post–Civil War era, a number of schools also began admitting women to their previously all-male institutions. While opposition to coeducation was still quite strong, particularly in the more tradition-laden eastern and southern colleges, between 1870 and 1890 the number of women in higher educational institutions increased from 11,000 (21 percent of all students enrolled) to approximately 56,000 (36 percent of all students enrolled), a fivefold expansion. While only 29 percent of all institutions were coeducational in 1870, by 1890 a full 43 percent were admitting both men and women. After the 1870s, the majority of college women, most coming from an expanding middle class, were enrolled in coeducational as opposed to female-only institutions.[74]

When Wishard began his crusade to expand the college YMCA movement on coeducational campuses, he found in many cases that men and women desired to form groups that allowed for joint membership. Wishard saw little problem with this, and in fact stated distinctly that he "assumed that it was better to have mixed associations in the colleges than none at all." Devoid of a national YMCA policy on female membership, local communities largely decided for themselves whether or not to admit women to association chapters.[75] In many associations formed in the West, in fact, mixed gender groups were widely accepted well into the 1880s. Although the Portland Convention had recommended a "sex test" for association membership and had also specified that convention representation be based solely upon male tallies, Wishard noted that such provisions were "lightly regarded" in Indiana where he received his training.[76] Wishard had little understanding of the seriousness with which these guidelines were upheld in the East because, despite periodic joking about the large number of bachelors on the membership rolls in New York, he had never been briefed on the need for strict gender segregation. On a number of campuses, women provided leadership, bolstered attendance rolls, and revitalized stagnant movements. At Lawrence University in 1882 and 1883, the president and corresponding secretaries for the movement were Miss Ann Wilson, prize mathematician, and Miss Carrie Althouse, "the best soprano singer on campus."[77] Wishard estimated that sixty to seventy-five mixed associations were initiated by 1882.[78]

When Wishard was told by association leaders that he would have to separate men and women within established groups—a task he compared to "unscrambling eggs"—he began to turn his attention to the women's religious groups that had been springing up slowly across the nation.[79] Initiated in 1866, a Women's Christian Association (WCA) movement, representing women of all ages and positions in life, had been gathering steam as a loosely federated grouping of local clubs. One of the leaders of the WCA, Mrs. H. Thane Miller, recommended in 1881 that the organization encourage the formation of local clubs in the "young ladies' colleges and seminaries" with all-female enrollment. When this work began to bear fruit, Wishard recognized a potential solution to his own problem. Meeting in 1883 with Miller, he noted the importance of such groups, but then suggested that these chapters also be formed in coeducational institutions. Since the name "Young Men's Christian Association" marginalized young women and since women hesitated to speak, pray, and contribute in "mixed assemblies," both argued that female students would secure more ownership in an organization that "bears their name."[80]

Together with female leaders from the WCA, Wishard in 1883 began vigorously promoting the separation of mixed-gender YMCAs and the formation of new female organizations. Within three years, nearly a hundred college "Young women's Christian Associations" chapters were formed nationwide with a collective membership of over 2,000 students. When they attempted to secure acceptance of these groups into the larger WCA, however, the organization balked at the evangelical church membership requirement, forcing the students and their state representatives to propose a new national organization. At Lake Geneva in 1886, the group formed the national YWCA, comprising chiefly student women's groups but open to city associations that "might care to unite with this evangelical movement." The birth of this new organization provided Wishard with a ready means of facilitating the transition to a male-only college organization.[81]

This did not, however, imply a straightforward process. While many institutions complied with these changes, Wishard's attempts to separate the genders met with more forceful resistance on campuses such as Cornell and Wisconsin.[82] Such battles were perhaps most vehement at the University of Michigan, where a strong mixed association had been in place since 1870, the first year of female enrollment at the institution. In fact, the SCA (Student Christian Association) enrolled some of the most influential female students of the day, including Miss Alice Freeman (later president of Wellesley College Alice Freeman Palmer). Wishard visited the campus in 1885 and informed leaders that they would be expelled from the association if they refused to separate by gender, and his appeal incited an immediate negative reaction. According to one historian, Wishard met with the female students privately and informed them that their presence would deny the men membership in the state and national movement and that "the boys wanted them out but were too gallant to say so." "The boys," however, upon hearing of this covert meeting, informed the female students that they would "fight to keep them in to the last ditch." When the principle of gender separation was formalized in 1886 in the state convention, the Michigan SCA was expelled from the movement. Student leaders felt that this had become a matter of principle, spurred by the growing equality of the genders and by the principle of coeducation in the university. "When the distinctions and restrictions that have separated the sexes in all branches of the world's work are being broken down and ground to dust beneath the wheels of progress," campus leaders asked in 1896, "why should not those same distinctions and restrictions cease to exist as regards God's work?" It was not until 1897 that college YMCA leader John Mott was able to convince a small number of men to leave the SCA and join the national YMCA.[83]

The process of separating the genders in campus religious work was in many ways antithetical to the trajectory of coeducational higher education in the United States. Professor Freeman of the University of Wisconsin, in fact, noted that "the plan of separating students from each other according to sex is contrary to the principles of coeducation, contrary to the spirit of modern society, inimical to progress, and is a relic of barbarism."[84] At the same time, the arguments utilized, rooted in the rhetoric of specialization and efficiency, were quite consistent with emerging national and collegiate ideals. By arguing that women could be better served in female groups that would provide them both opportunities for leadership and a focus on women's issues, there was an appeal to specialization that resonated with contemporary thought. Wishard, who actually met his wife while attempting to separate men's and women's groups at one college, was later convinced that this effort was critical to the long-term growth of the movement. "I shall ever believe," he noted, "that the disconnection of the men and women in the college Associations has resulted in vastly increasing the efficiency of organized Christian work in the colleges and cities for which these two kindred movements stand."[85]

The Expansion of a "National" Movement

Despite these challenges, Wishard's attentiveness to visitation and correspondence proved remarkably successful as the number of enrolled students and the number of campus chapters increased dramatically in his decade of leadership. While 1,320 students were involved in YMCA groups when Wishard assumed leadership in 1877, 14,193 students were involved by the time of his departure in 1887.[86] The rate of new chapter formation also escalated with each passing year. While associations multiplied at the rate of one every four weeks during Wishard's first year of leadership, by the fall of 1880, campus chapters were being birthed at the remarkable rate of one per week. By the end of the 1881–1882 academic year, there were 175 functioning college YMCA chapters, and by the time Wishard relinquished direction of the movement in 1887, 284 chapters dotted the American higher education landscape, leading him to predict that "the time will come when all colleges will be united except a few who think they can work better alone."[87] By 1883, the growth of the movement had reached such a critical mass that C.K. Ober, a graduate of Williams College with leadership experience in the YMCA, was selected to join Wishard as an assistant secretary.[88]

While the Midwest always had the greatest raw number of local chapters, Wishard noted in 1881 that the southern colleges had a larger percentage of associations than any other part of the country. The eastern colleges proved to be a bit more difficult to induce into membership, and only five such chapters existed east of the Allegheny Mountains in 1877. Blaming this on the class consciousness, conservatism, intellectual snobbery, and localism of these campuses, Wishard noted that the insularity of these locales prevented them from expanding beyond their boundaries. During the 1880s, however, many of these institutions did capitulate to Wishard's repeated appeals.[89] Understandably, the one region that remained largely outside of the YMCA's extensive reach was the West. In 1887, Wishard traveled to the Pacific Coast and found conditions quite challenging. Although he noted that "colleges are becoming as numerous on the Pacific Coast as presidential candidates," he also suggested that western schools maintained very low attendance figures and that the total number of Christians was "very small." In some schools, he thus found his work to be chiefly evangelistic, sowing seeds among the few "Sons of the Golden West" in order to produce a later harvest. The University of California at Berkeley formed a chapter in 1887, and during Wishard's tour he was able to secure six new organizations in California, six in Oregon, and three in Washington.[90]

Another gap in Wishard's desired saturation of the American academy was found in the domain of the professional school. Lacking conscious planning along these lines, YMCAs were formed over this 10-year period in only 16 of 135 medical schools, 10 of 250 normal schools, and not a single law school or commercial college.[91] Alternatively, they were far more successful at promoting the student movement in African American collegiate institutions, although this work from the outset was organized outside of the College Department. Leaders of the general African American YMCA work quite early on saw the formation of collegiate chapters as a strategic endeavor.[92] Representatives from Fisk, Howard, and Walden colleges were present at the 1877 Louisville convention, but further growth awaited the concentrated labor and visitation of YMCA staff members. In the years leading up to the late 1880s, much of that work was undertaken by Henry Edwards Brown, hired in 1879 as a traveling secretary for African American Association work. Though a northern white, Brown was an abolitionist who felt that the YMCA was well-equipped to facilitate the transition of recently freed slaves into their new responsibilities as American citizens. Having served as the founder and president of Talladega College after an undergraduate career at Oberlin, Brown was charged with cultivating grassroots support for black association work in the South. Although he started by looking within black churches and communities, by 1883 he found black colleges to be strategic centers for association influence. Able to reach large numbers of the young and educated black elite in these settings, Brown suggested that work among African American colleges would serve both to Christianize the future leaders of the black community and to raise up a potential secretarial force for black city YMCA work.[93]

There were possibly five YMCA chapters in black colleges in 1883, but by 1885 Brown had already visited most of the "leading colleges for colored students in the southern states," forming new associations in twenty-one of the schools and strengthening those already in existence.[94] According to yearbook statistics, by 1888 there were 27 associations in African American colleges and universities with a membership of 1,093 students. These associations, however, were wholly separated from the centralized college work in white institutions, segregated from Wishard's leadership, from state conventions and secretaries, and from the growing intercollegiate movement nationwide. Earlier in the 1870s, Weidensall stated that he was unable to meet with the collected representatives of white and black colleges during the Lowell International Convention because the southern white students refused to gather with the students from Howard University.[95] Such a posture of division was surely not unusual in the era immediately following the Civil War, but it did generate structural patterns that would be difficult to transcend. Wishard certainly applauded Brown's work and commended the enthusiasm of black students. At the same time, there was no attempt to coordinate chapters during these years, black student work existing within a segregated domain. Such separation would continue until well after World War I.[96]

Wishard's tenure as secretary of the intercollegiate movement, essentially from its origins in 1877 until 1888, was thus characterized by enormous growth and expansion. Despite the challenges faced on a number of local campuses stemming from issues related to local pride, membership requirements, and gender, by 1886, this leader could say with confidence that "The intercollegiate movement seems assured of success." By 1888, the YMCA was the de facto representative agency for college student Christianity across the nation. As Dr. Roswell Hitchcock of Union Theological Seminary noted in 1885, "The omnipresence, and I had almost said the omnipotence, of the Young Men's Christian Association, is the great fact in the religious life of our colleges today."[97]

Chapter Two

Saving Collegians to Save the World

One of the immediate ramifications of the creation of an intercollegiate YMCA movement between 1877 and 1888 was the development of a standardized program for local chapters across the country. Of course, the number and variety of such groups nationwide, representing a wide array of institutional types and religious proclivities, militated against complete homogenization. Volunteer organizations had to be attentive to campus concerns and traditions, and the appeal to student initiative could be satisfied only by allowing for some decision-making power to be exercised at the local level. While Wishard saw wisdom in permitting variety based upon local need, however, there was also a desire to construct a generalized set of guidelines that would represent a broad national curriculum. One of the essential flaws of the earlier religious societies, according to Wishard, was their fragmented localism, resulting in scattered efforts, diffused energy, and a paucity of standardized thinking about college ministry nationwide. One of his major tasks, then, was to cast a clear and replicable vision for these clubs so as to fashion a cohesive movement. "The success of the college work," Wishard suggested, was dependent upon "uniformity of method."[1]

While the larger YMCA had devoted much attention to the problems and opportunities afforded by burgeoning urban centers, higher education provided unique variables that shaped the college work in distinct ways. Speaking to students and leaders, Wishard claimed:

> There is a vast difference between the busy town and city life, with its thousand industries; its exciting struggle for living, wealth and position; its exposure to temptation in every form; its dearth of social and literary privileges; and the quiet retreats of college life, with its oneness of interest; its parental supervision of president and professors; its library, reading room and literary societies; its daily religious exercises and positive Christian atmosphere. Therefore, since the conditions of life in the college and the outside world are so widely different, there must necessarily be a difference in the

> methods of Christian work as great as between that of the pioneer missionary and the pastor of a New England congregation.[2]

While such images would be altered by the end of the century, the early contrast between urban squalor and college refinement spoke volumes about the perceptions that drove early curricular decisions in the movement. If colleges were anything like the New England congregations to which they were compared, some urban programmatic elements could be effectively ignored. While city YMCAs were called upon to direct the broad social, physical, intellectual, and spiritual needs of displaced urban youth, the college chapters could assume a degree of holistic institutional care that narrowed their necessary scope. Eschewing early pressures to develop social and recreational activities, Wishard suggested that the college YMCA was free to develop the uniquely religious dimensions of student life. "The College Association," he noted firmly in his first year of service, "is limited to spiritual work."[3]

The nature of that spiritual work, however, could only be determined with reference to the campus environment and the specific religious challenges posed by its influence in students' lives. From Wishard's perspective, the collegiate culture was characterized by two mutually reinforcing "problems." First, and most alarming, was the apparent decline in the number of professing Christians attending institutions of higher education. Citing a number of statistical surveys, Wishard noted that, of the 60,000 students enrolled in American colleges and universities in 1879, less than half were professing Christians.[4] Repeatedly in his writings, Wishard pronounced that fewer college students had been converted to personal Christian faith in the 1870s than at any time in the nation's history. "It is a fearful fact," he related to students in *The College Bulletin*, "that notwithstanding the multitudes of prayers offered and the revivals which have followed, fully one-half of our students leave College unconverted."[5] While earlier religious societies had spent time disputing theological dogma and raising awareness of missionary work overseas, he suggested, they had completely ignored the pressing reality that "a majority of the College students of America are out of Christ."[6] The more immediate mission field for students, it appeared, was located right on their own campuses.

The need for student salvation was particularly urgent, from Wishard's perspective, because he felt that the college years represented both the best and the final opportunity for many to hear and respond to the gospel message. Wishard saw within students a plasticity that left them open and malleable to persuasion. "Their minds," he suggested, were at a peak point of readiness to receive the "lofty truths, commanding motives, and inspiring hopes" of the gospel of Jesus Christ.[7] Because of the nature of collective living, college was envisioned as the last chance for men to develop warm relationships with hearts pliable to the touch of friendship. Once this opportunity had passed, the loss of such influences and the hardening of habits threatened to greatly diminish opportunities for spiritual rebirth. In 1879, Wishard suggested that "It is doubtless true that the vast majority of students who leave college out of Christ will never accept him. This is not necessarily the case, but it is a fact that the great body of Christians are converted before they reach the age of twenty-five. If this be true, it may be now or never for our unconverted college friends."[8] In light of the changes in higher education, leaders were sure the YMCA held the best potential for serving this evangelistic purpose. At the 1885 YMCA Convention in Atlanta, college workers were told by a local pastor that "in view of the present tendency to secularize education to a most dangerous degree, it needs to be impressed upon the minds of Christian men that the churches have no aggressive evangelizing agency which can without challenge do its earnest work in the State Universities and professional schools, save some such arm of power as the College Young Men's Christian Association."[9]

Yet while the dearth of professing collegiate Christians was unquestionably the core "problem" for national leaders, they were equally concerned that "saved" students were not being trained to save others through active and practical Christian evangelism.[10] In part, it was the academic bent of college students that was blamed for rendering Christian students unproductive in their religious work. Without denigrating the life of the mind, both Wishard and Ober contended that the collegiate setting promoted an unhealthy reliance upon thought rather than action. Consumed with intellectual debate and the study of arcane facts, students were deprived of the call upon their lives to engage in vibrant Christian activity in spreading the gospel. "To be sure by having saved them you have done them an inestimable service," Wishard reminded leaders, "but upon you rests the responsibilities of preparing them to save others. It rests with you to decide whether they shall be merely a garrison or an aggressive army."[11] Because students within the academy had a tendency to be "theoretical rather than practical," Ober noted, they required contact with a "practical, wide-awake, dead-in-earnest body of men" such as could be found in the YMCA.[12]

In addition to this academic bias, however, Wishard and other YMCA leaders were also concerned that practical religious work would be impeded by the selfishness and parochialism of the campus environment. If leaders witnessed a broad complacency about practical evangelistic work on the campus, in other words, they also recognized a devastating blindness to the religious needs of individuals residing outside of the ivory tower. Because of students' relative economic wealth and academic privilege, leaders of the intercollegiate movement were fearful that students exposed to such rare opportunities would be tempted to keep the benefits of their education to themselves, using academic elevation as a means to secure their own positions in life. More to the point, in their self-contained academic coterie removed from real life, students were left unprepared for the soul work among common people that would mark their postgraduate labors. While many cherished the isolated nature of collegiate life that was beginning to dissolve, Ober noted that students' cloistered existence possessed significant spiritual liabilities:

> College life is so different from the outside world; we are so much a little world to ourselves; have given so much of our time to history, abstract sciences and the dead languages; have inherited so many college traditions and customs of the middle ages—that the average college student knows very little about, and sympathizes still less with, the ways of the great world outside the college walls.[13]

It was for this reason, in fact, that Ober claimed that an organization external to the collegiate environment was so essential to the spiritual development of college students. While the older religious societies with "local name and exclusive methods" threatened to limit Christian work to the provincial college setting, membership in a cosmopolitan and intergenerational organization like the YMCA would convert the "easy and wasteful and purposeless lives" of "secluded students" into channels "where work is done for God."[14] Between the initial 1877 gathering at Louisville and the next biannual convention held in Baltimore in 1879, therefore, college YMCA leaders attempted to fashion a curriculum that would allow local chapters to embrace the larger association purpose "to save men in college and to send men out to save men."[15]

The Curriculum of Student Evangelism

The primary locus of the YMCA's evangelistic ministry was "personal work," purposeful and relational evangelistic contacts with fellow students. Unlike the event-oriented revivals of an earlier era, the YMCA sought primarily to mobilize individuals to serve as personal missionaries to their unbelieving peers. Though the goal of conversion was constant, methods varied based on local conditions. At Indiana Asbury University in 1879, the YMCA formed a committee consisting of one converted student for every ten unconverted students, the duty of each to bring those ten to salvation. At Lincoln University in 1884, the YMCA personal work committee held meetings for curious students and saw significant results. Of the forty-two collegians "out of Christ" at this time, forty were present to ask for prayer during the meetings, twenty-four professing salvation. At Ohio Wesleyan, a vigorous 1883 freshman YMCA evangelistic campaign resulted in over two hundred conversions during the first three weeks of the semester.[16] Leaders on the Princeton campus regularly articulated this desired end, noting that every student should "at his admittance become a missionary" with his greater ambition "the personal conversion of every man in his class." "If all were to do their duty in this particular," the leaders suggested, "membership in the college and in the society would be coordinate and coterminous."[17] If statistics were accurate, the number of students professing Christian faith through the personal work of association members was quite remarkable. Within the first two years of the movement's existence, 500 students had experienced Christian conversion under the auspices of association personal work. In each of the succeeding years, these numbers escalated even further. In 1879, 700 students were brought into the faith, followed by 1,100 students in 1880, 1,500 students in 1882, and 1,800 students in 1885. Such realities point to the fact that, while publicized campus revivals may indeed have diminished after the Civil War, Christian conversion through extracurricular relational evangelism was still a potent reality for college students in the late nineteenth century.[18]

Yet while such statistics might have generated celebratory remarks, Wishard was always dissatisfied with the tallies. "The gladness we feel in contemplating this wonderful result," he noted of the 1,500 students accepting Christ in 1883, "does not banish altogether the feeling of sorrow which comes as we think of the more than 30,000 students who reached the end of the college years unconverted, and far less liable to accept the Saviour than at the beginning of the year, every day of rejection only increasing the hardness of their heart."[19] One year later, looking at the movement since 1877, he noted that, "Over seven thousand students have professed Christ during the past seven years, but over seventy thousand have during this time left college unconverted. . . . Shall we continue to see such a mighty force of cultured men sent into the world to operate against Christ?" "There was," he concluded, "much land yet to be possessed."[20]

Campus chapters were directed to form student-led committees to facilitate personal work, local town evangelism, devotional meetings, prayer, Bible study, missionary meetings, intercollegiate correspondence, and provision for an association room. Yet while the personal and community evangelism committees were obviously attuned to the need for conversion-oriented work, it is equally clear that every aspect of the association program was dedicated to this soul-saving vision.[21] Prayer meetings, for example, were self-consciously directed to this end. Colleges had been celebrating, at least since the 1750s, annual Day of Prayer events, and class prayer meetings were common—if sometimes poorly attended—in most institutions. Yet by the 1880s, the traditional Day of Prayer, typically held on the last Thursday of January, was beginning to fall upon hard times. In some cases, the Day remained as a required activity, linked to class cancellations, prayer meetings, and sermons by faculty members or outside speakers. In other places, however, the event was either terminated or mocked at by student attendees. At Dartmouth, the student newspaper observed: "Among students, the 'Day of Prayer for Colleges' . . . is not observed by prayer, to any appreciable extent, but it furnishes an opportunity for the student to seek diversion in various ways, and consequently the day is welcomed and unanimously voted a good institution." The *Bowdoin Orient* similarly noted that "the meaning and sense of the day has long since been forgotten, and its only significance is that some well-known divine preaches in King's Chapel before about a dozen students and a couple hundred old ladies living around about Brunswick."[22] While maintained on many campuses as a vestigal token of divine affirmation and blessing, the vigor of the event had seemingly passed.

Rather than terminate this historic exercise completely, however, many administrators simply passed this event off to the campus YMCA chapters. Of 181 associations in 1884, for example, 107 helped organize the Day of Prayer activities for their campuses. YMCA leaders saw the Day as an opportunity to enhance both prayers for unconverted friends and personal work to speak to others about their souls.[23] While some were fearful that setting up this day would lessen the impulse for continual evangelism throughout the year, few could argue with the results. At Brown University, students described an "unusual amount of interest" in spiritual matters after the Day of Prayer, while at the University of Virginia leaders dated a massive outbreak of personal work to the Day. Oberlin dated the "best Christian work of the term" from the Day of Prayer, and in February of 1880, the associations at Dennison University and the University of Mississippi recorded major revivals. Taking part in "one full day's work for Jesus," the lion's share of conversions came in conjunction with this winter appeal. As Wishard contended, "To many of the unconverted, this day proves to be their 'spiritual birthday.' "[24]

While the Day of Prayer served to raise awareness about the gospel and provided a venue for enhanced personal work, however, leaders recognized that it was hardly effective as a long-term solution to the evangelistic gap on and off the campus. From the beginning of the movement, Wishard desired to make regular, systematic Bible study the cornerstone of preparation for active evangelism. On a more academic level, he suggested reading through the Bible from beginning to end each year to gain a broad sense of soteriological themes related to the Cross, the blood of Christ, and the Holy Spirit. He also recommended that students read the Bible devotionally, stating that daily enrichment through the biographical analysis of biblical characters would forestall the spiritual "dyspepsia" so common to the collegiate experience.[25] Yet despite the importance of such work, these elements were easily dwarfed by the evangelistic utilization of the biblical text. Sunday schools, he noted, were effectively communicating the intellectual and devotional components of Bible study. "The especial purpose of the College Young Men's Christian Association Bible study," however, was "to so familiarize the student with the Word and its use as to render him skillful in impressing upon his unconverted fellow students the necessity and the way of salvation."[26] Rather than replicating what Sunday schools were already teaching, in other words, Wishard wanted to relate biblical knowledge to the personal work theme of the movement as a whole. At Amherst and Yale in the 1880s, for example, approximately a hundred students met once each week in groups of five "to study the Word, with the view of ascertaining its answers to the various objections offered for not accepting Christ." It was the practical use of the "sword of the Spirit" for evangelistic competency that became the central purpose of association Bible study.[27]

In order to formalize this purpose, Wishard, together with Ober and YMCA leader Richard Morse, traveled to evangelist D.L. Moody's home to prepare a series of Bible study outlines that could be used by the college movement. The main purpose of the thirty-three-lesson undertaking, Wishard pointed out, was to teach students "how to use the Scriptures in winning unconverted men, one by one, for Christ." Along these lines, the group included a series of lessons on "fundamental Scripture Truth," including issues such as "Christ our Savior," "The Holy Spirit," "Confession of Sin," and "Repentance."[28] Yet they were also clear that biblical and doctrinal knowledge alone did not guarantee evangelistic success. Therefore, a second series of lessons was devoted to the study of "individual cases." The lessons were to prepare students to deal with unique "types" who rejected the message of Christianity, in this case represented by the ignorant, the doubting, the self-satisfied, the fearful, the discouraged, the willful, and backsliders, among others. While Wishard noted that the variety of such cases was infinite, he also suggested that the class focus upon actual, rather than imaginary, individuals during these sessions. In the last ten minutes, therefore, students reported on evangelistic encounters, detailing difficulties engendered by each interaction that could be discussed the following week. Using the Scriptures to diagnose spiritual maladies, such practical apologetics converted the Bible from a theological and devotional text into a sourcebook for personal evangelistic work. Offering the curriculum to groups around the country, about 13,000 copies were sold within two years.[29]

While such programs surely dominated the association vision in these years, the provision of attractive physical space for such activities was also envisioned as a means of fostering evangelistic outreach. Wishard was convinced, in fact, that students would judge Christianity in part on the basis of the physical beauty of YMCA facilities, associating a dignified physical setting with an equally dignified faith. "Don't expect men who have little interest in spiritual things," he warned, "to find anything especially attractive in an old lecture room, where the fire has gone out, the only light a lamp whose chimney is rarely cleaned, hard benches and a blackboard; every spot in the room suggesting some well executed College joke or a 'flunk' in recitation, and a severe and well deserved reprimand flung from behind the Professorial desk."[30] To this end, Wishard suggested that a student committee be formed to raise funds, deal with administrators, and locate and furnish a room that would be aesthetically pleasing and functional. In attempting to secure adequate facilities, students often appealed to administrators by noting that religious organizations should be placed on par with other campus extracurricular activities. At Princeton in 1878, the organization asked for a room that would provide a central gathering place for discussion, prayer, and Bible study by citing the elaborate facilities afforded to the campus literary society. "If meeting for the reading of Shakespeare and Milton are beneficial," they questioned, "why not those for the study of the Bible and for prayer?"[31] Such attempts seemed to be quite effective. At the University of Illinois in the 1880s, the campus YMCA meetings took place in a "cold room" on the top floor of the chemistry building, and at least one leader blamed the small attendance on this factor. When Association members found a new room with a stove and more pleasing appearance, they saw attendance soar.[32] While classrooms were often still used for meetings on many campuses, Wishard was clear that such facilities were inadequate for soul work.

Exporting the Vision

In keeping with Wishard's desire to equip students for evangelistic work beyond the campus, by the latter half of the 1880s, students were increasingly encouraged to export their vision of "personal work" both to other campuses and to the surrounding communities. From the very beginning of the movement, Wishard promoted the idea of "neighborhood work," local evangelistic outreach in jails, Sunday schools, and hospitals.[33] While chapters engaged this work in a variety of ways, perhaps the chief means of operationalizing this evangelistic vision came through deputations, teams of student evangelists commissioned to Christian work off campus. The idea for such efforts came most directly through the visit of an influential delegation led by Henry Drummond from the University of Edinburgh. In the wake of crusade evangelistic work with D.L. Moody, Drummond initiated groups of "holiday missioners," faculty and students who traveled to local towns in the Scottish lowlands in order to take part in evangelistic preaching during college breaks. This scholar soon began sending similar groups of students to other colleges and universities—notably Oxford—in order to share the gospel message with interested peers. Fresh off a series of religious revivals, this team hoped to spark similar enthusiasm across the Atlantic.[34]

In the fall of 1887, Yale, in response to Drummond's appeal, sent out a number of deputation teams consisting of three or four students to communicate the gospel in other colleges, city missions, Sunday schools, young people's meetings, and preparatory schools. The object, they noted, was not "exhortation or the discussion of theological problems" but rather a personal explanation of the change in their own lives through the work of Christ.[35] Within the next two years, many YMCA chapters nationwide sponsored similar work among other colleges and in neighboring communities. In 1889, Harvard YMCA leaders noted that they had sent out "several deputations" during the fall semester, while Princeton had seen groups go to Elizabeth, New Jersey, and New York City. Nor was this simply a movement among elite New England colleges. In 1889, Knox College students initiated a movement not unlike Drummond's holiday mission by traveling during the summer to twenty-eight towns and giving evangelistic talks in churches, Sunday schools, and young people's societies. Such work surely enthralled Wishard, who saw this as a tremendous illustration, not only of active religious work, but also of students' ability to move beyond the provincial "ivory tower" walls of the colleges.[36]

The excitement over local evangelistic deputations was largely paralleled in the 1880s by a growing passion for global evangelism. Despite the fact that many YMCA chapters emerged directly from Societies of Missionary Inquiry, explicit missionary training and education was not represented in the initial curricular planning at Louisville. Wishard's own interest in career missionary work notwithstanding, the early leaders of the movement saw missionary activity as critical but perhaps relevant only to those preparing for the foreign field. Leaders of the movement at the University of Michigan, in fact, noted in 1882 that "Our organization was born just twenty-five years ago of a dying parent known as the 'Society of Missionary Inquiry,' but as far as is known to the present generation of students, very little of that parent's own proper spirit has manifested itself in the offspring prior to its twenty-fourth year."[37] By 1880, several within the association were speaking of the "corporate indifference" to foreign missions within the college movement, noting that in their enthusiasm for campus and local evangelism they had seemingly neglected the last recorded words of Jesus to "go into all the world" to preach the gospel.[38] Perhaps a reflexive reaction of those attempting to highlight the distinctions between the new movement and former missionary societies, foreign work was viewed initially as a movement elective.

This exclusion, however, was quickly overturned. During his study of church history at Union Theological Seminary, Wishard came across the account of the origins of collegiate missionary enthusiasm through the Haystack meetings held at Williams and Amherst, and he read a host of other books detailing the missionary exploits of former collegians. "It is easy enough to imagine how one who was directing a Student Movement in the last quarter of the century would be impressed by the deeds of the Christian students of the first quarter," Wishard later noted of his historical studies. "We were their lineal spiritual descendants and successors; what they had begun it was ours to complete; they had sown in the century's early springtime; we were to reap what they had sown in the century's autumn."[39] For Wishard, the accounts of early collegiate missionary efforts sparked a sense of obligation for rekindling the dying embers of this initial enthusiasm. While their forebears had lacked nothing in passion and ambition, they did lack the organizational machinery and the intercollegiate bond that now existed in the YMCA movement to make their visions a reality.[40] "The work of each association," Wishard now noted, was to be "local, then neighborhood, then national, international, and world wide," or in biblical terms, "first to Jerusalem, then Judea, then Samaria, then the ends of the earth."[41]

In addition to the influences at Union, Wishard was spurred on by former college roommate Robert M. Mateer, one of the authors of the original Louisville circular in 1877. A student at Princeton Seminary, Mateer and other like-minded seminarians began planning in 1879 for a gathering of theological students interested in foreign missionary labor. Two hundred and fifty students from thirty seminaries gathered in New Brunswick, New Jersey in 1880 to give birth to the Inter-Seminary Missionary Alliance (ISMA), and by 1883 fifty seminaries were allied in this work. While Mateer and other ISMA leaders urged the YMCA staff to funnel missionary interest through the new organization, Wishard was uncomfortable with such surrender. Together with Ober, Wishard did attend many of the early ISMA conferences as a "corresponding member," and he urged interested students to attend with them. However, both Wishard and Ober refused to form a partnership with the ISMA because they feared a dilution of missionary interest in the local YMCA chapters. Wishard instead heralded the movement as a clarion call to the colleges to follow their example. "The relation of our colleges to Missions is really more important than that of our Theological Seminaries from the fact that the men who fill the Seminaries pass through the Colleges," he noted. For Wishard, missionary enthusiasm sparked in college would bear fruit in the seminary for those attending these institutions. However, since an estimated nineteen out of every twenty college students would not pursue advanced theological education, the college YMCA likely provided the best opportunity to offer this challenge to a broad constituency.[42]

Wishard used a number of approaches to fan the flame of missionary interest. He invited missionary spokespersons such as J.E.K. Studd to tour the American colleges in order to spark enthusiasm.[43] In addition, he advocated for a vigorous program of missionary education on each campus. Association reading rooms were therefore to be well stocked with missionary papers, maps, magazines, history, and biography.[44] Furthermore, students took part in monthly missionary meetings, by 1880 a staple of local chapters nationwide. Held either in place of one of the weekly religious meetings or as a separate event, each missionary meeting was devoted to a particular field, taking up three central questions: "What are the peculiar characteristics of the field?" "What methods are employed by the missionary in prosecuting the work?" and "What are the results?"[45] In 1881, for example, chapters were urged to follow the prescription of the missionary periodical *The Gospel in All Lands* in analyzing the characteristics of American Indians, Roman Catholics of Europe, Africans, Roman Catholics of America, the Chinese, the Oriental Churches, the Japanese, the Jews, the races of India, the Mohammedans, the Malays, and "unoccupied mission fields" in order to enhance missionary efforts to reach such groups. Wishard also recommended missionary hymns, including such traditional stand-bys as "From Greenland's Icy Mountains," "The Morning Light is Breaking," and "There's a Cry from Macedonia." Such "war songs," he noted, "never grow old."[46]

Spread throughout the YMCA's evangelistic curriculum was a broader concern that the vocational aspirations of collegians be aligned with the soul saving proclivities of the movement. Leaders regularly urged students to pursue careers as pastors, missionaries, and YMCA general secretaries, commending these vocations above competing "secular" alternatives. Even when other vocations were selected, Wishard was certain that they could be directed towards evangelistic work. He saw medical students, for example, as a potentially critical force for world missions. By 1883, leaders sponsored a YMCA conference in Lake Forest, Illinois among Chicago-area medical colleges for the expressed purpose of fostering medical students' interest in missionary fields. By December of that year, 600 had gathered from the seven medical colleges in and around Chicago, one hundred declaring their intent to go to the mission field "anywhere, anytime, to do anything for Jesus." Similar meetings were held in other urban centers—notably in New York in December of 1884 and Philadelphia in 1885—with similar numbers and similar results.[47] Attempting to secure students for such positions, YMCA leaders revealed that their emphasis was to equip students for explicitly religious careers, specifically those aimed at gospel proclamation. While they did not denigrate alternative careers, the association existed in part, YMCA leaders contended, to raise up among educated Americans the future evangelistic leadership of the Protestant church.

Sharing the Vision

Wishard recognized that the essential genius of the evangelistic YMCA program could be secured and extended only through inspiration provided by intercollegiate relationships and contact with religious leaders outside the college. The *College Bulletin*, of course, was one means of providing timely news regarding work on other campuses and throughout the YMCA. Wishard regularly reported local campus YMCA events, conversion statistics, and prayer needs, and he never neglected to provide descriptions of international and state YMCA conventions on the pages of this monthly periodical. At the same time, he desired local chapters to take initiative in maintaining relationships with other campus chapters, sharing needs and praying for other clubs during regular prayer meetings.[48] Wishard also recommended letter writing, noting repeatedly that, like fraternity correspondence, these letters should be "full of encouragement," not formal or stiff but "warm-hearted, familiar letters, as are exchanged between friends."[49] Also encouraged were visits between local associations in close proximity, the leaders at Amherst noting, "We cannot overestimate the importance of systematic Inter-Collegiate Visitation. . . . We see no reason why the Associations of neighboring colleges should not visit each other as often as the baseball nines."[50]

Because isolation was in part a factor of college life, however, Wishard suggested that correspondence and visitation would never suffice. He realized early on, in fact, that students would require more explicit connections with religious leaders outside of the college in order to transcend parochial tendencies. In order to enhance both intercollegiate relationships and contact with practical workers, each active member was entitled to a "College Vacation Ticket," priced at about five cents in the mid-1880s, which allowed the student to enjoy the privileges of local town YMCA facilities and programs during college breaks.[51] In addition, Wishard vigorously championed the importance of student attendance at state and international YMCA conferences. Such meetings, leaders proposed, would place students in contact with Christian leaders who were experts in evangelism and personal work among young men. Blending practical training in such work with intercollegiate communion, conferences were envisioned within the movement as critical sources of inspiration, teaching, and the personal networking essential for sustained enthusiasm. During the first ten years of the movement, biannual YMCA conferences in Baltimore (1879), Cleveland (1881), Milwaukee (1883), and Atlanta (1885) proved to be important locations for delegates to discuss critical topics and make programmatic decisions. While most of the time was taken up attending sessions with other YMCA-affiliated groups, there was always time devoted to meetings designed exclusively for collegians. State conferences functioned similarly, with time divided between general YMCA discussions and those specifically devoted to the collegiate work.[52]

Despite Wishard's initial enthusiasm, however, these attempts did not satisfy the students. The source of discontent was the lack of sufficient time for college-specific concerns in the hectic agenda of a multifaceted organization. At the state conference of Massachusetts in 1882, students were forced to devote an extra day just to the college movement because they were unable to cover necessary topics in the allotted time. In 1883, they decided to start their own conference for New England college YMCAs. Reminiscing four years later, De'W.C. Huntington, student president of the Yale YMCA, noted that "there is such a wide difference between the interest of college men and the interest of young men at large, that they felt in a great measure shut out from a close participation in the discussions and interests of those meetings and looked forward to some opportunity for conference upon topics pertaining to the peculiar circumstances of college men."[53] From this time onward, state college conferences independent of other YMCA groups grew increasingly common, with Iowa and Wisconsin also moving in this direction in 1883 and many other states adopting the model in 1884 and 1885. The success of such gatherings quickly became evident when the numbers of students participating began exceeding figures for previous YMCA-wide state and national conventions. The New England conference alone attracted 120 students in 1885, and that year approximately a thousand students attended college state conferences around the country. It appeared that while students desired to be intercollegiate, they were less inclined to join those in the YMCA outside the college realm.[54]

The excitement generated by these exclusively collegiate state conventions was not lost upon Wishard. Despite his desire to see college students break free of exclusivity and make connections with lay leaders beyond the ivory tower, he also recognized that the bonds created in these more focused meetings were critical for extending and deepening the college movement. Revivals were frequent within the college-specific contexts. Extended time was provided for in-depth conversations on college problems and program solutions. And while those participating in the college state conferences had lost some of the interaction with those taking part in other dimensions of the YMCA, they were still privileged to hear from lay speakers who would "transmit the enthusiasm, the well-tried methods, the wide-awake aggressiveness of business men to the somewhat exclusive monotonous routine of student life." Thus, he reasoned, the benefits of "non-provincial" messages might still be gained even as the liabilities of all-inclusive conferences were surmounted.[55] While many state leaders criticized the college YMCA for its new separatism, Wishard conceded that college students would require intercollegiate fellowship on a more exclusive basis to maintain interest in the work.

In the wake of these successful state efforts, therefore, Wishard began to wonder about the potential for a similar move at the national level. While students had continued to faithfully send delegates to the biannual international YMCA conferences, the timing of the May events, almost always near the end of the academic year, limited student attendance. While the eighty-three delegates at the 1881 International YMCA Convention in Cleveland represented a threefold increase over Louisville in 1877, this total was proportionately quite small for a growing national movement that now boasted a student membership of 8,491. In addition, critiques regarding the lack of focused time for collegiate fellowship and discussion were growing in number and force. Wishard's attempts to expand the time allotted to college work within the international conferences had been largely rebuffed. At the 1881 gathering, for example, Wishard submitted a proposal asking for the convention to devote a single day to the work in the colleges, but the idea was rejected "because of the feeling that such an arrangement might create criticism by giving undue and disproportionate emphasis to the college work."[56] While the rationale was understandable, Wishard began to question the adequacy of these settings for stimulating enthusiasm for the collegiate movement nationally.[57]

Wishard's nascent thoughts regarding a separate national college conference were stimulated during his time preparing Bible study materials with Ober and Morse at Moody's home in Northfield, Massachusetts. During their stay, he remarked to Ober that the sprawling landscape of this New England homestead might provide an ideal setting for a national collegiate gathering. Moody was no stranger to work with collegians. Throughout the 1870s and 1880s, he drew large student crowds during crusades in Great Britain, Scotland, and elite New England colleges such as Yale, Princeton, Harvard, and Dartmouth. In addition, Moody proved to be an ardent supporter of YMCA college work, soliciting funds for Ober's salary and for the missionary deputations of J.E.K. Studd.[58] Despite these links, Wishard's initial request for a student summer conference was "crossly" rejected by the evangelist. Moody was doubtful that college men would freely give up their vacation time for a conference on Bible study and religious work. In addition, despite his previous successes, Moody conveyed a profound feeling of inadequacy in working with academically oriented collegians, largely stemming from his own lack of educational background. Wishard attempted to reassure him on both counts. He noted that Moody's home at Mt. Hermon provided an ideal setting for a combination of recreation and religious training, stating that "the river and the hills and such ball games and races as we could arrange" would give students a true vacation even as they were trained in life and thought. On the second count of Moody's supposed inadequacy, Wishard reminded him not only of his previous success with students but also of the fact that students were in need, not of intellectual acumen, but of inspiring lay religious activism. Devoid of such influences on the campus, he assured Moody that there was nobody better equipped to address collegians on these themes.[59]

Upon the evangelist's agreement in 1886, Wishard and Ober actively recruited delegates from institutions across the country.[60] Fueled by the more widespread availability of rail travel, approximately 235 students from 89 colleges attended the initial conference, spending 26 days together studying the Bible and evangelistic work. In subsequent years, Wishard spent a great deal of effort attempting to stimulate the colleges to send larger delegations. To accommodate more students, the conference was moved from Mt. Hermon to Moody's 300-acre property at Northfield. The conference time was also cut in half (to two weeks) and scheduled immediately after commencement.[61] Recognizing financial limitations, Wishard orchestrated deals with the railroads to secure discounts and, since the price of one dollar per day for room and board proved to be exorbitant for some, made available 10 × 12 tents so that students could sleep along Northfield's grassy slopes. Such efforts appeared to pay off. While various campaigns to gather a thousand students for the summer conference fell short of realization, by the time of the 1887 conference, 269 students from 82 colleges were in attendance at the new Northfield campus, and the numbers then escalated to 400 in 1888 (representing 100 colleges) and 500 in 1889 (representing 126 colleges). After the 1886 conference, gatherings increasingly drew internationally as well. In 1887, the conference played host to two students from England and fifteen from Canada. By 1889, there were fifty-six foreign students including delegations of fourteen students from six British universities, nine representing Scotland and Ireland, and twenty-two from Japan (recruited by Wishard himself) in addition to the eleven attending from Canadian schools. In many ways, Northfield became an important locus of transatlantic and global student religious impulse that would have far-reaching implications in the coming years.[62]

Selling the Vision at Northfield

More than any other dimension of the association program, the summer conferences encapsulated Wishard's programmatic ideals. In fact, it was in this venue that prayer, Bible study, missions, intercollegiate correspondence, and training for personal work all came together in a single "event." Though Moody's desire for flexibility militated against a rigid daily schedule, the 8 a.m. meetings were taken up with discussions, led by Wishard and others, designed to train students in methods of evangelistic Bible study, personal evangelism, and general YMCA college work, analyzing such topics as "How to Reach and Hold New Students," "The Evangelistic Meeting," "Systematic Personal Work," "Deputation Work," and "The Bible Training Class."[63] At 10 a.m. and in select evening sessions, invited speakers addressed students on biblical exposition and learned defenses of Christian doctrines such as the veracity and verbal inspiration of Scripture, the nature of sin, the exclusivity of Christ, and the bodily resurrection of the dead. Seeking speakers who could provide inspiration and yet also relate well to the academic bent of collegians, Moody tapped such luminaries as James H. Brookes, A.J. Gordon, Arthur T. Pierson, William Rainey Harper, Henry Clay Trumbull, Hudson Taylor, S.M. Sayford, and James McCormick in the first three years of the conferences. Many of these were Moody's friends, and most shared his theological predilections for premillenialism and dispensationalism.[64] For the students, the messages served as vigorous appeals to maintain orthodoxy amidst the emerging bromides of liberal theology just beginning to exert significant influence in the American colleges.

While such themes were highlighted, Moody and several of his invited guests also focused upon the inadequacy of human knowledge and effort for achieving real spiritual results. One of the pervasive themes of the conference in the early years was the need for consecration, prayer, and the power of the Holy Spirit in preparing one for Christian work. Students were urged, individually and corporately, to pray and wait for the Holy Spirit to fully claim their hearts and minds. While the Spirit dwelled in each Christian, Moody noted, some lacked the power and "unction" of the Spirit in their lives, resembling machinery devoid of steam.[65] Frequently utilizing examples from the Old Testament, the evangelist warned students against dependence upon their own intellectual prowess, reminding them that only God's power could help bring student peers to salvation. In response, a number of students met during the final four days of the initial conference at six o'clock in the morning to "wait upon the Holy Spirit" in prayer. Linked to Moody's own experience of an "enduement of power" through the Holy Spirit and to the larger holiness emphases of the Keswick movement and its passion for the "higher Christian life," this spiritual tone was one of the distinct marks of the Northfield experience. Leaders sought, and students pursued, a "power from on high."[66]

Perhaps the greatest legacy of the Northfield conferences related to their nourishment of the foreign missionary emphasis of the movement. The impetus for discussion on such topics came from Robert Wilder, a recent Princeton graduate who had been invited by Wishard to attend despite this alumni status. Wilder's father, Royal, had been a member of the Andover "Society of Inquiry on the Subject of Missions" and had served as a missionary to India before assuming editorial leadership of the new *Missionary Review*. Preparing for career missionary work, Robert entered Princeton in 1881, where he excelled as a Phi Beta Kappa student in Greek and Philosophy, as president and secretary of the YMCA, and as an officer of the Prohibition Society that successfully closed every saloon in Princeton. Stemming from an 1883 conference of the ISMA, Wilder and a few like-minded students initiated the Princeton Foreign Missionary Society, an organization devoted to the study of missionary fields, prayer for global evangelism, and personal work on campus. While the organization had become well established by the time of Wilder's graduation, he was not content with a local missionary society. In his recollections, he pointed out that a global missionary uprising in Demark, Norway, Sweden, and Great Britain in the 1880s had elevated his hopes of collective efforts.[67] Wilder envisioned an American missionary uprising, joining with these other agencies to generate a global student missionary movement to redeem the world "in this generation." He hoped that Mt. Hermon would be the launching pad for such a movement.[68]

While discouraged by the lack of attention to foreign missions in the stated conference agenda, Wilder asked Moody if the students might be able to take one of the evenings to address missionary topics. Three sons of missionaries and seven foreign students representing different countries presented information regarding the needs and opportunities for missionary labor. The students were then asked if they were willing to sign the Princeton Foreign Missionary Society declaration to affirm a life commitment, barring unforeseen circumstances, to career missionary work. Throughout the remaining days, a total of ninety-nine students signed the declaration. When another signed on the last day, the "Mt. Hermon Hundred" had become the seed bed of Wilder's desired national movement.[69] Together with John Forman, another Princeton alumnus, Wilder visited 162 institutions after the conference in order to recruit students for careers in missionary work.[70] Some critics ridiculed the pressured pledge techniques and suggested that they were forcing young students into premature vocational decisions. Others accused them of denigrating alternative "secular" pursuits as less "Christian" than missionary work. Yet this deputation was remarkably successful, securing 2,106 volunteers (including 500 women) in 1886–1887 alone. Princeton President James McCosh, reflecting upon the work of these two Princeton alumni, effused praise: "I am amazed at their success. . . . Has any such offering of living men and women been presented in this age—in this country—in any age, or any country, since the day of Pentecost?"[71]

Not all students honored their pledges. Some rescinded their promises after receiving more information about the extensive requirements of denominational mission boards. Others experienced parental opposition or simply succumbed to the temptations of more lucrative professions. Wilder acknowledged from the beginning that there would be "some chaff among the wheat," but he also recognized that local missionary bands would be essential to sustaining the missionary fire for those who were serious about this work.[72] Some campuses fashioned their own committees in the wake of the missionary revival, but none was connected to the larger YMCA. College association leaders grew fearful that this missionary "gusher," as it was often called, would generate a cleavage in collegiate religious work, reifying the fragmented efforts of an earlier era. For Ober, the great fear was that missionary agencies would "de-missionaryize" the general associations, sapping global zeal from the collegiate mainstream. It was advised that those joining missionary bands instead comprise the missionary committee of the college YMCA and thus avoid becoming a segregated organization.[73] "The hope of making the movement a success," Ober concluded, "was to make the college Associations responsible for it."[74]

The resulting Student Volunteer Movement for Foreign Missions (SVM), a joint effort of the YMCA, YWCA, and the ISMA, was designed as a means of maintaining and standardizing the work of this rapidly growing movement.[75] Locally, volunteers worked through missionary committees of the YMCA in all campus-wide missionary education and recruiting, but they were also free to have their own meetings for encouragement and fellowship. By 1891, 6,000 students had signed the pledge of commitment and 320 recent graduates had already "sailed" to missionary lands. In addition, college associations had begun to initiate missionary fund drives for the financial support of alumni on the field. Princeton students pledged $3,800 within twenty-eight hours in order to support John Forman for missionary work in India, and between September of 1888 and June of 1889, over $30,000 was raised for missionary causes through Wilder's appeals. With such an optimistic response, it is easy to see why the SVM leadership chose to embrace as their "watchword" the Princeton Foreign Missionary Society appeal: "the evangelization of the world in this generation."[76]

The missionary enthusiasm spawned by campus missionary meetings, medical missionary conferences, the Mt. Hermon and Northfield gatherings, and the SVM surely confirmed in Wishard's own mind the wisdom of his decision to multiply evangelistic work through the college students of the nation. Anchored by model constitutions, joint conferences, an intercollegiate journal, and prolific visitation, he was able to produce a highly systematized and reproducible model of an association "curriculum" that could absorb local variables even as it assimilated groups within a common structure. Through prayer, Bible study, deputations, the development of rooms, missionary meetings, and direct personal work, the organization between 1877 and 1888 was deeply committed to save the souls of students and to train collegians to save the souls of those inside and outside the college.[77]

In this sense, the college YMCA represented the mainstream vision of Protestant evangelicalism on a number of levels. For example, the college YMCA was a perfect expression of the primacy of personal salvation in the religious life. While such themes were indebted to the Protestant revivalism of the Second Great Awakening, they bore the characteristic stamp of evangelical ideals heralded by the revivals of 1858. As Kathryn Long has argued with regard to this era, the spirit of revivalism was conservative and individualistic in nature. Unlike the earlier New England revival tradition, which combined "conversionist piety" and social reform, the revival of 1857–1858 marked a shift to the importance of individual evangelistic conversion and "shared devotional piety."[78] True to this reality, the focus on the soul and divine regeneration in the YMCA clearly overshadowed social involvement. Even when students were involved in "neighborhood work" and deputations outside the colleges, it was clear that the emphasis was not on social service or social justice but personal heart salvation. If there was a vision common to the early intercollegiate YMCA, it was the conviction that social betterment would be fostered by the conversion of individuals. This form of evangelical consciousness—described by historian Mark Noll as activistic, immediatistic, and individualistic—was the primary framework for the association program in these years.[79]

In addition, the practical and activist thrust of the college associations clearly identified the movement with evangelical patterns of this era. Wishard's call to practical "religious work" among students was fueled by a sense that the academic culture of the colleges would promote stagnation in the Christian life, threatening cloistered youth with an enervating intellectualism. Wishard hoped to demonstrate to students that Christianity was an active religion, that it found its primary calling in outward expressions of evangelistic fervor rather than in the theoretical confines of the life of the mind. Without question, this was a movement away from the practices of the earlier religious societies. As the historian of the YMCA at the University of Wisconsin put it, the organization "took the Christian religion from the theoretical shelf it had occupied in the religious societies and brought it to the level of everyday life" through practical evangelism.[80] The collegiate movement, if not wholly anti-intellectual, was certainly designed as an antidote to the stultifying intellectualistic culture of the colleges. As leaders of the University of Michigan association noted in 1883, they were attempting to convert "intellectual societies marked by discussion to devotional societies marked by spiritual conviction and practical action."[81]

While remaining firmly entrenched within the evangelical context during the next thirty years, some of this orientation would begin to change as the colleges responded to the broader cultural shifts and social crises of the Progressive era. By the 1890s, initial changes in the culture of higher education, coupled with new visions of Christianity, would begin to reorient the fundamental purposes and perspectives of the college YMCA movement. The practical activism that was directed toward evangelism in the 1870s and 1880s would soon be extended to other domains, and it was a newly minted Cornell University graduate, one of the original Mt. Hermon hundred, who would direct the association into its new era.

Section 2

Character and Service, 1888–1915

Chapter Three
Expanding the Boundaries

Among the many prominent students whose lives were changed through the first Northfield conference, Cornell sophomore John R. Mott stood as a singular character. Mott had been a charter member of the YMCA formed in 1883 at Upper Iowa University, where he excelled in debate and oratory but felt confined by the denominational rigidity of the school.[1] Seeking academic excellence and a more direct route to the professions, Mott decided to transfer after his freshman year to Cornell, enamored with this university's elective system and its reputation as a "free-thinking institution." Entering as a sophomore, Mott was greeted at the train station by members of the Cornell University Christian Association (CUCA) who assisted him in finding housing, registering for classes, and making contact with upperclassmen and faculty members. Won over by their personal attention, Mott immediately joined the organization and quickly rose within the ranks. By the end of his first semester on campus, he had already been elected vice president and noted in a letter home to his parents that it was his "determination to make our Association the best of all the college YMCA's in the United States."[2]

In addition to furnishing a context for leadership development and social standing, the Cornell YMCA also provided a fertile ground for the growth of Mott's vocational vision. His parents desired that he return home after graduation to run his father's business as a lumber merchant. At the same time, both by inclination and talent, Mott felt drawn to the professions, particularly law and politics. "On the other hand," he noted, "there is a still small voice saying, 'go ye into all the world and preach the gospel to every creature.' . . . Can I in the face of such promptings turn from them?"[3] While Mott later recalled receiving guidance by opening randomly to a Bible passage that directed him to "turn many to righteousness," additional confirmation came through his discussions with English cricketer J.E.K. Studd. Studd, who had come to Cornell as an emissary of the college YMCA to address students about missionary work, had himself given up professional status, reputation, and wealth to proclaim the gospel as an itinerant missionary. Tortured by his own desire for a prestigious and lucrative career, Mott found in Studd's appeal to "seek first the Kingdom of God" the directed challenge to obey the "still small voice" leading him to vocational ministry. "When I gave up to God in this way I settled forever the question what I am to do in life," he wrote to his parents. "It is to be soul saving."[4] Mott was in many ways the model product of Luther Wishard's early YMCA program.

Mott's presence at Moody's first summer school at Mt. Hermon in 1886 proved to be of signal importance in directing his ministerial career aspirations.[5] As one of the "Mt. Hermon hundred," Mott was dedicated to the task of missions, though he was unsure what form it might take vocationally. Selected to travel along with Robert Wilder to recruit students for foreign missions, he eventually decided against this deputation in order to return to Cornell for his junior year as the CUCA president. A "hindered" Volunteer, Mott immersed himself in YMCA activities during his final two years as a student and saw the enrollment of the CUCA explode. The active membership during the 1886–1887 academic year had increased from 130 to 330, a fact Mott attributed to the vigorous personal work of his staff and fellow students. *The Association Bulletin*, the official organ of the movement at Cornell, was expanded from an eight-page paper to a twenty-page publication with cover and four pages of advertisements. Mott doubled membership in the White Cross Army, a program promoting personal holiness and sexual purity among university men. He greatly expanded student Bible study, added service work in the slums of Ithaca, and began a vigorous building campaign to improve facilities. By the end of his tenure in 1887, he could tell those critics who saw Cornell as a godless institution that the university was "no longer an infidel hot bed . . . its Christianity is more on the alert and more active than in many of the largest religious colleges in the East."[6] "I am often led to think," he concluded, "that this is the main reason that I was ever brought here."[7]

Still unclear about a precise ministry vocational direction, Mott soon found himself surrounded by multiple possibilities. The Cornell YMCA leadership asked him to come back to campus as the first paid YMCA secretary in the school's history. George L. Burr, librarian of the President White Library, asked Mott to join him for research in German and Latin manuscripts. At the same time, philosophy Professor Jacob Schurman asked him to apply for a fellowship. In light of Mott's love for Cornell and Ithaca, these offers must have held deep appeal. However, C.K. Ober, in the wake of Wishard's 1888 resignation, also offered Mott a position as an associate intercollegiate secretary.[8] Despite the difficulty of the decision and the apparent opposition of his parents and professors, Mott agreed to this national role for a one-year term, noting that experience in "practical work—dealing with all kinds of men, problems, and things" would be "a most helpful change after six years spent largely in book-work."[9] Graduating Phi Beta Kappa with honors in History, Mott left Cornell with a sense that he was leaving behind potential prestige in the academic world. When Ober left the movement just two years later, Mott's "one-year" agreement stretched into a twenty-four year tenure as the senior secretary of the college YMCA. As he noted years later, he "got in so deep" that he had "never been able to get out of it."[10]

A Growing Need

Anchored by his combined emphases on holiness pietism and entrepreneurial business acumen, Mott brought to his new post much of the energy and expansionist vision that had characterized his local efforts at Cornell. He was certainly impressed with the significant progress that had been made under Wishard and Ober, celebrating the consolidation of multiple independent societies into one "grand agency."[11] The growth within the YMCA had been quite remarkable throughout the 1880s. By 1890, Pennsylvania had more college chapters than the entire country possessed in 1877. In 1893, Yale alone had as many student members as could be found in the entire movement in 1877. Mott noted that, through the YMCA, there had been 20,000 conversions, 45,000 trained Christian laymen, 20,000 students entering the ministry, and 5,000 committed to foreign missions in the 6 years since Mt. Hermon.[12] Yet Mott was equally convinced that there was more ground to cover. He was displeased with surveys revealing that only 27,000 of over 80,000 men in the colleges were association members. Even more disturbing, however, were statistics demonstrating that only 21,000 of 40,000 Christians in the colleges belonged to the YMCA. Remarking that such figures divulged inefficiency, he asked rhetorically, "Should there be nearly as many Christians outside the Association as within?"[13]

If there had been compelling reasons for the YMCA to initiate college work in the 1870s and 1880s, Mott and his colleagues were quite certain that the dynamics of college life at the turn of the century made such a presence absolutely essential. Mott's leadership corresponded with the beginnings of a four-decade enrollment explosion in higher education. As historians Lawrence Veysey and Roger Geiger have suggested, 1890 marked a critical turning point in the move from the experiments of "idealist pioneers" to the rapid expansion, standardization, and prestige of university structures.[14] Taken as a whole, undergraduate enrollments in higher educational institutions grew from 62,839 in 1870 to 237,592 in 1900 and 355,430 by 1910. While only 3 percent of the eighteen to twenty-one-year-old population attended college in 1890, 4 percent attended in 1900 and about 5.12 percent by 1910. Many institutions were flourishing in this era, but the emerging state universities showed the most dramatic increases. Between 1890 and 1910, the population at state universities increased from 22,816 to 101,285, and in the relatively compressed time frame between 1895 and 1906, these universities grew 112 percent as compared to a 55 percent increase in other types of institutions. While it is true that students attending colleges and universities still represented a fairly small percentage of the college-age population (an even greater enrollment boom still lay in the post–World War I future), it was clear that higher education was beginning to emerge as a legitimate component of an increasingly articulated American educational ladder.[15]

Of particular importance was the growing number of middle-class collegians. While schools had previously attracted a polarized mix of wealthy students seeking to confirm status and the aspiring poor, by the 1890s the new American middle class began attending *en masse*. On the one hand, families with greater financial resources were now better able to forestall the income-earning potential of their adolescent sons. Additionally, these institutions were increasingly envisioned as means of ensuring a social mobility that would differentiate middle-class individuals from newly arrived immigrants. Propelled by the increasing importance of technical competence and scientific expertise, a growing complement of professions sought to generate objective credentialing systems that would elevate standardized, skilled performance.[16] In addition, the social world of the college provided a proving ground for those who desired to demonstrate traits of character and personality commensurate with professional success. As Helen Horowitz has contended regarding this era, the colleges provided a "trial run" for the business of the real world, settings where the student would "slough off provincialisms and learn an ease of manner and a style designed to ensure his acceptance in society."[17] While previously the colleges were seen as appropriate only for the formal professions of law, medicine, and divinity, the number of vocations requiring such education soon expanded greatly. Alongside more traditional professions now stood students preparing for jobs as merchants, manufacturers, chemists, farmers, and engineers. In addition, the advent of widespread coeducation highlighted the vocation of teaching, and colleges in the early twentieth century were graduating more teachers than any other profession.[18] The practical courses commended by earlier critics of the classical curriculum now had a market upon which to draw.

Beyond the importance of mere numerical increases, college YMCA leaders also pointed to the increasingly ambiguous moral and religious influence of college and university faculty. For example, some leaders recognized that the moral qualifications of faculty members had been lost as purposeful evaluative criteria for hiring. Writing for *The Intercollegian* (successor to Wishard's original *College Bulletin*), Professor Herbert Lyman Clark from Olivet College noted that professors should be blamed for "a very large proportion of the religious indifference in college" because they were no longer chosen "with reference to their power of molding the characters of their students" but only for their "academic knowledge and specialized expertise." Because of such hiring criteria, in many cases faculty were actually envisioned as morally corrupting, demonstrating personal habits sure to antagonize morally sensitive Christian students. Clark, in fact, suggested that students were frequently obliged to ask professors of high intellect but questionable character "not to use so much slang in lectures, to treat the young women with ordinary courtesy, and to show some respect for the Christian religion."[19] SVM leader Robert Speer noted to Mott in 1902 that one of the single most critical religious problems in universities was not the conversion of students, but rather the conversion of the "beer drinkers and the German specialists on our faculties." "There are," he suggested, "too many Sunday golf players, low standard men on the faculties now."[20]

Many others accused faculty members of turning students from the faith of their homes, using intellectual arguments to foster doubt and "rationalism" without providing credible religious alternatives. While YMCA leaders never attempted to discredit academic pursuits, they warned that certain professors, particularly those in departments of science and philosophy, were apt to lure impressionable students to embrace atheistic philosophies and reject the "simple faith" of their homes. As one leader put it,

> Particularly in some of our congregational colleges and state universities, the department of philosophy, in a spirit of intellectual conceit, is encouraging a type of liberalism and criticism which is little short of the rankest infidelity. The moral and religious teachings of the home and church are ruthlessly set aside and frequently nothing constructive given in their places. Great question marks are placed around such great fundamentals as prayer, the Bible, and even the Savior himself. The result is that consciences are deadened; the distinction between right and wrong made less vivid and real. Those who are already indifferent become more so, while sincere and earnest Christians are often put at sea in their religious thinking, and not infrequently their faith overthrown altogether. Not a small item in the seriousness of this situation is the fact that the men in charge of said departments of philosophy are often among the most popular men in the institution.[21]

Not all teachers, of course, were so destructive. By all accounts, faculty of state universities in the late nineteenth and early twentieth centuries were still broadly "Christian" in affiliation. In 1898–1899, for example, a University of Illinois religious survey revealed that among the 125 faculty members at the institution, there were only eight professing no religious affiliation, the report further noting that "in every case those who are not members of any church are yet believers in the Christian faith."[22] However, YMCA leaders felt that even those faculty who might have exerted a positive moral influence were denied opportunities in a changing higher education setting that decimated the contexts necessary for student moral "learning" from these superiors. On the one hand, it was becoming more common for faculty to change institutions, reducing both the development of long-term relationships and the potential for extended influence in a single institutional context.[23] On the other hand, as college YMCA Secretary J.D. Dadisman of the University of Kansas suggested, failure was linked to the practical exigencies of the expanding university. Because of increasing class sizes, it was "practically impossible for the instructor even to know the name of more than a small proportion of those in his classes," thus decimating the potential for the knowledge required to sustain in loco parentis relationships.[24] When combined with hiring criteria that denied character as a prerequisite, these forces represented dangerous harbingers of dwindling professorial guidance.

In addition, YMCA leaders also blamed German models of specialization and professionalism for rendering university faculty morally inert.[25] As Princeton President Francis Patton suggested to a group of YMCA leaders, specialization of function had divorced religious vitality from the academic component of professorial work. Called to pursue a very specific research agenda, religious considerations were now viewed by faculty members as, at best irrelevant, and at worst harmful to scholarly focus and productivity. As historians Jon H. Roberts and James Turner have pointed out, one of the most important results of specialization in the academy was the "tendency of university professors who believed in God, as most did, to keep their religious beliefs in one box and their academic ones in another."[26] The classroom, therefore, was increasingly stripped of explicit religious content, defined as a place to grapple with scientific facts rather than spiritual values linked to personal belief. Patton also informed YMCA leaders that the pressures of scholarly expectations were diminishing faculty opportunities to actively invest in student religious life. "This increasing specialization of function whereby the professors feel under obligation to make some specific contribution to the literature of their respective departments," he suggested, "is more and more divorcing them from active interest in the life of the undergraduate, and particularly in the religious life of the undergraduate."[27] The fact/value distinction that privatized religious devotion as a matter of subjective expression was effectively pushing matters of faith to the extracurricular periphery.

Rather than lamenting the loss of the in loco parentis ideal, Mott chose to view such realities as indicative of the providential destiny of the college YMCA. Just as the forces of moral and religious righteousness were retreating from their central position at the core of professorial life, the YMCA was poised to fill this gap. In 1902, Robert Speer put the matter poignantly, noting that the YMCA "stepped into the breach just at a time when, through specialization and the rise of the ultra professionalism of pedagogy, the course of the influence of college faculties was changing greatly." "Whatever its origin," he stated, "the change came by which undergraduates were left to go to the devil if they wanted to, so far as the faculties were concerned. . . . However that may be, the defect of the colleges in the care of the religious interest of their students has been provided against by the Association movement."[28] Mott himself suggested in 1903 that since "many Christian professors have not taken so active a part in promoting the moral and religious life of students as formerly," the association was growing "just in time to help meet the need occasioned" by such neglect.[29]

Mott's evaluation of the nonparticipation of faculty in student religious life did not dissuade association leaders from procuring the involvement of worthy teachers in this extracurricular venue. In fact, as specialization splintered appropriate academic and devotional contexts, the YMCA became one of the primary channels within which religiously devout faculty members could serve as mentors to undergraduates. Select faculty invested heavily in the association, acting as sponsors of the organization, joining as Bible study leaders or speakers at weekly meetings. Alternatively, some faculty also served as members of the local YMCA Board of Directors, providing administrative assistance for organization policies and procedures. Those with previous YMCA experience as students were obviously the most likely candidates for such involvement. At the University of Illinois, for example, Samuel Parr and Ira O. Baker were both presidents of the campus YMCA during their tenures as students at the institution. When they assumed roles as faculty, both became chairmen of the YMCA Board and were actively involved in the day-to-day operations of the campus chapter. Such involvement clearly revealed a sea change in university thought. Previously called to link Christian ideals to scholarly classroom inquiry, interested professors now assisted students in forming extracurricular religious values segregated from curricular concerns.[30]

Science, Voluntarism, and the YMCA

In addition to the rapid decline of professorial moral and religious influence, compulsory religious activities were also on the wane. While many of the colleges leading into the 1880s still maintained a battery of required religious practices, ranging from chapels to church attendance and days of prayer, by the 1890s and early twentieth century these were becoming increasingly rare in public institutions. This shift was closely related to broader changes in perceptions of the proper relationship between religion and education in American colleges and universities. As noted previously, opposition to sectarianism had a long nineteenth-century history, rooted both in the practical considerations of attracting a diverse student market and in more theoretical ways related to Scottish Common Sense "Enlightened Christianity."[31] Some continued to defend the nonsectarian vision for these reasons. Yet by the 1890s, the argument against explicit religious sentiment within the curriculum grew even more vigorous as an expression of commitment to the necessary "freedom" of scientific processes. As a number of scholars have argued, the 1870s and 1880s witnessed a growing departure from the Baconian sense of deriving academic principles from an externally "given" religious worldview. Increasing reliance upon a Comtean positivism that described a broader evolution from dependence upon religious authority to scientific empiricism pushed many to denigrate appeals to particularistic a priori religious interpretations. In this sense, the call to nonsectarianism increasingly took on a different meaning than it had when previous leaders eschewed the overbearing influence of single denominational perspectives. Nonsectarianism now constituted a rejection of any religious perspective in academia marked by specific theological grounding in revealed truth.[32]

While scientific courses had been offered long before, the privileging of free scientific inquiry as the fundamental epistemological grounding for higher education represented a new dominant paradigm. Labeled by George Marsden as "methodological secularization" and by Jon Roberts and James Turner as "methodological naturalism," such a move was perhaps not surprising in these heady days of scientific advance.[33] Science had clearly generated significant results in medicine and various technological fields, and most predicted a sunny future for continued growth in these domains. Perhaps more importantly, many began to argue that the history of science demonstrated that processes and patterns previously explained by supernatural causation were now attributable to scientifically explainable natural forces. Cornell President Andrew Dickson White, for example, spoke of how religion had been used to posit divine causes of human sickness and both flat earth and geocentric theories in astronomy. Since medical science, physiology, and astronomy through the scientific method had overturned such faulty explanatory hypotheses, the logic of such historical examples was for many irrefutable. Given time, all phenomena would eventually succumb to naturalistic explanations, eliminating the need for reliance upon miraculous supernatural causation. Darwin's evolutionary theory, a live issue by the 1870s and 1880s, became for many a flash point topic along these lines.[34]

Pious Christians, particularly those tinged with Protestant liberal ideals, seldom found reason to attack this evolving dominance of free scientific inquiry. The moral habits of science—honesty, perseverance, and truth-seeking—appeared quite compatible with Christian aims and emphases. Many contended that Christians had little to fear from free and honest inquiry since most assumed that the results of such work would harmonize with cherished biblical and theological constructs. College leaders thus did not see such positions as in any way antagonistic to Christianity, especially if linked to biblical injunctions to love one's neighbor and steward the created order.

Even more to the point, whatever may have been said about religion in the curriculum, college and university leaders were still deeply concerned about the religious welfare of students, particularly undergraduates. While value-laden religious ideals might be seen as either irrelevant or dangerous to the freedom of inquiry within the curriculum, they were still viewed as essential components of the personal moral and faith development of undergraduates. The precedent of clergy presidents was beginning to wane by the 1870s and 1880s, but leaders in higher education still demonstrated an active interest in the religious life of the campus and frequently addressed extracurricular gatherings related to moral and religious living. While such concerns may have been peripheral to what George Marsden has called "heart of the enterprise," they demonstrated continuity in the legacy of care for undergraduate moral welfare.[35]

Such care, however, was marked by a pervasive spirit of voluntarism that accompanied the new ideals of free inquiry on the campuses. For many scholars in this era, freedom of inquiry required the elimination of compulsion in both curricular and extracurricular venues. Appeals for elective course offerings were typically rooted in this sense that academic progress could best be facilitated by giving students freedom to explore areas of interest.[36] More importantly for the YMCA, many contended that a free and "chosen" faith engagement was far more authentic than forced religious practice. When YMCAs first appeared on college campuses in the 1860s and 1870s, students were in most cases still required to attend some combination of chapel, morning and evening prayers, and morning and evening church services on Sundays. By the final decade of the century, many of these practices were either abandoned or placed upon a voluntary basis. Evening prayers and the second Sunday church service were often the first to go, followed by the requirement for any church attendance. The elimination of compulsory chapel began gradually throughout the 1860s and 1870s but became more generalized by the end of the century, particularly in state institutions. New universities, such as Johns Hopkins and the University of Chicago, made chapels voluntary from the outset. Midwestern schools, less anchored by weighty traditions or ties of denominational heritage, also led the way, with the University of Wisconsin abandoning compulsory chapel as early as 1868 and the University of Michigan moving to voluntary prayers in 1871. Early attempts to secure such voluntarism were met by sharp criticism in many more tradition-bound schools, but by 1886 Harvard had given up compulsory chapel, and Columbia followed suit upon the chaplain's retirement in 1891. A study by Michigan President James B. Angell conducted in 1890 revealed that, among twenty-four state institutions, twenty-two still conducted chapel services but only twelve were compulsory by that time.[37]

Yet the proper nature of this voluntary student religion was a source of considerable tension in this era, and the YMCA found itself in the middle of rancorous debates along these lines. Many administrators, particularly those with strong liberal Protestant proclivities, desired that the colleges and universities provide voluntary religious options that highlighted modern scientific ideals. College-sponsored voluntary chapels, for example, were often led by a rotating group of visiting ministers, now including occasional representatives of Unitarian, Catholic, and Jewish congregations.[38] A visible manifestation of the nonsectarian (and increasingly interfaith) stance of these institutions, universities typically insisted that such invited speakers downplay theological distinctives, focusing instead upon singing, short prayers, and generalized Christian ethical teachings. Such themes, unlike previous forays into more explicitly doctrinal issues of Jesus' death, the nature of eternal life, and the theology of sin, could be considered without any recourse to religious antecedents, and some argued that they could be empirically and scientifically validated independent of text-based appeals.[39] Many administrators hoped that such voluntary experiences would introduce students to a faith anchored in modern scientific scholarship infused with notions of free inquiry.

For others, including Mott, the genius of scientific freedom was not to be found in the creation of a voluntary religion marked by free inquiry but rather in the provision of student freedom to select self-chosen religious activities. Students, he argued, should be able to develop and attend religious groups in accordance with their own desires, consistent with their needs and reflective of their religious passion and zeal. While such voluntary choice might be exercised in favor of a naturalistic and scientific faith, Mott contended that freedom would also allow for groups anchored by a more traditional authoritative religious grounding. Such a distinction was critical for Mott because he believed that a broad scientific and ethical religion would blunt the potency of a strong Christian witness and render the faith a bland collection of ethical bromides.[40] Citing the Harvard Religious Union, the Fortnightly Club at Cornell, the University Christian Association at Stanford, and the Students Christian Association at the University of Minnesota, he noted that generic Christian belief without the specificity of Protestant evangelicalism "could not be harmonized with aggressive, deep, progressive spiritual work in the name and for the sake of Jesus Christ."[41]

Debating Voluntarism at the University of Chicago

The tension between these two views was clearly evident in the lengthy formation of a YMCA at the new University of Chicago in the early 1890s. From the time of the founding of this institution in 1891, Mott saw the growing university as a strategic location to establish a YMCA chapter that would be a model nationwide. While the organization had obviously been successful in eastern institutions such as Yale, Cornell, and Princeton, Chicago presented an opportunity to demonstrate the compatibility of the YMCA with the new brand of emerging research university in the late nineteenth century. Without question, Mott had every reason to anticipate ready acceptance of the YMCA platform in this new setting. He had a long-standing friendship with President William Rainey Harper. In addition, Amos Alonzo Stagg, a former Yale YMCA leader, had just arrived at the university as football coach and associate professor of Physical Culture. With such networks already formed, Mott was confident that Chicago would become a center for association influence in the world of higher education.

Mott visited the campus in 1892 to discuss the possibility of forming a YMCA, but it quickly became evident that the task would not be straightforward. While many in attendance desired a YMCA formed along evangelical lines, a significant faction desired a religious organization that would be more inclusive, inviting members of all faiths to pursue a scientific religious vision. When Harper heard of the dispute, he called a meeting to discuss possible organizational frameworks for university religious work. Clearly dissatisfied with an organization that would not be "Christian," he was also hesitant to give full endorsement to the YMCA as the representative agency of University of Chicago religious expression. The president instead expressed a desire for an organization that would reflect modern scientific religious scholarship and the breadth of Christian opinion across the theological spectrum.[42] Mott was grieved at Harper's position and carefully pleaded his case throughout the fall of 1892, informing him that no college president had ever opposed the formation of a YMCA chapter.[43] The very spirit of open scientific freedom in the universities, Mott contended, mandated an institutional tolerance for student-initiated religious activities along the lines that would best meet their needs. "Should not the very spirit of liberality which some of the professors were urging," he suggested, "admit that if a small body of young men believe they can do a certain important work for Christ better by coming together on an evangelical platform, they should be accorded the freedom and privilege of doing so?"[44] For Mott, the rhetoric of scientific freedom implied a tolerance for a variety of religious forms, including an unabashed evangelical conviction.

Mott pressed Stagg to move forward in the formation of the organization on campus, sending him a sample constitution based upon the national model and encouraging him to begin the process of selecting officers for committee work. Stagg faithfully assembled students, but Harper was displeased that Mott had decided to move forward without permission.[45] Reluctantly, the president allowed for the formation of the YMCA on campus, but only by placing it within the province of an umbrella "Christian Union," formed to promote the "harmonious organization on the basis of those elements of religious faith which are held in common." Describing this federated structure, Chaplain Charles Henderson noted that the University sponsored a Christian Union, "for all who profess and call themselves Christian" and a YMCA and YWCA "for the more evangelical persons who feel the need of the means of more direct Christian work."[46]

By all accounts, the YMCA flourished within this system and, to Harper's dismay, became the vital spiritual force on the campus. By 1899, the organization had 161 members, including 83 enrolled in YMCA-sponsored Bible study work. They sent twelve students to the summer conference in Lake Geneva, the "banner delegation among the colleges and universities of Illinois." Perhaps more alarming to Harper was the degree to which the most esteemed faculty of the University and Divinity School immersed themselves in association work. Speakers to YMCA gatherings in 1900–1901 included such luminaries as Edgar J. Goodspeed, James Hayden Tufts, Shailer Mathews, and Gerald B. Smith. One of his most influential professors, botanist John M. Coulter, served on and later chaired the YMCA advisory board, working with the organization into the 1920s.[47] At the same time, the more scientific religious perspective offered in voluntary chapels had not generated the predicted student interest. One student, writing for the *University of Chicago Weekly*, noted that the voluntary chapel services had been deeply disappointing because they consisted only of a prayer, a song by the choir, some perfunctory announcements, and "an entertaining address by some professor, upon a topic with little bearing on religion."[48] Chaplain Henderson noted to Harper that attendance was discouragingly small, injuring the reputation of the University around the country. "We are humiliated before ourselves, our students and visitors from abroad," he told the president, "deepening the impression that very few people, even of the religious leaders, care for devotional life."[49] Scientific religion was not faring well at Chicago.

Harper was irked by the success and rising status of the YMCA and implied that such work served as a distraction from scientific religious acumen.[50] Henderson, however, disagreed with the president's assessment, stating that "As to the success of the YMCA and YWCA being a cause for discouragement, that depends on the point of view. In my judgment their meetings are the hope of real devotional life in the University. The Chapel service, the Christian Union, and nearly every form of religious activity find in these persons who believe in the YMCA the most patient and zealous workers. Instead of discouraging them I believe we should do all in our favor to cheer them."[51] Henderson found that the movements' foundation in evangelicalism was a powerful source of its potency, proclaiming that "There is no power of cohesion outside those who are certain to work by 'evangelical' means. This does not mean that we should cease to build the Christian Union, but it does mean that the others should not be discouraged."[52]

This dispute clarified the critical tensions implicit in the scientific notions of "freedom" that consumed higher education discourse in this era. For Harper, allegiance to free inquiry necessitated the highlighting of a form of religion imbued with modern scientific ideals, commensurate with the curricular aspirations of these schools. Alternatively, Mott and Henderson both interpreted scientific freedom more in terms of student "election" of the religious models reflective of their desires and needs, even if those models were deemed oppositional to the scientific ideals of free inquiry. As it turns out, it was Henderson's (and therefore Mott's) argument that won the day, and it was this sentiment that largely dictated the course of religious expression in the universities through World War I. Increasingly, educational common sense in colleges and universities demonstrated a Jeffersonian commitment to a religion of broad, ethical ideals within the curriculum, and an allowance for voluntary religion of any variety in the extracurriculum. As Julie Reuben has contended, while commitment to free inquiry had set some educators upon a journey to discover a modernistic religion, that same scientific ideal became the goad for others to sponsor an unrestricted tolerance for diverse religious activities.[53] While Harper saw much at stake in securing a scientific religion, Henderson's posture of "cheering" the independent work of students had secured a widespread acceptance.

When given the opportunity, students seemed to "freely" select the YMCA. Despite administrators' early enthusiasm for religious voluntarism, student nonattendance at campus-sponsored voluntary religious activities demonstrated a pervasive apathy along these lines. By 1896, only fourteen students on average were attending chapel at Johns Hopkins, and about thirty were attending regularly at Columbia, belying President Nicholas Murray Butler's bold predictions of increased student interest.[54] On many campuses, embarrassed administrators reduced the number of days such services were held. In other cases, they relinquished leadership of voluntary chapels to the YMCA. By the early 1900s, in fact, many local YMCA chapters were given responsibility for securing chapel speakers and taking charge of logistical arrangements for these events. The vast success of the YMCA in this era provided a visible contrast to these ineffective agencies. Between 1890 and 1915, Mott's restless entrepreneurial vision set in motion an ambitious attempt to saturate the American collegiate landscape with YMCA programs. Convinced that these institutions were ripe for a vigorous association presence in an era marked by limited faculty involvement and new ideals of voluntarism, Mott worked to build this movement into a higher education staple. By the time he was finished in 1915, the YMCA was the dominant force in collegiate religion across the nation and around the world.

Expansion of the College Movement

Even taking the expansion of higher education enrollments in this era into consideration, the membership explosion within the movement during Mott's tenure was astounding. While the 284 chapters extant in 1887 boasted 14,193 members, by 1892 there were 425 associations with 27,334 members. Just one decade later, in 1902, 681 chapters sustained a membership of 41,800, and by 1912, 2 years before Mott's position as college secretary was complete, there were 800 chapters with approximately 65,000 members nationwide. While the escalating number of chapters surely bolstered enrollments, growth in existing organizations contributed significantly to the expansion of the movement. Iowa State saw membership rolls jump from 90 in 1890 to 719 in 1909. The Oberlin YMCA grew from 162 in 1903 to 523 in 1909, while the University of Illinois saw growth from 90 in 1889 to 1,066 in 1913, making it the largest collegiate chapter in the world. By 1901, one in four male students nationwide was a member of the YMCA, making Princeton President Francis Patton assert, "The Young Men's Christian Association has well nigh the monopoly of the religious culture of our universities and colleges."[55]

Mott was consumed with the prospect of expanding the reach of the college YMCA to regions and institutions previously unaffected by its influence. Within the United States context, Mott, like Wishard, recognized that the West was the only region of the country that lacked a significant intercollegiate YMCA presence. Cognizant of the rapid growth of institutions in this region and noting that no representative of the college movement had visited the Pacific Coast since 1887, Mott left in 1892 to embark upon a thorough tour of the western colleges in order to initiate new chapters and strengthen the few already existing. Finding only six functioning associations on the entire coast (most of which he noted required "reconstruction"), he described a pervasive absence of Bible study, prayer, evangelism, and missionary interest.[56] Mott suggested that the West was at a disadvantage because of its isolation, shifting population, and "the special activity of the devil," producing "more Sabbath desecration, gambling, profanity, vulgarity, drinking, impurity, infidelity, and indifference concerning things spiritual . . . than in any other section of the country."[57] Mott's visit, however, was transformative along these lines. While only six associations existed prior to his visit, twenty-nine chapters, including a new sixty-member club at Stanford, had been initiated by the end of Mott's tour.[58]

In order to enhance the geographical range of the movement and secure comparable intercollegiate consciousness in all regions of the country, Mott arranged to expand the summer student conferences. By 1889, Mott had discussed the possibility of an alternative to Northfield for students of the midwest, and in 1890 a summer conference modeled precisely after Moody's gathering was offered at Lake Geneva, Wisconsin. By the end of the century, conferences were also held at Pacific Grove, California and Ashville, North Carolina. By 1912, nine conferences were held each year, including one for African American college students at Kings Mountain, Tennessee. While such offerings converted a single national conference into multiple regional conferences, they allowed more students to reap the benefits of such experiences. The four 1901 conferences entertained 1,404 students, and by 1913, 3,000 students attended these gatherings.[59]

In addition to this national growth, one of Mott's signal contributions to the college YMCA movement was to expand its global reach. As early as 1884, a student YMCA chapter had been originated at Jaffna College in Ceylon. Planted by Frank Sanders, a recent alumnus of Ripon College, this association was the first recorded college YMCA chapter formed outside of a North American context. Soon after, college associations were formed in Tungchow, China and Berlin, Germany by recent graduates of the American YMCA program. The Northfield conferences, always attended by sizable delegations from foreign countries, cemented relationships between students from various nations and therefore served to generate a growing international Christian student consciousness.[60] Mott himself traveled extensively overseas beginning in 1894, responding to invitations to assist professors and students in developing religious organizations in their colleges. Considering a unified worldwide movement under the auspices of the YMCA, which had been Wishard's initial vision, Mott chose instead to construct a federation of national movements, each completely independent and adapted to local context. In 1895, the World's Student Christian Federation (WSCF) was formed, anchored by a minimalist organizational purpose of bringing students of the world to Christ, building them up in Christ, and sending them out for Christ in practical religious work. Biannual conferences were held in various member nations, and by 1913, the WSCF was uniting together national movements representing 2,305 local associations with a membership of 156,071 students and professors in 40 nations.[61]

While geographical growth was a critical factor in Mott's expansionist vision, he also prioritized a more inclusive penetration of diverse institutions of higher education. Among neglected institutional types, Mott in his early years emphasized the growing number of professional schools, particularly those in urban centers such as Chicago and New York. He correctly recognized here a growing market that was largely untapped by student organizations. Estimating in 1890 that there were 175 medical schools, 49 law schools, 177 normal schools, 100 "schools of science," and 222 "commercial colleges," Mott lamented the fact that in "not more than 100" of these institutions was there "any special effort being put forth to bring the students to Christ, and to enlist them in work for him." Reflecting the ideals of the parent YMCA, such students, he claimed, were subject to intense temptations, many coming from small towns and rural areas to face for the first time the evils of intemperance, gambling, and impurity in the cities.[62] These students typically worked harder than their other collegiate peers, focusing more concentrated energy on academic endeavors. Enrolled for shorter durations and more widely geographically scattered than those living on residential campuses, the shifting professional school population presented significant programmatic challenges to an organization that depended upon frequent interpersonal contact. Furthermore, because many lived at home, these students were more likely to express their religious involvement in churches rather than campus organizations. With significant pressure from Mott, YMCA chapters were planted in a majority of such institutions, but student memberships never approached the numbers in residential collegiate settings. In 1905, for example, when an estimated 31 percent of traditional male college and university students were enrolled in the YMCA, a mere 17 percent of professional and normal students were so enrolled. With the exception of a few schools where the YMCA proved to be enormously successful, professional schools remained on the periphery of the YMCA's extensive organizational reach.[63]

Mott was far more successful in his attempt to assimilate theological seminaries into the college YMCA fold. Unlike the case in many other nations, where seminaries and colleges were part of the YMCA from the beginning, in the United States the colleges and seminaries began as constituent members of disparate, and at times competing, organizations. In most seminaries, student-initiated religious expression typically took the form of missionary societies, either independent or integrally connected to the ISMA. Mott applauded the missionary sensitivities of theological students, but he also noted that the separation of these groups from the work of the college YMCA had generated a dangerous "hiatus between the future leaders of the Church and the educated laity" of the colleges.[64] Mott was convinced that seminary students and college students had much to gain from an interdependent relationship. Theological students could benefit, he surmised, from contacts with future professionals because collective efforts would facilitate their future ministry to such individuals and break down clergy/lay walls. By joining this practical movement, such students would also avoid overintellectualizing the gospel message and would be called upon to engage in practical work that would sensitize them to surrounding need. On the other hand, collegians could benefit greatly from the theological expertise of seminarians, allowing these advanced students, many of whom had at one time been members of the college YMCA themselves, to "bring corrective influence and wise guidance to the undergraduates who are facing real religious difficulties." In addition, Mott hoped that contact would reduce college students' naive prejudice against ministers and open their eyes to career possibilities in such work.[65]

While such arguments might indeed have been persuasive to seminary leaders, ultimately the decision to affiliate with the college YMCA was related more explicitly to discontent within the ISMA. Those working under its auspices felt isolated from the significant work taking place within the SVM, and they deeply desired the global connections to facilitate international relationships with other seminaries. However, since the newly created WSCF would not allow multiple agencies to join from each nation, the only means to capitalize on such enthusiasm was to join the college YMCA. Thus, in 1898, the ISMA was dissolved and seminary groups were reorganized as the "theological section" of the intercollegiate YMCA, committing to a broad purpose of deepening the spiritual life of the seminaries, extending interest in worldwide missions, and making contact with other theological students around the globe. While in 1898 only twenty seminaries were official members of the college YMCA's theological section, by 1913, eighty seminaries had joined the movement with participation rates of close to 70 percent. Robert Wilder, tapped to be the first general secretary over these theological schools, was convinced that membership in the YMCA would "widen the horizons" of seminary students, giving them a "contact with life" that was often lacking in their isolated and cerebral contexts.[66]

In addition to these professional schools, YMCA work in black colleges and universities also increased dramatically in this era, although this had little to do with the organized efforts of Mott or others in the white college movement. While Mott was informed about such work and regularly reported advances in black colleges in *The Intercollegian*, student work in African American institutions in this era was conducted from within the YMCA's Colored Work Department rather than within the general Student Department. By the time Mott assumed his position in the association, the early work of Henry Edwards Brown had already generated twenty-seven chapters in African American institutions with a membership of over a thousand students. However, it was William A. Hunton, the first African American YMCA secretary, who developed a more comprehensive program. Hunton was a native of Canada, the son of an ex-slave whose home was a station on the underground railroad. After serving for a time as director of black YMCA work in Norfolk, Virginia, he was placed in charge of the larger Colored Work Department, newly establish in 1890. Jesse Moorland, a graduate of Howard University, joined him in 1898 to assist both with the formation of black city associations and with the college work as well.[67]

Hunton quickly recognized that he had his work cut out for him, acknowledging that African American institutions were not privy to many of the same advantages that characterized white colleges and universities. In the 1901–1902 school year, Hunton noted to Mott that, while some white campuses had their own YMCA general secretaries, no black colleges could afford such a luxury. Travel expenses and the requirement of summer employment prohibited the development of student conferences along the lines of Northfield, and many local chapters could not afford the Bible and mission study materials of the movement. In some cases, he noted, such resources were also too advanced for black students, and few institutions could find leaders qualified to direct such groups. In addition, race prejudice was also a factor, witnessed most notably in the refusal of white state and regional secretaries to visit black institutions. Denied access to white conferences and deprived of a more robust secretarial force, black colleges were isolated from the resources of the general intercollegiate movement.[68]

Despite such difficult factors, the progress of the college YMCA movement in black institutions continued apace. By 1899, "colored chapters" of the YMCA existed in 45 out of the 100 African American colleges in the nation, boasting a membership of 2,200 students.[69] By 1904, when the number of black students involved crested over 5,000, 50 percent of the students in these institutions had joined the YMCA. Spurred by this growth, national leaders in 1905 supplied another secretary, George E. Haynes, to assist with the work. Such minimal leadership, however, was not nearly enough to sustain a growing movement. Traveling regularly to seventy-five or more campuses per year, Haynes in 1908 noted that the YMCA chapters in five schools had never been visited since their origin years earlier, while thirteen had not been visited in over three years. Relying largely on students and the indefatigable efforts of Hunton, Haynes, and Moorland, by 1902 there were 103 associations in black colleges. These chapters actually dominated the activity of the Colored Work Department in these years. Between 1896 and 1912, about two-thirds of all African American YMCAs were located in black colleges.[70]

On a similar but more limited scale, college association leaders also worked to initiate YMCA chapters in the Native American boarding schools then arising in various pockets around the country. Wishard had actually visited a functioning college YMCA chapter at the Santee Agency Normal School in 1885, immediately accepting this group into the larger national movement. One year later, Ober established a chapter at the Carlisle Indian Training School in Pennsylvania, which Mott listed just a few years later as one of the most "aggressive and progressive in the state." By 1900, there were forty-four American Indian Associations, including nine in Indian boarding schools. By 1911, associations had been formed in such institutions as the Salem Training School in Oregon, Bacone, Tulsa, and Chilocco in Oklahoma, Haskell in Kansas, and the Sherman Institute in California, prompting the YMCA to hire R.D. Hall as a secretary in charge of the YMCA work among Indian boarding school students. One Chippewa delegate to a college YMCA Bible study conference noted that "the Indians of our School are engaged in a great uprising, no longer for the scalp of the white man, but for making Jesus Christ known among their fellows."[71]

One of the immediate by-products of this vast proliferation and diversification of student YMCA chapters was a crisis of supervision. Mott quickly recognized that even his own significant energy and managerial expertise was insufficient for managing a rapidly multiplying movement. In the 1890s, only one in twenty-five local college YMCA chapters had permanent secretaries, and while he was continually calling upon state YMCA organizations to develop college subcommittees, only about twelve of thirty-six had done so by 1895. In 1891, recognizing his inability to cover the field, Mott began forming small groups of undergraduate students to work among the colleges of their states, and by 1892 seventy students were so engaged in twenty-five states. Designed to uphold the organization's voluntary character, these student pairs encouraged local chapters and assisted with ideas for growth and standardization. Mott also promoted a series of president's conferences, events designed to train campus student leaders in a way that would alleviate extensive demands for national and state staff intervention.[72]

Yet despite the theoretical commitment to student leadership, the sheer size of the movement mandated a radical expansion of the collegiate YMCA employed staff at the national level. Attempting to account for regional, institutional, and programmatic variety, Mott expanded the secretarial force in ways that provided leadership in each of these domains. Between 1897 and 1899, the number of paid national college secretaries was doubled from five to ten, representing leadership over the East, West, and South regions, medical, theological, and preparatory schools, Bible study work, and administrative tasks. By 1914, there were seventeen national secretaries, new positions having been created for work in social service, sex education, and Indian training schools, among others. Mott had an excellent eye for potential leaders, and many of the individuals who assumed these roles, such as Fletcher Brockman and Gilbert Beaver, became Mott's close personal friends and associates, later moving on to significant careers in religious work.[73]

"The Most Religious Communities in the Country"

For Mott, the fact that the YMCA was now present in "practically every prominent institution of higher learning on the continent" and had developed more extensive intercollegiate bonds than any other college organization was not only a source of pride but also a clear indication of the enduring potency of collegiate religious enthusiasm in an age of diversifying activities and institutional secularization.[74] The remarkable expansion of the college YMCA, in fact, explains why, despite the loss of formal religious content in curriculum and chapel, educators and YMCA leaders could remain unabashedly enthusiastic about the Christian character of institutions of higher education at the turn of the century. There was almost complete unanimity among prominent administrators in the belief that student religious interest and expression at this historical juncture was far greater than it had been in the 1870s and 1880s.[75] Surveys of the religious practices of students seemed to bolster such sentiments. University of Michigan President James Angell conducted an analysis in 1896 of nearly five thousand students in five state universities, the results indicating that 55 percent were church members while 90 percent attended or were at least affiliated with some local congregation. Such statistics were confirmed ten years later when an independent national study of state universities estimated an even higher 60 percent of university students as church members. That same study noted that 20 percent of college men and 50 percent of college women belonged to the YMCA and YWCA, respectively. Surely bolstered by increasing female enrollments—figures for female church attendance and voluntary religious participation were always higher than for males—such statistics nonetheless provided comfort for those fearing widespread secularism on the campus and ammunition for those seeking to validate the direction and Christian character of these institutions. Educators could argue indeed that, despite the many changes in university education, students, at least by all measures of personal practice, were more religious than ever before.[76]

Not surprisingly, such measures were utilized liberally by college YMCA leaders to document the successes of the movement. Writing on "The College Man's Religion at the Beginning of the Twentieth Century," Mott suggested that the colleges were "the most religious communities in the country." Comparing collegians in 1903 with "the young men of America as a whole," Mott found that, while about one in twelve (between 10 and 15 percent) young men outside of the colleges were members of evangelical churches, more than half of the male collegians were so affiliated.[77] Looking back historically, he also estimated that only one in three college students had been a church member prior to the advent of a national college YMCA movement. Taking full credit for such statistics, Mott could state unflinchingly that "The fact that the colleges and universities to-day are the most Christian communities of the United States and Canada is a fact that we must place to the credit of this movement more than to any one other force."[78] If there had been a trend toward secularization in the university, Mott contended, that trend had been decidedly rebuffed by the YMCA. What is so interesting about all of these enthusiastic appeals is that they each cite evidence in terms of personal and extracurricular criteria rather than institutional or curricular progress. Higher educational institutions were more Christian because voluntary Christianity had exploded, not because the universities themselves had become more devoted to religious aims.[79]

While it would be wrong to see such optimistic evaluations as purely self-serving—both administrators and YMCA leaders were undoubtedly sincere in their enthusiasm for these developments—it is clear that they did not choose to acknowledge the potential concomitants of these shifts. What Mott, in his dramatic vision of providential destiny, failed to recognize was the possibility that the YMCA might actually have facilitated faculty and administrative separation from the moral and religious lives of undergraduates. With great insight, Robert Speer admitted during his discussion of the decline in faculty religious interest in American colleges that "Possibly the growth of the YMCA movement might have contributed to this change through relieving the faculties of some sense of responsibility."[80] This recognition that the growth of the YMCA might have actually contributed to faculty religious apathy was rare in this period of movement triumphalism.

Yet it is absolutely critical to a full understanding of the nature of secularization during the "age of the university" in American higher education. In an era characterized by specialization and division of labor, it was quite natural indeed for faculty members to embrace their roles as "pure" scholars while leaving religious influence to this extracurricular agency. Many faculty members, driven by visions of specialization and positivist ideals of the separation of facts and values, were apt to view their religious commitments as outside the ken of scholarly life, more appropriate in the context of private rather than public settings. Just as significant, however, was the fact that the religious culture of undergraduates was now in the hands of the YMCA, making it quite easy for these individuals either to shed their responsibilities for such work or to make their contributions by investing time within this extracurricular venue.

In the end, it is clear that the YMCA was a response to the changes taking place in higher education in the latter decades of the nineteenth century. As strict faculty oversight, religious compulsion, and in loco parentis ideals grew anachronistic, the YMCA became a student-initiated force for spiritual guidance and moral responsibility on the campus. As the value of free inquiry promoted a move to voluntarism and student freedom, the YMCA became an active expression of student self-selected religion. As specialization of function became the model for scientific higher education, the YMCA assumed its place within the structure of the university as the agency that would attend to the religious welfare of undergraduate students. By the early twentieth century, higher educational institutions had solidified a pattern in which the religious life of undergraduates was expressed chiefly within the voluntary extracurriculum. Securing a permanent place within that extracurriculum became the new central task of the college YMCA.

Chapter Four
Building a Campus Presence

In his popular reminiscences of college life around the turn of the twentieth century, Yale alumnus Henry Seidel Canby described a campus pervaded by a student-defined romantic activism. The strenuous life of college activities in this era, he claimed, defined achievement in a way that fostered the bustling pace of extracurricular participation. Students were encouraged to join a host of clubs, fraternities, and athletic teams in order to furnish a campus identity and secure prestige in the competitive and hierarchical world of college honors. In addition, Canby suggested that this world was one of intense loyalty and insularity. Reflective of the progressive homogenization of student age and stage of life, campuses of this era generated a focused and intensive peer culture. Driven by its own values and inspiring a passionate devotion to those ideals, the college was, he suggested, a "community that defined its own success, pursued it constantly, and was arrogantly indifferent to the ideals of others, asking its members for complete and whole-hearted allegiance." In contexts where "the young made a world to suit themselves," that world cultivated a love for alma mater that eschewed external loyalties in favor of an active, campus-centered existence.[1]

Certainly, by the late nineteenth century, the desire to obtain "success" during the undergraduate years was measured more in relation to the competitive extracurriculum than to classroom brilliance. Intellectual aspiration was decidedly downplayed—even denigrated—within such settings as students ostracized academic leaders and heralded those who excelled in other areas. Content with a "gentleman's C," many students worked hard enough to ensure a passing grade while spending the bulk of their energy on campus social climbing. Indeed, the importance of such extracurricular "proving grounds" had provided the impetus for Princeton President Woodrow Wilson's famous sentiment that "The side shows are so numerous, so diverting—so important, if you will—that they have swallowed up the circus, and those who perform in the main tent must often whistle for their audiences, discouraged and humiliated."[2]

This mentality created a significant mood of social ambition and frenetic activity on the campuses. Bishop Sherrill of Yale asserted that "The emphasis in the college of those days was on doing. Every social pressure was exerted to make a man try for a team, or a paper. The basis of judgment was largely how much he did."[3] This larger spirit was captured quite well in Owen Johnson's 1912 fictional classic, *Stover at Yale*. In the end a critique of the intellectual superficiality of the activity-oriented campus, the novel depicts the collegiate career of Dink Stover, a successful student who achieved notoriety through his social and athletic exploits. Describing the campus, Johnson noted:

> Everyone was busy, working with a dogged persistence along some line of ambition. The long, lazy afternoons and pleasant evenings were not there. Instead there was a grinding of the mills and the turning of the wheels. . . . He felt it in the open afternoon, in the quiet passage of candidates for the baseball teams, the track, and the crew; in the evening, in the strumming of instruments from Alumni Hall and the practicing of musical organizations, and most of all in the flitting, breathless passage of the News heelers—in snow or sleet, running in and out of buildings, frantically chasing down a tip.[4]

Such a mentality possessed a degree of self-abnegation, to be sure. The effort expended in campus activities had as its larger purpose the greater prestige of the institution. At the same time, one's ability to "do something" for the college was a clear path to personal success.

College YMCA leaders quickly recognized the potential fragility of the organization's existence in such a campus environment. On the one hand, the insular world of the college complicated the presence of a club with ties to an external organization. Luther Wishard denigrated collegiate insularity, but the reality of the campus in this era made it immediately evident that intense campus loyalty was essential to any organization's long-term viability. In other words, failure would be assured if students saw local chapters as externally imposed, structures foisted upon them by an expansion-hungry religious organization with national rather than campus sensitivities.[5]

On the other hand, there was a growing concern that the YMCA would be eclipsed by other campus activities, serving only as an alternative area of involvement for those unable to achieve success in more legitimate campus channels. For some, insecurity about the positioning of the organization on the campus was a function of the type of student attracted to YMCA membership. Leaders within the movement were somewhat concerned that, despite the overall growth of the campus chapters, the more popular students were bypassing the association and choosing instead the more prestige-affording options of fraternities, athletics, student journalism, and various music groups.[6] Professor Herman Horne, a faculty sponsor of the YMCA at Columbia University, described the chief weakness of the organization in terms of its failure to attract the most prominent student leaders of the campus social hierarchy. "[We] have some, but not as many as we ought to have or should have, of the representative college men—those men that lead their fellows in the various legitimate college diversions," he noted. "Sometimes it occurs that students entering college with ambition for leadership are rather advised against joining our association number, as though we constituted an unrepresentative college body."[7]

To be "unrepresentative" within a campus culture defined by social prestige and college loyalty was, of course, an organizational death knell. One campus YMCA leader suggested that too many organizations were simply taking the men who naturally came to the chapters rather than "leaving those men to come of themselves and going after the *big* men." "Any fool can catch a cotton-tail," he concluded, "but it takes a hunter to catch a grizzly"(emphasis in the original).[8]

To remedy this failure, many chapters did indeed make direct plans to recruit popular men to validate their moral crusade. At the University of Michigan in 1913, campus leaders were urged to "use prominent athletic and campus men" to lead YMCA meetings early in the year to demonstrate the prestige of the association.[9] At the University of Wisconsin in 1910, leaders of a dwindling chapter set their sights on appointing the captain of the football team to the association presidency, despite the fact that he had no connection to the YMCA. Securing his loyalty, the organization grew dramatically over the next two years, achieving mainstream status within the University.[10] For most chapters, the men who ruled the collegiate social world possessed the greatest potential for elevating the status of the movement and drawing people to its ennobling influences. "No class cares for a leader who is good for nothing else," one local secretary noted when speaking about YMCA Bible study groups. "He must be a representative college man. It is worthwhile for him to be a winner in scholarship, athletics, or other student contests, just for the vantage-ground it gives him in extending Christ's Kingdom."[11]

The extent to which local chapters could embrace this approach was keenly demonstrated at Amherst College in these years. Bruce Barton, future marketing genius and influential author of a number of popular religious texts, claimed that he was advised, as a transfer sophomore in 1904, to "keep away from the Y.M.C.A" because of its placement outside of the social-status affording groups. He described the organization at the time of his matriculation as consisting of a group of men who had not been invited to join any fraternity and therefore "found a certain spiritual revenge in conducting the Y.M.C.A. as a kind of fraternity of their own, untainted by contact with the rest of the college. . . . They were in the college world but not of it."[12] In 1904, students and alumni concerned about this negative reputation pressured the captain of the basketball team, a prominent fraternity man, to run for the presidency of the organization. When he won, he fought to end all creedal prerequisites for membership and proposed instead a new common creed, "I believe in God and Alma Mater." Although the international YMCA condemned this move, Barton was convinced that the presence of a fraternity member and the absence of exclusivity helped students to see that one no longer needed to be a "mollycoddle" to be religious.[13] A popular leader, combined with a creed that blurred the line between love of God and college, ensured compatibility with an activist and insular student culture.

Commensurate with this desire for prestige within the campus social hierarchy was a strong push to recruit fraternity members. On one level, YMCA leaders stated that the organization should seek out fraternity men because of the potentially destructive influence of the houses. A report on college fraternities issued by the national college YMCA movement in 1908 stressed that the "segregated life" of these groups promoted "habits of living" marked by "utter religious indifference" and "positive moral degradation." There were, the report suggested, "comparatively few chapters without their leprous spots, in the form of one or more men whose example is positively harmful morally."[14] One report from Purdue University in 1905 stressed the fact that fraternity members were often pressured to avoid religiosity, disdain for religious activities serving as a badge of loyalty to the fraternal bonds. C.V. Hibbard of the University of Wisconsin recalled, in fact, that fraternity men regularly gathered to "heckle" YMCA members as they walked up the hill for their religious meetings in the law building. As one author summarized, young fraternity members, influenced by house traditions, often embarked on a collegiate career that was nothing less than a "religious vacation."[15]

While such descriptions implied a sense of discord with a Christian organization, national and local YMCA leaders seemed unabashedly enamored with the concept of attracting such students to membership and potential leadership in the movement. The 1908 report described fraternity members as "men of enterprise" who were "capable of exceptional service." "Forming a disproportionate part of the most progressive and representative element," the document noted, "they dominate the college activities and are frequently men of exceptionally large and clear conceptions." While the report condemned fraternity exclusivity, it nonetheless implored chapters to seek out such individuals for what they could contribute to the association. "In a number of institutions," the document suggested, "the association greatly needs men of the virility and social culture which characterize many fraternity men. Their acquisition puts the Association in a more dignified and commanding position." Betraying a sense of inferiority, the fraternity report concluded that YMCA chapters failed to secure such men chiefly because of the weakness of the religious program. "The type of Christian life exhibited by many Association leaders, perfectly sincere, perhaps, but narrow-gauged and unpleasantly solemn, has repelled many," the report suggested. "The Association is sometimes too narrow in outlook and too ineffective in management to command the respect of energetic men of wide vision."[16]

Not content with attracting a solid core of religious-minded students, leaders between 1890 and 1915 began to understand that the very survival of the YMCA within the burgeoning structure of American higher education depended upon a greater immersion within the networks of student life. The solution, it appeared, was straightforward. Chapters would need to become indispensable, tightly integrated with the larger student culture and enmeshed within the daily lives of undergraduates. In addition, they would need to develop an inclusive means of relating their religious ideals to the prestige-generating domains of the campus. In this era of increasing campus pride linked largely to extracurricular exploits, YMCA leaders recognized that forging a campus presence was the only path to success. Through new structures, personnel, and programs, they hoped to generate an all-student organization in American higher education.

The Building Movement

One of the most critical factors in enhancing the centrality of the YMCA on local campuses came through the construction of buildings devoted to association work. Prior to the 1890s, many of the chapters utilized campus rooms to hold religious and prayer meetings, often moving from year to year depending on availability, group size, and administrative whim. Wishard's appeals notwithstanding, many of the rooms utilized for YMCA purposes, typically classrooms, were poorly equipped for religious meetings and other functions. Even before Mott assumed leadership in the organization, therefore, local leaders were discussing the need for a stronger physical presence on campus, citing comparisons with other college organizations. As a provisional response, some associations began renting houses close to campus that provided both rooming options and space for gatherings. Mott, however, was clear that this was only a temporary solution. Soon after assuming leadership, he began pressing hard for a national campaign that would provide adequate facilities for the expanding college YMCA.[17]

The rationale for such a project was derived from many sources. Doubtless Mott remembered well his own campaign at Cornell. In addition, city YMCAs across the country were experiencing a building boom that impressed Mott greatly. Two hundred and ninety YMCA buildings were erected nationwide between 1900 and 1916, and financial investments in building projects doubled from 1906 to 1912 alone. However, the most pressing reasons came from the setting of higher education itself. Mott noted that the rapid construction of "handsomely appointed" dormitories, science buildings, libraries, and gymnasia across the country had raised student expectations. Even more to the point, the rapid construction of fraternity houses offered a challenge to the expanding YMCA. As structures housing various organizations set down footprints on a rapidly crowding campus landscape, Mott realized that there was emerging a vigorous contest in which groups would compete for the physical space of student extracurricular life. As campuses developed self-sustaining communities, Mott surmised, those organizations that established a physical presence on the campus would continue to draw student interest. Those that failed to do so would be marginalized as peripheral agencies, at best, and external impositions, at worst.[18]

Mott quickly recognized that such structures were critical to fulfilling the movement's passion to achieve a centralized presence on the American campus. A well-designed building, first of all, would enhance the popularity and dignity of the association in the eyes of students. Mott believed strongly that collegians, particularly the "unconverted," would evaluate a campus organization primarily by material criteria and physical attractiveness. "If (the YMCA) meets in some dusty classroom, or in a cold roomy chapel, or in a poorly furnished room on the third floor," he noted, "they will judge it accordingly. As long as it is true that 'the Lord looketh on the heart,' it will be equally true that 'man looketh on the outward appearance.' "[19] While leaders were clear about the need to gain credibility in the eyes of all students, they seemed to be particularly concerned with the opinion of popular students. Because a building would allow for the organization to attend to the "social side" of student life, Mott reasoned that the YMCA would be far more likely to attract the "athletic and society men" of the campus with such facilities in place. Since universities still provided little space for recreation, a building could provide a place for that "very valuable class of men" to join in YMCA activities.[20]

Of equal importance, Mott and other YMCA and campus leaders viewed these buildings as important sources of unified student life and school spirit on the campuses. In an era where elective course offerings had begun to erode "class spirit" and where the removal of student residence from campus had encouraged fragmentation, YMCA buildings were advertised as hopeful forces for common life. President Frank Strong of the University of Oregon, in his speech commending the YMCA building, noted that such a structure would create a place for unified, nonsectarian religious fellowship, but he also felt that it would "supply one very marked deficiency in the University of Oregon . . . a center for the general student life of the university. Such a building would add immensely to the cohesiveness of the institution, to its unity." At Rutgers in 1912, one enthusiastic student leader spoke of the YMCA building as "the central point in college life; a place, other than the chapel, where fraternity and sectional spirit may broaden into college spirit; or in other words, make our fireplace the cozy corner of a University Club."[21] While fraternities catered to their own memberships, Mott hoped that the YMCA building could become the democratic center of campus social life.

Finally, and perhaps most importantly, Mott envisioned the YMCA building as a home, a place where students, far removed from the refining influence of parents, could be nurtured in an environment of moral sanctity. Appeals for buildings, in fact, were often couched in terms of preventing student participation in more unseemly activities. At Bates College in 1912, the general secretary noted that because there was no adequate provision for a YMCA building, students were seeking forms of recreation in the city which, "if not in themselves dissipating and degrading, are at least so by their associations." Students, he noted, were "increasingly looking to the moving picture, the dance hall, the theater, pool rooms, bowling alleys, and the street for their recreation and amusement." At Georgia Tech, a school bulletin assured parents that students would find at the YMCA "a wholesome atmosphere and adequate amusement, making it unnecessary for a boy to go to the city to spend his idle hours." Here was the architectural equivalent of the in loco parentis ideal that was gradually being eliminated from higher education. As faculty members became less willing to offer guidance to students under their care, many looked to YMCA buildings as "homes" that would secure a neglected familial ethos and promote student moral development. "In a large degree," Mott noted, "the Associations must make up to the students what college life necessarily takes from them, viz: the influence of home life." In loco parentis was for Mott, at least in part, a "bricks and mortar" affair.[22]

In this sense, it is perhaps not surprising that campus leaders saw YMCA buildings as powerful public relations tools against claims of university godlessness, assuaging fears that religion and morality had been marginalized on the campus. Mott, for example, noted that a building campaign transformed the opinion of many who had viewed Cornell as a "godless institution." At the University of Iowa, President Charles Schaeffer suggested that the building served as "vindication from the defamation, must I say malice, that has showered upon our fair city," ensuring that outside observers could "no longer point to the State University of Iowa as an institution of infidels." "The charge has never been true," he noted, "but today Close Hall refutes every argument of that kind." Similarly, President W.M. Beardshear of Iowa State College served on the committee that commended a YMCA building campaign in part because, "The presence upon the campus of such a building would go a long way toward removing a widespread but erroneous idea about the college, viz: that Christianity and Christian things are discounted." For many presidents, a structure devoted to student religion was precisely what was needed to deflect fears regarding the secularization of the American university.[23]

Buildings had been constructed on a few campuses before Mott's rise to leadership, but early efforts had been local and sporadic at best. Princeton's Murray Hall, erected in 1879, was followed by structures at Yale (1887) and Cornell (1890). These early projects were widely heralded by association leaders as portents of an expanding future, one optimistic spokesperson at the Northfield conference in 1889 suggesting that it was "only a matter of time when every prominent college in our land will have upon its campus a handsomely appointed Association building."[24] Indeed, despite delays caused by the Depression of 1893 and its aftermath, the construction of YMCA buildings did become a veritable "movement" between 1890 and 1915. Thirty-six structures on local campuses devoted exclusively to the use of the YMCA had been erected by 1915. Of the thirty-six, twenty-three were built on the larger campuses with over one thousand students. Only eight were constructed on campuses with fewer than six hundred students, and twenty-five of these edifices were located on state university campuses. With rapid construction occurring throughout the country, it was no exaggeration for Mott to proclaim in the 1890s that "The day of experiment has passed. . . . The building movement has begun."[25]

Buildings were sometimes supplied through the contributions of a single donor, and the rolls included some of the most influential philanthropists of this era. At Princeton, the first college YMCA building in the nation, Murray Hall, was erected in 1879 when alumnus Hamilton Murray donated $20,000 for the construction of a structure that would benefit the "Christian work of the students."[26] Later, philanthropist William E. Dodge provided the funding for Dodge Hall (which was attached to Murray Hall) at Princeton and Earl Hall at Columbia University. Madison Hall at the University of Virginia, valued at $80,000 at the time of its construction in 1905, was the gift of Dodge as well, bestowed by his wife after his death. While Mott provided the energy for a building at Cornell University, it was trustee and New York publisher Alfred S. Barnes who donated the money for Cornell's Barnes Hall. Noted philanthropist Horace B. Sillman donated a building to Union College in Schenectady, New York, and also gave the largest contribution for the building at the University of North Carolina.

Not to be outdone, John D. Rockefeller provided the funding for the YMCA's Rockefeller Hall at Georgia Tech and contributed over half the amount for the association's Lane Hall at the University of Michigan.[27]

At times, donors attached conditions to their contributions. At Johns Hopkins, for example, Eugene Levering contributed $20,000 toward the construction of a "suitable Hall" but mandated that the building be used to promote "study of the historical and philosophical foundations of Christian belief." "Believing as I do in the Holy Scriptures as the inspired Word of God, that these furnish the ultimate criterion of an authority for religious truth, that Jesus was divine, and in the spiritual change described in the New Testament as the new birth," he noted, "I desire that no teaching or lecture inconsistent with these truths be allowed in said building."[28] While still nonsectarian with regard to the evangelical Protestant denominations, such a request for Levering Hall demonstrated an attempt to use these facilities to restrain theological wandering and ensure the propagation of a solid and orthodox Protestant faith.

In cases where single large donors could not be secured, attempts to raise money usually involved canvassing work with students, faculty members, administrators, alumni, and members of the community. College and university leaders, recognizing the importance of a social center on campus and eager to allay public fears over secularization, often contributed significant funds to YMCA buildings. At Dartmouth, for example, the majority of funds for Bartlett Hall were secured through the personal canvassing efforts of the religiously zealous president of the university, Samuel C. Bartlett. At the state University of Iowa, it was the Alumni Association that came to the rescue of a faltering building campaign. Devoid of meeting space after a fire destroyed another building on campus, the Alumni Association agreed to canvass for the association provided that they be given an office in the building and name the structure "Alumni Hall."[29] Quite frequently, assistance was also given from individuals off the campus. The building campaigns at the University of Michigan and University of Wisconsin, for example, included significant contributions from the Presbyterian churches of these states, collected during church services held on the Day of Prayer for colleges.[30]

While the proportions given by each group varied depending on locale, the stories of personal sacrifice were quite remarkable. At Virginia Tech, undergraduate student Lawrence Priddy left his college commencement exercises and immediately began a personal canvass seeking funding for a YMCA building. Over a 2-year period in which he devoted himself to this cause, Priddy traveled 36,796 miles by rail, 518 miles by horseback, 582 miles by buggy, wrote 3,419 personal letters, and called on 18,864 people to raise the $25,000 necessary for completion. Faculty at DePauw

University donated the proceeds of the winter university lecture course for the construction of the building on their campus, and at Wesleyan University, students "on the fringes of poverty" subscribed $6,000 for the construction of their YMCA building, this total averaging about $60 per student at a time when tuition was only about $75 per year. At little-known Maryville College in East Tennessee, poor students spent their summer vacation burning the 300,000 bricks necessary to begin YMCA building construction.[31]

By the early 1900s, a fairly stable pattern of architectural design could be found. Nearly all buildings housed offices for campus YMCA leaders, rooms for Bible study and missionary meetings, and rooms for general campus club activities. Most had a single auditorium, often the largest on the campus, to provide space for mass meetings, lectures, and entertainment. The better endowed buildings possessed bowling alleys, barber shops, gymnasium facilities, swimming pools, and restaurants. Many had game rooms, complete with billiard and pool tables, and some had "smoking rooms" as well beginning in the early twentieth century. To appease administrators, many buildings also reserved rooms for a variety of student activities. At the University of North Carolina, one of the upstairs rooms was used for signal practice for the football team. At the University of Virginia, Madison Hall housed lyceum lectures, music rehearsals, college publications, and rooms for visiting athletic teams. To attract faculty and reveal the links to the academic components of college life, campus YMCA leaders were also encouraged to add libraries, reading rooms, and space for tutoring deficient students. The libraries and reading rooms, in fact, were often the treasures of the association building, many organizations spending thousands of dollars on reading materials in biblical studies and missions. As Mott could state truthfully by 1903, these buildings had something for everyone.[32]

Increasingly, college YMCA buildings were also constructed to include dormitory space for student residence. The advantages of such an arrangement were obvious. Since a majority of universities in this era lacked sufficient building space for student residence, providing such facilities increased the buildings' usefulness on campus and therefore earned the organization favor in the eyes of administrators. Dormitory space also expanded the home-like quality of the buildings and secured worthy peer influences on a full-time basis, serving as an antidote to moral degradation off campus. Finally, and perhaps most importantly, this arrangement provided a permanent source of income for the buildings, thus allaying the need for a sizable endowment fund. Associations, in fact, often rushed to finish these areas of the buildings first so that the income could be generated quickly. At the University of Wisconsin, fifty-nine students in the 1905–1906 school year chose to live at the new YMCA building but had to use ladders to get to their rooms because of the unfinished state of the building.[33]

Because they were so well endowed, and many colleges and universities themselves did not yet provide comparable facilities, YMCA buildings often became the "hub," not only of religious life, but also of the general social life of the campus. By assuming a physical presence as a social center, YMCA buildings solidified the position of the chapters, drawing in students who would not otherwise have come into contact with the organization and raising the prestige value of the association. As W.D. Weatherford noted, "the Young Men's Christian Association, by having its own building, becomes the host of the other organizations and is constantly helping them in their work. It therefore puts the Christian activities of the college in the forefront and makes the Young Men's Christian Association a broad organization interested in every phase of student life."[34] The secretary at the University of North Carolina suggested that the building cemented the always tenuous relationship between the association and the campus. "Twelve months ago the Young Men's Christian Association meant nothing to the students of the university," he noted. "Now they would not know how to get along without it."[35] Without question, the construction of YMCA buildings on college and university campuses was an important variable in the attempt to secure a centralized, insider, "all-campus" presence.

Building a Secretarial Force

In addition to their powerful role in entrenching the YMCA within the physical space of university campuses, the construction of YMCA buildings also indirectly spawned another significant innovation for enhancing campus presence—the hiring of YMCA campus general secretaries. Because of the financial, administrative, and interpersonal expertise necessary to orchestrate the operations of full-service facilities, the national college YMCA leadership soon recognized that schools would require full-time staff. While volunteers were adequate for organizing Bible studies, personal work, and delegations for summer conferences, the acumen necessary for building operations far outstripped both the technical and the experiential resumes of undergraduates.[36] Models for such a position were not lacking. Because city YMCAs entered the building phase somewhat earlier than the student chapters, they had begun hiring secretaries for the work as early as the late 1860s. By 1895, the YMCA *Yearbook* documented a total of 1,159 secretaries nationwide, and from 1900 to 1920 the number of secretaries increased by approximately 323 percent nationally.[37] Yet while this influence was obviously enormous in terms of setting a YMCA precedent for such hiring, many local campus leaders actually pointed to campus athletic organizations as the exemplar paradigms for this new position. Taking a cue from teams that recruited top seniors to come back as "graduate coaches," local campus YMCAs began to tap accomplished student leaders to remain for a time as general secretaries for the progress of the work. Because of their link to the college and proven loyalty to the YMCA, these recent graduates were considered ideal agents for leading the chapters forward.[38]

Secretaries were generally hired by local campus YMCA Boards of Directors or Advisory Boards, committees responsible for chapter direction and increasingly for the administrative, financial, and legal affairs of the associations. While such committees existed on nearly every campus, their importance was accentuated and their role enhanced when campuses possessed buildings. Most often, boards consisted of a combination of faculty members, alumni, local community "men of affairs," members of the institution's board of trustees, student YMCA leaders, and the local general secretary. At Georgia Tech in 1910, for example, the board consisted of two–five students, five–seven businessmen, and three–five faculty. The University of Chicago advisory board possessed an equal number of faculty, alumni, businessmen, and students, while the University of Illinois staff included four faculty members, two students, and a single college alumnus. While constantly reminded of the importance of student initiative, boards provided stability and leadership that addressed present issues in terms of the longer history of an individual chapter and in light of the experience of age.[39] The general secretary was therefore to serve as the liaison between the administrative boards and the student body.

Perhaps not surprisingly, the first campus secretaries in the United States were hired at Yale, Princeton, and Cornell between 1886 and 1890, corresponding roughly with the construction of YMCA buildings in those locales. Still, this represented only 3 of 276 local chapters in existence at the time. After a very slow beginning, however, the numbers escalated sharply. While the intercollegiate movement could boast no student secretaries at its origin in 1877 and only one a decade later (at Yale), by 1900 thirty-eight leaders had been hired. By 1903, Mott could broadcast the fact that there were 60 secretaries giving full or "near full time" to this work, and by 1908 there were YMCA general secretaries at 103 colleges and universities in North America. While this still represented only about 15 percent of all local chapters, the explosion of paid campus ministry directors demonstrated that, while the movement was still a voluntary college organization, it had expanded to the point of requiring, in larger settings, a permanent leadership position outside of the student body.[40]

The qualifications for the secretary position were many and varied. Particularly in the early years of the new position, Mott and his colleagues made clear that the YMCA secretary was to think expansively about his role as that of a "campus pastor," the spiritual leader of the college community. In reality, this was increasingly true. The rise of the YMCA student secretary on American campuses corresponded quite closely with the decline of traditional religious leadership positions on the campus. While it was seldom true that the presence of a YMCA secretary led to the dismissal of a long-placed chaplain or campus pastor, it was often the case that such individuals were not replaced because the YMCA secretary was viewed as filling that role on the campus. When the campus chaplain at the University of Virginia retired in 1896, for example, the presence of an active YMCA secretary rendered the position duplicative and the administration therefore decided to leave it unfilled.[41] Here was yet another sense in which administrators were willing to allow a voluntary organization to assume religious functions that had once been the formalized domain of the institutions themselves.

Because of this, YMCA general secretaries were assumed to have responsibility, not only for administering programs, but also for more general campus pastoral care. At the University of Wisconsin, the general secretary noted that he often received letters from concerned parents and friends urging him to "look after" certain students. One mother in 1901, for example, introduced her son in a letter to the secretary, proclaiming that "He is inclined to be a little wayward and I would take it as a personal favor if you would kindly look after him and try to get him interested in YMCA work."[42] In 1908, when it was clear that the hiring of secretaries was becoming more generalized across the movement, YMCA national Secretary Charles Hurrey wrote that "Far transcending in importance the office and financial work or public activities is the calling on men in their rooms, mingling in their athletic and social life, sharing in their joys and sorrows, and courageously leading them into the deepest Christian experience."[43] In the absence of parental influence, and recognizing the decline of faculty oversight, YMCA secretaries were considered the spiritual shepherds of the campus and perhaps the best representatives of in loco parentis in this era.

Such spirituality was also to be accompanied by a robust social presence in all areas of campus life. Paralleling the YMCA desire to attract representative men to group membership, many recommended that secretaries distinguish themselves within the social world of the college. H.L. Heinzman, himself a general secretary at the University of Iowa, suggested in 1910 that every student religious leader was to be an "ideal college man." "We have come to think and to lead students to expect," he noted, "that every YMCA Secretary has been a leading fraternity man of the college, an athletic hero, a 'number one' student, and everything else that we can put into the college hero."[44] Far from pious recluses, secretaries were to be flooded with school spirit, freely entering into the full complement of campus activities in order to make connections and network with well-positioned students. "If he is not large enough to enter every phase of student activity, he will fail in serving some men of his school," one secretary noted. "If the Student Secretary is a loyal and enthusiastic college man (and if he is not, let us hope that he will soon be forced from our ranks) . . . his interest will become one of the deepest concern for the welfare of the school."[45] To form an organization inclusive of the campus and demonstrative of college pride, the secretary, really the public face of the association, had to excel in making himself a vibrant presence on the campus.

When speaking of such criteria, Mott often used Henry B. Wright, secretary of the Yale association from 1898 to 1901, as his exemplar. The son of an acclaimed Yale professor and Dean, Wright had been voted third "most admired" man in the class, of 1898. He was Salutatorian of his graduating class and a member of Phi Beta Kappa, Delta Kappa Epsilon, and the prestigious Skull and Bones. While in office, Wright expanded Bible study, recruited Moody and Mott for evangelistic talks, built up the Northfield delegation, and initiated a building campaign for the YMCA chapter at Yale's Sheffield Scientific School. Despite these significant accomplishments, however, Wright was most highly regarded as the "friend at large of the Yale undergraduate." He occupied the role of "older brother" to Yale men of this era, giving wise counsel from his office in Dwight Hall and demonstrating a ubiquitous presence at Yale social events. According to his biographer, Wright was frequently approached by students with questions about issues as diverse as pledging fraternities, relationships, career, and the moral status of dancing. Combining social success and spiritual vigor, Mott proclaimed Wright to be "the ideal student secretary."[46]

On many campuses, in fact, the secretary was viewed as a kind of bridge figure who might serve to restore the fragmented relationships between faculty/staff and students by standing between these two poles. Because they were recent graduates and spent a great deal of time in social settings with undergraduates, they could identify closely with the student body. Yet in many cases, the colleges and universities themselves paid portions or all of the YMCA secretarial salaries, making such individuals de facto members of the college staff. While salaries were often supplemented by alumni and parents, the fact that colleges invested financially in these individuals meant that secretaries had allegiances not only to students but to the institutions. Looking at the general secretary position from their vantage point in 1900, Yale scholars James B. Reynolds, Samuel Fisher, and Henry B. Wright suggested:

> He was not one of the members of the faculty, but he promoted a very happy kind of intercourse between them and the students, commanding the

> confidence of both. He was not an undergraduate, but the great majority of the students had known him in recent years as one of their own number and as one whose Christian character commanded their respect. . . . He has thus served as a happy medium of personal Christian intercourse between the students and their undergraduate officers.[47]

Southern Secretary W.D. Weatherford stated that, while faculty and students were often divided from each other, the secretary would be able, "because he is an older man and knows the faculty and pastors better, and at the same time because he is nonofficial and knows the students better," to "bring all together into a more perfect whole." A clear harbinger of the "dean" positions that would rise up in the following decades, the college YMCA secretary was what one president called the "discovered missing link" between the faculty/administration and the student body.[48]

Between 1887, when the first campus secretary was hired, and 1905, the trajectory of the secretary position was largely a function of individual personality and local context. Like secretaries in the YMCA more broadly, these men frequently lacked advanced (or in some cases any) theological education. For most secretaries, especially prior to 1910, this position was viewed as a temporary prospect, a helpful stepping stone to pastoral work, YMCA leadership, or an expanded role on the campus. There were many advantages, of course, to remaining employed on the college campus after graduation. Friendships could be fostered, campus projects could be continued, and the social world of the campus remained accessible. An extra year or two to contemplate a future career was a welcome possibility for many, and the fact that secretaries could engage in graduate training (most took courses in Divinity) made the option that much more compelling. It was not, however, viewed as a "profession." There existed no external agency to centralize and standardize training for these individuals, and while they surely made contact with other leaders at state and national conferences, they lacked resources providing for their own specific vocational expertise. Despite the fact that many national leaders pushed for commitments to stay for a full collegiate "generation," few remained longer than three years in these settings. The position, in this sense, was a means for individuals to bridge the gap between a successful college experience and the uncertainties of permanent vocation.[49]

Beginning in about 1905, however, as a critical mass of such agents developed across the country, the move to generate a secretarial profession began to pick up steam. By 1908, college secretaries started to meet in annual conferences to discuss matters of association policy, program, and effective campus methodology. Several leaders began to stay on in their positions for five or more years, demonstrating that one could choose such work as a legitimate long-term vocation. As campuses witnessed the growth of standardization that inevitably accompanied the arrival of a general secretary, the permanence of such men was heralded as a boon both for the associations and for the colleges as a whole. These combined elements—the development of professional gatherings, the prolonged duration of service, and the confirmation of college and university leaders—conspired to heighten the sense of the YMCA secretaryship as a legitimate career. Such a sense would only grow in the coming years.[50]

Especially in those settings where secretaries stayed for longer durations, this position served as a critical means of enhancing the presence and reputation of the YMCA on American campuses. These individuals made connections with large numbers of undergraduate students, in most cases serving as the only collegiate staff members paid to exercise moral and religious oversight for students. They served as role models for the vision of the association, putting on display an attractive form of Christian faith. They served as personal bridge figures between religion and student life, crossing boundaries between disparate social groups to enhance the penetration of Christianity on the campus. For many undergraduates, in fact, they were the most representative nonstudent faces in the institution. Together with the value of YMCA buildings, secretaries cemented the position of the organization in American higher education.

Serving the Campus

For YMCA secretaries, one of the most critical means of developing a more inclusive campus presence was through social and service activities directed squarely at the student body. While local chapters had engaged minimally in such work even before 1890, by the time of World War I the YMCA had quite literally become the de facto center for student services and activities on campuses nationwide. As enrollments grew rapidly, institutions were often unprepared to meet the student housing, employment, and entertainment needs created by this deluge. Much as they did with other dimensions of student life before World War I, administrators allowed such matters to be handled by voluntary means, and in the early years of the twentieth century it was the YMCA that filled this sizeable gap. Until administrators assumed responsibility for these services after World War I through the hiring of deans and other officers of "student development," the YMCA played a critical role in caring for student needs, making themselves increasingly indispensable to their institutions.

The first foray into campus social service came through the publication of the university freshman student handbook, sent out to students even before they arrived on campus. Designed to fit "conveniently in a man's vest pocket," the handbook, funded by local advertisements, was a resource that typically contained valuable orientation information (campus maps, lists of area churches, and brief sketches of student organizations), a complete student/faculty directory, and schedules of campus events. For example, the Indiana University "Red Book" contained the names of all students and faculty, a description of YMCA activities, and an agenda for the freshman student orientation. The Northwestern University "N Book" followed a similar format but also included brief sermons (from 1880 to 1894), lists of available service opportunities (from 1895 to 1915), and a comprehensive guide to the city of Evanston. Several "freshman Bibles," as they were often called, also served to educate incoming students on the "yells, songs, and traditions" of the institutions, many of which were practiced jointly during the YMCA-sponsored reception at the beginning of the academic year. The book, one YMCA leader noted, was to "detail those facts about the college the Freshmen need to know, but dislike to inquire about too frequently."[51] While the University of Illinois, the University of Virginia, and Ohio State had all initiated handbooks as early as 1884, by 1900 they had become a nearly universalized symbol of the YMCA's connection to student life and culture.[52]

YMCA chapters also provided strong programs for orienting students to the campus. In the years before the war, almost all local chapters maintained a policy of sending the most "socially attractive" members of the group to meet freshman students at the train station upon their arrival. Assisting with luggage, they escorted new students to the YMCA information bureau, where the cabinet was waiting to provide a variety of new student services. Those requesting information about the campus were given tours by upper-class students. Those in need of housing were provided with a compiled listing of all available rooms and boarding houses. At Cornell in 1902–1903, 400 students were provided with rooms through the YMCA. At Yale, leaders noted that all houses and rooms listed had been personally inspected by association members. In the era preceding the widespread construction of dormitories, the YMCA was thus responsible for placing many incoming students into appropriate places of residence, a monumental task. YMCA chapters nationwide assisted new students during class registration, a job made more pressing by the increasing availability of elective classes. They also sponsored freshman receptions, attended by professors, administrators, and student leaders. As one state secretary put it, "They had a most Christian style of 'hazing' newcomers."[53]

Beyond freshmen tasks, one of the most important arenas of YMCA campus service in these years related to student employment. Most local chapters in the years preceding World War I maintained "employment bureaus" in order to assist students in finding jobs on campus or within the local community to defray collegiate expenses. Each association chapter maintained a card catalogue of students requiring work and a record of positions in which men were needed. The types of jobs secured by students through association auspices were quite diverse, ranging from permanent yearly employment to odd jobs designed to provide quick sources of income. Clerical work was typical in both permanent and temporary settings, but manual labor was also quite common. At Bowdoin in 1912, students took up jobs such as addressing envelopes, canvassing, caring for horses, tending furnaces, lawn care, tutoring and teaching, typewriting, waiting on tables, and washing dishes. At Rutgers, among 278 students who secured jobs at this same time, 100 were engaged in farming tasks, while 32 students served as waiters, 28 in clerical positions, 14 in factories, and the balance in various odd jobs. The YMCA at the University of Southern California started a lawn mowing company to enable students to generate weekly income.[54]

The number of students securing employment through the association, coupled with the raw income figures generated through student labor, demonstrated the immense value of these efforts. At Northwestern, where this bureau was created in 1898, more than a hundred students (of a total student body of 377) found jobs through the organization that year. Four years later, when the total number of students had increased by only 30, the number of students assisted by the YMCA jumped to 190. In 1905–1906, students at Columbia University earned $39,660 with the assistance of the YMCA, approximately one-third of the total amount raised by all students that year. During a period in which many students were working, the YMCA employment bureau surely served as a financial lifeline for those possessing "more brains than means."[55]

Furthermore, the association was involved in setting up social activities for the institution. YMCA chapters regularly sponsored receptions, banquets, postexam parties, football pep rallies, and dances, attempting to place such events on an all-campus basis. YMCA buildings, of course, also provided much entertainment in the bowling alleys, gymnasiums, and billiard and pool game rooms. The addition of dances and billiard playing revealed a subtle departure from more conservative Christian opinion on popular amusements at the time. In fact, University of Illinois Professor Thomas Burrill was forced to defend the addition of billiard and pool tables against the outcry of select local ministers. While he acknowledged that such activities were often "associated with bad and demoralizing conditions," he also noted that the YMCA's provision of such amusements prevented students from pursuing such enjoyment in more unseemly locations. This was the YMCA perspective in most locales in the early twentieth century. With the exception of more specifically taboo activities related to drinking and sexual exploration, the movement largely allowed for social activities that had been corrupted only by their associations. As long as dancing, billiard playing, and bowling were incorporated into the YMCA building rather than other degrading contexts, they encouraged a participation in such activities.[56]

College association leaders also attempted to develop an organizational infrastructure that would allow them to minister to international students, fashioning the Committee on Friendly Relations among Foreign Students in 1911. Charles Hurrey, Mott's former executive assistant, took over as general secretary of this new organization, which eventually supported not only the YMCA, but also the YWCA, the German Students Cooperative Association, the Hindustan Association, and the various campus Cosmopolitan Clubs. The Committee established programs that bore the unmistakable mark of the prewar YMCA. Students were met by members of the Committee upon arrival in the country and were given assistance in finding accomodations, purchasing railway tickets, selecting courses, securing employment, and locating campus fellowship. They helped those who were sick and provided loans for needy students. In select locations, they even established "international houses" to support the service work with this population. YMCA students initiated English classes for those struggling with the language, and they also invited these students to the summer YMCA conferences. By 1919, in fact, 455 international students were attending the six national conferences, individual nations sending large delegations.[57]

Such work, which also included biannual book exchange programs, nonmedical care for ill students, and academic tutoring, surely enhanced the organization's exposure even as it solidified its reputation among students and administrators. In 1900 alone, the University of Minnesota YMCA sponsored large university socials, postexam celebrations, a boarding house bureau, an employment bureau, tutoring services, a student loan fund, and a loan library for textbooks. "These agencies place the Young Men's Christian Association in a position commanding the respect of the non-Christian element of the student body and faculty," the general secretary noted. "In this way is also gained an avenue to the heart of many a student who will never forget the hand of good cheer and helpfulness that was extended to him by a band of young men who own as their Master the loving Son of God . . . "[58] As with the construction of buildings and the hiring of secretaries, the development of a vigorous campus service program had the effect of linking the organization more closely to the life of the institution and therefore making it essential to the functioning of the campus. If the association chapters were formerly envisioned as popular extracurricular expressions of student voluntary religious activity, they were increasingly regarded as integral aspects of general campus life.

A Popular Movement

The attempt to develop a strong campus presence through the construction of buildings, the hiring of secretaries, and the development of a campus service program paid obvious dividends for local chapters. Administrators were quite pleased with this expansive work. On the one hand, the YMCA served for many as powerful evidence against accusations of "rampant secularism" and godlessness in the university.[59] When asked what would be lost if the YMCA were no longer functional on campus, University of Oklahoma President Buchanan stated that the "advertising value" of the organization would be missed, noting that he "Gave talk on religious life of the university and I stressed the activity of the Y. It makes men and women feel safer in sending sons here." The registrar at the University of Denver stated similarly that "Many students who might go to a church school in order to be in a religious environment will come to the university because it has a YMCA." With active YMCAs, school leaders could point to a vibrant religiosity at the center of student life.[60] On the other hand, administrators also recognized that in a time of expanding enrollments and limited staff and facilities, the YMCA was indispensable to student care. President Cyrus Northrop of the University of Minnesota, for example, was effusive in his praise for association service work:

> With us in the University of Minnesota, where not many of the students are wealthy and many are dependent on their own efforts for support in a greater or less degree, the association welcomes the new student when he comes a stranger to the university. It helps him to find a home suited to his means and proper for him to live in. . . . It finds employment for those who must do something for their own support. It furnishes free classes to those students who are deficient in preparation and must be trained without expense. . . . Dealing with material somewhat different from the blue blooded sons of a long line of college ancestors in Harvard, or Yale, or Princeton, it ultimately makes out of this material true men who will fear God.[61]

In addition, between 1890 and 1915, the YMCA assumed a place within the campus social hierarchy that made it a potential locus of prestige and school spirit. The YMCA in many locales succeeded in attracting representative men to association leadership. In 1909, one of the newly elected presidents of the Princeton association was president of the junior class, captain of the basketball team, and voted most popular in his class. The Princeton *Alumni Weekly* boasted that in the previous few years the president's position had been filled by a varsity football player, the editor of the *Princetonian*, and other "well-known men" who "commanded the respect of the whole college." Arthur Reed Kimball at Yale noted in 1900 that the leaders of the campus YMCA were "representative men, those who achieve its highest social honors." "In other words," another concluded, "the Young Men's Christian Association at Yale has ordinarily been in the hands, not of sanctimonious-looking, narrow minded Christians, but of manly, straightforward undergraduates, interested in the broader aspects of college life, yet convinced that without the religion of Christ all life must be incomplete." By the early 1900s, students lobbied for roles in the YMCA much as they would for positions in fraternities and other campus organizations.[62] Even more to the point, YMCA membership by the turn of the century was also heralded as an indicator of school spirit. Canby remarked that students were often encouraged to join the YMCA not because of religious growth but because it was a means of supporting the college. Students were told, "You fellows that aren't out for the teams or the musical clubs ought to see whether you can't do something there. It's a good thing, anyhow, and religious and all that; but what I'm saying is that it's a college activity and ought to be supported. Where's your spirit, anyhow?"[63]

At times, this close connection between religious and campus prestige caused some to question the motives of eager recruits. Describing the tendency for student social and religious aspirations to blur, Yale observers commented that "The danger is surely not that outspoken Christian men be cut off from playing a prominent part in the social life of the college, but perhaps it is that some men may be drawn into the religious work without sincere conviction, in the hope of social gains."[64] At Yale, the organization grew so influential that many brought accusations that students were using the chapter for purely political and social reasons. Widely publicized rumors circulated around campus suggesting that the YMCA had an underground link to one of the senior societies, largely controlling the outcome of the selection process. Students rankled by this "admixture of politics and religion" assembled an anti-YMCA club called the "Holy Pokers" to dismantle what one called the Dwight Hall "Machine." Though the group was largely unsuccessful (likely due to the fact that two out of three members of the Yale community were YMCA members in 1899) and chaos subsided the following year, the social life on the campus was deeply splintered by this "ungodly schism." "Fortunately," Henry Wright suggested, most undergraduates would not allow the YMCA to become an agency of feigned religious enthusiasm. "Yale despises the hypocrite," he quipped, "despises him with a more genuine feeling than she despises any other species, unless it be the snob."[65]

Wright's optimism, however, may have been somewhat overstated. In a national survey conducted by Edwin Starbuck addressing the factors contributing to a deepening of the spiritual life on American campuses, a number of students suggested that YMCA student leaders were chosen on the basis of social and athletic rather than spiritual qualifications. One student noted that "leaders in the Associations habitually do things that, according to my standard, are not in keeping with the character they ought to have." Another commented, "I have noticed that our religious association has always been controlled by a type that seems to me to gravitate there by selection. The athletic fellows (members of the Association) are anything but religious." While only five of fifty-four described the YMCA as a "dynamic spiritual force," the majority indicated that the organization was quite popular, drawing social leaders to its meetings and activities. For some, a growing campus presence occurred concomitantly with a dwindling spiritual identity.[66]

To their credit, many YMCA leaders recognized the potential incompatibility between heightened prestige and the movement's ongoing spiritual mission. The state student secretary of Illinois commented in 1903 that "It cannot be too often emphasized that the Association is not a mere ethical or philanthropic organization. Its strength does not lie in gymnasiums or buildings or employment bureaus, or in any secular enterprise, however worthy that may be." Likewise, the YMCA general secretary at Syracuse University suggested in 1905 that chapters seeking to secure status among the most popular students of the campus were making "Judas bodies" of the Associations, pushing them away from their true purpose. Mott was clear that the reliance upon popularity and campus influence would be the greatest temptation faced by the movement. "Here and there may be noted a perilous tendency of trusting for spiritual results apparently more attractive buildings, social influence, perfection of organization, able secretaries, the increasing prestige of the Association movement, than to God himself with whom alone resides the ability to vitalize and energize our work," he stated. More than most, in fact, he saw that secularization was a likely concomitant of a prestigious campus presence. "To keep a large and commanding organization pure in its influence requires special watchfulness and prayer," he suggested, because "the prestige of the Association coming from the building and large membership, will lead some to join for the influence it may bring. Right here we cannot be too vigilant. . . . The one and only way of safety is to emphasize the spiritual side of our work."[67]

Interestingly, however, the "spiritual side" of the work was shifting. Between 1890 and 1915, sparked in many ways by this activist student culture, the college YMCA movement experienced a substantive revision of purpose and program that generated a less distinctly theological perspective. As both leaders and participants began to move from the evangelistic themes of the early movement to a vision of practical activism through character building and service, the place of Christian belief within the association began to erode. The desire for movement prestige surely exacerbated this shift, but theological changes were also at play. The vehicle for this more explicit secularization was muscular Christianity, a perspective that highlighted the "strenuous life" of purity and active service rather than regenerative soul work.

Chapter Five
A Muscular Faith

In his 1920s novel *Elmer Gantry*, Sinclair Lewis provided a memorable portrait—indeed a parody—of a YMCA Week of Prayer gathering at fictional Terwillinger College. At the center of his depiction was state college YMCA Secretary Judson Roberts, a former standout in football, baseball, and debate at the University of Chicago. Nicknamed "the praying fullback," Roberts applied his aggressive personality to campus revival meetings. While he could bring audiences to tears with his descriptions of "the Christchild," Lewis noted, he could also have them "stretching with admiration as he arched his big shoulder-muscles and observed that he would knock the block off any sneering, sneaking, lying, beer-bloated bully who should dare to come up to him in a meeting and try to throw a monkey-wrench into the machinery by a lot of quibbling, atheistic, smart-aleck doubts." He was, Lewis editorialized, "a real red-blooded regular fellow!" Roberts urged admiring onlookers to consider career missionary work and then to "think how clean and pure and manly you'd want to be if you were going to carry the joys of Christianity to a lot of poor gazebos that are under the evil spell of Buddhism."[1] Irrevocably linking physical and spiritual commitment, he urged undergraduates to embark on a regimen of exercise and cold baths, noting that bodily strength was a necessary prerequisite for spiritual labor. In the presence of this man, Elmer Gantry was "allowed to swell with hero-worship." As the impressionable student fell under Roberts' spell, the "praying fullback," it appeared, had won another recruit.[2]

Himself a member of the Oberlin College YMCA during his undergraduate years in the first decade of the twentieth century, Lewis had personal experience with these kinds of events. He was active in the campus chapter and noted that the organization "impressed me very highly as an indication of positive earnest muscular Christianity," quite distinct from the "hypocritical watery sort of thing" he had found in other religious organizations.[3] Though a fictional account, therefore, his portrayal resonated deeply with the emerging philosophy of the college YMCA in the era between 1890 and 1915. As Clifford Putney has recently described, the desire to secure a manly faith was a pervasive theme in American Protestantism at the turn of the century, reflecting in many cases an anxiety regarding the identification of Christianity with feminine interests.[4] In a 1910 YMCA survey, statistics revealed that church membership was about two-thirds female, and representatives of the Men and Religion Forward Movement in 1911 estimated that there were "3,000,000 more girls and women in the churches of America than men and boys."[5] Other studies demonstrated that male Sunday school attendance dropped precipitously after age twelve, a frightening prospect since adolescence constituted the ideal incubation period for youthful conversions.[6] Tethered to this reality, college YMCA leaders also recognized that vocational ministry was losing its attractiveness as a career for college men. Of the nearly 1,200 graduates from Yale, Harvard, Columbia, and Princeton in 1904, less than 30 stated that they were planning to enter the ministry, and by 1915, students preparing for pastoral work included a disappointing 3 at Cornell, 2 at the University of Illinois, 9 at Indiana University, 4 at the University of Oklahoma, and 8 at the University of Pennsylvania. According to many critics, Christianity was progressively losing its masculine influence.[7]

Interestingly, the loss of men from church and pulpit was frequently described, not in terms of masculine irreligion, but rather as a clear sign of the incompatibility between typical religious forms and the masculine character. A number of books critiqued Protestant middle-class Christianity as sentimental rather than virile, a faith of rest, repose, and abstract thought rather than forceful action. Typical religiosity, they noted, was too otherworldly, calling young men to lives of personal piety and pallid virtue rather than to active engagement in the affairs of the world. Popular Congregationalist Minister Josiah Strong claimed that "There is not enough of effort, of struggle in the typical church life of to-day to win young men to the church," adding that a "flowery bed of ease does not appeal to a fellow who has any manhood in him."[8] Emphases on love and tolerance, coupled with portrayals of Jesus that highlighted his meek and peaceful demeanor, seemed to alienate men from institutional religion. While few denied the importance of values such as love, peace, and moral refinement, many were unequivocal in noting that the masculine, practical, and activist components of Christianity had been largely and inappropriately eclipsed by these softer themes. John Childs, assistant YMCA secretary at the University of Wisconsin, claimed that Christian institutions were producing "men who consciously, or unconsciously, are becoming established in the belief that religion is for women and children, and, perhaps, a few sentimental men."[9]

According to Mott, such religious proclivities made pastoral ministry unappealing to many young men. His 1908 book, *The Future Leadership of the Church*, in fact, suggested that, while the "allurements of material progress and success" contributed to the attractiveness of other professions, the incompatibility of ministry and manhood was chiefly to blame:

> The absence of men from certain churches, and the thought that these churches have to deal mostly with women and children, have helped to cause, in some minds, the misconception that the ministry does not present sufficient opportunity to exercise the talents of strong men. Again, the young man notices that some city churches are made up of people of wealth, culture, and refinement, and decides that work with such people would be too easy and along too soft lines. . . . The dullness, lack of variety, parochial outlook, and sectarian spirit which characterize the ministry in many a community do not appeal to the aggressive young man as offering vent for his spirit of Christian enterprise. Then it is necessary to bear in mind the prevalent ambition of young men "to do things," as they express it. To build a bridge, to organize a corporation, to frame a law, to discover some new way to relieve physical suffering—all these seem to be achievements, but the work of the ministry does not seem to them in any real sense the achieving of tangible and important results.[10]

In a series of books commissioned by the college YMCA attempting to pique student interest in vocational ministry, many authors decried the absence of manly exemplars in ministry settings.[11] *The Claims of the Ministry on Strong Men* by George Angier Gordon suggested that the churches were filled with "wooden priests" who lacked the personal vitality necessary to command attention and exert influence beyond the church walls.[12] In a similar vein, Oberlin's Edward Bosworth noted that, in order to open themselves to the possibility of a ministry career, most college men would have to "shake off the memory of some church, whose minister has not been alive to (outside) opportunity, and whose contribution to the welfare of the community is so meager and indistinct as to make his ministry decidedly unattractive to an earnest man."[13] It was the negative example of such men, these authors suggested, that had dissuaded male students from looking to the ministry as a potential venue of active service to the kingdom. Men were fleeing the church because ministers were unfamiliar with the "rough and tumble" realities of life and too immersed within the domains of home and parish to understand the masculine world.

In a related sense, many felt that cultural shifts were threatening the very nature of masculinity, leaving men docile in the face of great social need. Boys were growing up in home and school contexts increasingly dominated by female influences. The oft-cited closing of the American frontier had supposedly blunted the self-assertive wanderlust impulse, diminishing the need for bodily strength and courage while generating a sense of enervating confinement. Urbanization and industrialization had shifted the source of male identity from production to consumption, and many began to contend that urban luxury had enfeebled muscular ardor in the male population. Distanced from manual labor and softened by the comforts and conveniences of consumer wealth, many spoke of middle and upper class "overcivilization," the decline of physical and moral strength stemming from lives of easy self-indulgence. Corrupted by their own economic success, the middle and upper classes were therefore threatened with a loss of both the Puritan work ethic and the Puritan moral will, generating an "overrefined" sensitivity lacking strenuous effort. For some, this trend could only presage a form of gradual "race suicide." Statistics revealed a declining birth rate among Anglo-Americans, setting the stage for the replacement of effete native stock with a new wave of muscular immigrants. Combining moral, class, and racial fears, the attack on feminization was broad-based in its rhetorical power.[14]

Shifting economic realities in the urbanizing and industrializing fin-de-siecle United States also complicated the development of masculinity. This concern certainly included the lack of physical exertion awaiting young men in sedentary, white-collar professions, but the critique was more inclusive. In a broader sense, men were purportedly losing the initiative and self-reliance that had blossomed under the rubric of mid-nineteenth century entrepreneurial capitalism. Corporate capitalism, by contrast, seemed to constrain manly self-assertion within webs of bureaucratic norms. Many found career ceilings within the layers of middle management or clerical work, blunting aspirations for business leadership. With fewer men either owning their own farms and small businesses or possessing first-hand contact with the products of their labor, the perception of individual potency was greatly curtailed. The self-made man, previously a staple of masculine self-identification, was rendered by industrialization a convenient vocational fiction.[15] Such economic realities pointed to a collapse of the Victorian gender system that had linked and coordinated female religious piety with male economic competition. Such a system had worked in part because it allowed for both unimpeded economic progress and also a sense of feminine restraint on the potential abuses of such labor. Once the small-scale entrepreneurial capitalism of the nineteenth century began to give way to corporate models, however, a feminized Protestantism was no longer needed to balance an unmitigated male aggressive competition. For a male identity cramped and emasculated by economic factors, feminine religion looked enfeebling rather than helpful.[16]

While responses varied, many Protestant leaders between 1890 and 1915 confronted the masculinity crisis of church and culture by highlighting the "strenuous" components of the Christian faith over and against its more refined elements. Through a number of books and organizations designed to address these themes, a picture of the ideal muscular Christian emerged with some degree of clarity.[17] Such a man was to energetically fight against sin and for righteousness, heroically bringing Christian principles to bear on the practical issues of life. Gentlemanly virtues, anchored in graciousness and gentility, no longer seemed appropriate as antidotes to overrefinement. As Joseph Kett has suggested, the middle-class spirit of activism at the end of the nineteenth century tended to downplay "maturity and intellectuality" while highlighting "physical prowess and perpetual becoming." In such a culture, Kett argues, the word "manliness" came increasingly to represent, not the opposite of "childishness" but rather the opposite of "femininity." Action was valued over reflection, aggression over restraint, and heroism over placid goodness. In this sense, American Protestants drank deeply of Theodore Roosevelt's appeals to the "strenuous life" of masculine initiative, exemplified in his own aggressive and imperialistic "Rough Rider" image. Attempting to restore the masculine potency lost through cultural, religious, and intellectual changes, muscular Christianity was poised to reinvigorate the cause of Christ and attract men back into the fold.[18]

Making Muscular Christians

Recognizing that many saw Christianity as weak and effeminate because of associations with intellectualism, introspection, and other-worldliness, members of the college YMCA began to suggest that campus chapters could serve an important role in helping students experience the manly nature of the faith. The movement's position on these issues was perhaps best articulated by Henry Churchill King, who, as president of Oberlin College, had molded this institution into a virtual factory for the production of muscular Christians. An ardent supporter of the college YMCA, King delivered an address at Northfield, later published by Association Press, that for many became the clarion call of the new emphasis. Entitled *How to Make a Rational Fight for Character*, this book documented the ways in which early-twentieth-century life had generated a "vacillating, flabby, self-indulgent generation." Physically and spiritually emasculated, Christian men, he proposed, had lost the vigor of their Puritan ancestors who had devoted themselves not only to taxing agricultural pursuits but also to a fierce spiritual battle against the forces of evil. What was most needed in this new cultural setting, King suggested, was a "new Puritanism" rooted not in conversion or theological acumen but in the development of strong and manly Christian "character."[19]

Such masculine character, in King's estimation, was to be defined by self-control and the ability to pursue a "clean life," defeating the temptations of the world through the vigorous exercise of willpower and personal "attention." "The moral life," he said, "is made up of a series of volitions that involve the definite choice of definite means to definite ends," and the key to success was therefore linked to the ability to resist the lure of sin and work instead for righteousness in each "moment of decision." What was needed was a decisive commitment to confront evil with boldness, pursuing righteous character in every aspect of what the YMCA now called the "fourfold" (intellectual, social, physical, and spiritual) life. "Christian brethren, this is life, temptation, trial, struggle, conflict, possible victory—the strenuous life," he told his audience of young men. "You cannot cowardly give it up."[20] As T.J. Jackson Lears has suggested, while theological constraints were softened during this era, there was no parallel "loosening of the bourgeois morality of self-control" among American Protestants. The strenuous life of bold resistance to sin continued as a cherished muscular ideal.[21]

The fact that character was linked so powerfully to self-control, willpower, and attention meant that the body was a critical factor in moral and religious growth. King reminded readers that they were not "disembodied spirits" but rather bodily creatures with physical natures that either helped or hindered character development. The strengthening of will and attention, King posited, could only come through a supply of "surplus nervous energy" available to fortify resistance to the urgings of human impulse. As the basic source of surplus energy to bolster the will, the body was for King the critical reservoir of moral power. In fact, he claimed that a lack of concern for health and vigor was a chief source of much collegiate immorality, fueled not by an evil spirit but rather by enfeebled flesh. Sleep, exercise, and proper diet were of primary importance in furnishing the capacity to battle against sin and temptation. If college men would attend to such needs, he reasoned, they would then have "a margin of capital, with power to attend, with power to will, with power, therefore, of self-control."[22]

In addition to bodily discipline, however, the other key component to forming masculine character was a life of active service to others. Moral failure, King felt, was too often the result of a sedentary Christian life rooted in passive listening rather than active devotion to helping others. Because men were "made for action," a life dedicated to working for human betterment was not only a divine command but also the best means of buttressing the moral will. In other words, while service was a good in itself, it was also a fundamental means of providing new challenges and desires that would replace sinful proclivities with righteous passions. "You can," King concluded, "make no hopeful fight for your own character, without beginning at once a service for others."[23] For King, character and service were therefore inextricably intertwined. The pure and healthy body was a potent tool of social betterment. Similarly, active service was both a reflection of personal goodness and a means to actively forge a righteous life. In King's own Oberlin College, the 1908–1909 YMCA Annual Report stated unabashedly that "The whole aim of the Association is to help develop the boys that come to Oberlin into manly Christian men . . . with strong Christian characters, broad minds and healthy bodies . . . men who believe that to serve God best they must serve their fellow men."[24] Thus combined, character building and service emerged as the sine qua non of muscular Christianity and of the college YMCA.

In order to help develop collegiate muscular Christians anchored by King's "new Puritanism," YMCA leaders were concerned that students possess human exemplars of this ideal. Some leaders felt that recruiting athletes to the movement would be one of the central vehicles for demonstrating to the campus at large the essential compatibility between Christianity and masculine vigor.[25] A Christian athlete, one leader commented, was of significant "usefulness" for the gospel because "too long already has an unbelieving world refused to see that Christianity is good for others than boys and old people, and that to be a manly man is to be a Christian." While young Christian men had previously been "laid upon the shelf as useless in His service," they were now poised to prove to the world that "in addition to faithful church service, public praying and conducting devotional meetings, they conscientiously play ball, jump, run, and in every way contend on an equal footing with most amateur non-Christian athletes." College leaders were eager to counter the conception that Christianity softened men by removing vigor and forceful drive. "If then, the Christian college athlete accomplishes nothing beyond the rooting out of these prejudices and false beliefs," one leader suggested, "he has served a very noble purpose."[26]

Since church leaders and missionaries seemed to fail in this regard, the important role of "mentoring" was assigned chiefly to YMCA secretaries. Perhaps not surprisingly, secretaries on local campuses were often chosen because of their demonstration of fourfold masculine Christian character. Student Clay Shepard of Oregon State University noted regarding his secretary, E.F. Colton, "We are told that a perfect man must be three-sided: he must be filled out mentally, physically, and spiritually, and Colton certainly comes nearer being that man than any I have ever seen. He is an old football player and he is one of the best built men I ever saw. You can see that he is a college man out and out, and when he talks to students he knows how to touch them; he is very sharp and quick and far sighted and sensible and he is a man that does things, not only says them."[27] On one occasion, southern Secretary W.D. Weatherford pulled together some of the best athletes at the college and challenged them to a competitive exercise regimen. Demonstrating his prowess on the giant swings and bar vaults, "he found a ready audience for his evening sermons." Student leaders at the

University of South Carolina spoke approvingly of Weatherford's visits because his intellect, physical strength, winsome personality, and spiritual passion "dispelled the idea, which might have existed in the mind of any man, that Christianity was not a manly thing."[28] If church leaders were increasingly described as devoid of masculine vigor, YMCA secretaries could fill the void.

Perhaps the most famous leader along these lines was Amos Alonzo Stagg, campus secretary at Yale in the 1890s and later head football coach at the University of Chicago. Stagg's meteoric rise within the college YMCA spoke volumes about the leadership ideals of the movement in these years. A central figure in the cult of Yale athletics, he pitched for the "Yale Nine" and was a member of five consecutive championship teams, regularly defeating arch rivals Harvard and Princeton. He was also a regular contributor to the school paper, took a post as financial editor of the *Yale News*, and sang tenor for the prestigious Yale Glee Club. Because of these many avenues of campus influence, Stagg was elected to Skull and Bones, the celebrated senior society reserved for the most influential students. Recognizing his potential for influence, Moody tapped this campus hero to organize athletic activities at the Northfield conferences. Then, at Mott's request, Stagg stayed at Yale to serve as the third general secretary of the association. Mott noted the power such an athletic presence could have both for the prestige of the association and the masculine reputation of the faith, suggesting that there was "no man in New England whose words will carry greater weight with college students" than Stagg. When he became secretary, he helped initiate the Yale Mission, an outreach to the "poor wretched men" in the community, and also spearheaded the work to send deputations to high schools to communicate the "Yale Christian ideal" of "moral excellence" and "service to others." While serving as secretary, he joined the football team at the urging of YMCA General Secretary Richard Morse "on the ground that it would increase his influence as a worker." He later confessed that he felt the YMCA was "capitalizing on his athletic prowess" for the benefit of the organization.[29]

With such leaders serving as showcases of fourfold muscular Christian ideals, the college YMCA program in these years was designed to develop similar qualities in students. By the turn of the century, discussions of doctrinal, theological, and evangelistic issues were increasingly rare in religious meetings, replaced by addresses on morality and social service. Within the first decade of the twentieth century, appeals to clean character and service coexisted in nearly equal measure. The 1903 and 1904 editions of *The Intercollegian*, for example, saw recommendations for local chapters to deal with "the battle with temptation," "fighting for the clean life," and "honesty in examinations," while also addressing "what is a call to service?" "the philanthropic needs of the community," and "opportunities for service in vacation." Local campuses appeared to follow these directives. At Rutgers, topics related to evangelism were gradually replaced between 1900 and 1910 with appeals to "the battle for purity," "clean athletics," "Christian service," and "serving Christ," while at Oberlin, leaders substituted traditional theological themes with such issues as "the war of character," "boys club work," and "opportunities for service." At least at the rhetorical level, the character and service language of muscular Christianity, replete with military metaphors, had become mainstream.[30]

A Muscular Bible

One of the most significant forces for communicating the new muscular vision, however, came through Bible study. Mott lamented in 1897 that less than half the number of active members of the movement were involved in Bible study, but the growth after this time, even taking into consideration the tremendous enrollment spike in higher education, was dramatic. Concerted rallies to enhance Bible study enrollment resulted in an expansion of participation from 15,990 in 1903 to 45,340 in 1907. Both private and state campuses showed similar renewed enthusiasm for campus Bible study in these years. While Yale boasted a healthy 200 enrolled in 1901, by 1908–1909 under General Secretary Kenneth Scott Latourette this number had escalated to 1,000. At the University of Illinois, which sustained one of the largest Bible study enrollments nationwide throughout this era, the 242 students involved in 1901 had mushroomed to 1,049 in 1907, 802 of whom persevered for more than two months in the program.

Not surprisingly, statistical summaries regularly boasted about how the associations had attracted "representative men." In 1908, statistical surveys revealed pointedly that 755 class presidents, 1,402 baseball players, 1,522 football players, 1,454 glee club members, 712 basketball players, 1,053 track stars, and 653 newspaper editors were involved in YMCA-sponsored studies. The involvement of fraternity men was also widely heralded as an indication of the mainstream status of Bible study. While 53 institutions possessed fraternity YMCA Bible classes with 1,909 enrolled in 1905, 125 institutions sustained such groups for 5,000 fraternity members by 1908.[31] While 3.8 percent of the total male student population enrolled in groups in 1886, 24.5 percent enrolled in 1908. With about a quarter of all college men taking part in YMCA Bible study groups and about 42,000 study manuals sold in 1910, it was little exaggeration for religious educator George Albert Coe to describe the movement as a "renaissance of popular Bible study" in American higher education.[32]

At first blush, it may appear ironic that a movement now driven by a philosophy of muscular faith would devote so much attention to book study. In reality, however, both the content and the method of such study were significant factors in the transition to the new paradigm. Mott proposed in the first year of his new leadership position that every association should make provision for three kinds of Bible study. "Personal work" study, as the name implied, sustained the legacy of Wishard's earlier emphasis by recommending the Bible as a practical ministry tool. Developing a series of "workers' Bible training classes," interested students gathered weekly to study both the biblical text and specialized evangelistic works such as L.W. Messer's *Christ as a Personal Worker* (1891) and H.C. Trumbull's *Individual Work for Individuals* (1901).[33] Yet while such study had been the dominant feature of association work, by the early 1890s many leaders were increasingly convinced that evangelistic Bible study was inadequate for the collegian. In an address to the assembled students of the YMCAs of New England, William Rainey Harper, then at Yale, claimed that in an atmosphere where scientific rigor was applied to religious topics it was unconscionable to neglect the academic study of the Bible.[34] Harper was convinced that students should engage in inductive textual study, employing rigorous scientific and analytical investigatory processes of observation and interpretation before discerning practical implications. A good friend of Harper's, Mott took the scholar's words to heart. Encouraging students to pursue so-called intellectual study, Mott and Ober published a monthly series of Harper's Old Testament studies in *The Intercollegian*. In addition, by the early 1890s, leaders had introduced a series of inductive study texts produced specifically for the college YMCA movement, including Robert Speer's *Studies in the Gospel of Luke* (1892) and *Studies in the Book of Acts* (1892) and Wilber W. White's *Thirty Studies in Jeremiah* (1896) and *Inductive Studies in the Twelve Minor Prophets* (1892).[35]

For Mott, the importance attributed to such academic study in no way distracted from the final "devotional" aspect of biblical analysis. Devotional studies took a variety of forms in this era, but the most common and widely publicized was the "morning watch," a program of daily Bible study aimed at personal spiritual growth. Mott challenged students to spend the first thirty minutes of each day in focused meditation on the Bible, providing a host of resources to aid in this task. Such systematic and prayerful reading, he suggested, was the best means of fostering a life of obedience, shoring up resistance to temptation, and heightening awareness of spiritual battles throughout the day. "By putting Bible study the first thing in the day a man prepares himself for his daily fight with self and sin and Satan," he noted. "He does not wait until the battle is on before he goes to put on the armor and gird up his sword." Mott repeatedly mentioned how the great Christian leaders of the past were observers of the morning watch, beginning their days with prayer and Bible reading. The weakness of the contemporary collegiate generation, he proposed, could be directly related to their failure to maintain such discipline. "Without doubt our failure to prevail with man and against evil in the world during the day is too often due to our more fundamental failure to prevail with God at the beginning of the day."[36] The muscular battle against sin proved to be one of the Scripture's fundamental contributions to college men.

By 1901, in fact, Mott had overturned his earlier notion that YMCA Bible study should serve evangelistic, intellectual, and devotional aims, claiming instead that the core efforts of the association should focus upon devotional study alone.[37] This dramatic shift was a function of two interrelated forces. The first was the rise of academic biblical studies within the colleges and universities. Spearheaded in large part by Harper, then president of the University of Chicago, the desire for academic coursework in Bible reflected this leader's fear that a lack of collegiate biblical sophistication would leave students in a position either of uncritically embracing simplistic notions of religious teaching or of discarding religion altogether because of its inconsistency with a growing intellectual acumen.[38] While he recommended classes in religious psychology and pedagogy, history and sociology of religion, and comparative religion, it was the academic study of the Bible that most excited the reformer. With unabashed optimism, he predicted that students would recognize the value of such classes if they were conducted on a strictly scientific plane, eschewing devotional content in order to promote intellectual credibility.[39] As Bible "chairs" and departments of biblical literature were organized around the country, Mott was convinced that the YMCA could best serve the interests of academic Bible study, not by pursuing such work themselves, but rather by "agitating" on local campuses for the hiring of Bible professors and for the place of biblical studies in the college curriculum.[40]

The growth of curricular biblical studies essentially created an institutional division of labor when it came to the Bible, with the colleges assuming purely academic study and the YMCA leading students in personal devotional work bent on character and service. In actuality, Mott hoped that curriculum and association Bible study would be complementary and mutually reinforcing factors on the campuses. On the one hand, devotional study in the YMCA would spark a love for the text that would elicit an interest in intellectual biblical studies within the curriculum. Likewise, Mott felt that students studying the text within the college curriculum would desire to gain a deeper experiential knowledge of its precepts and so join the Bible studies provided by the association.

Some, however, were not so pleased with this proposed dual system. University of Chicago Professor Ernest D. Burton felt that the division of intellectual and devotional purposes sent a negative message to students. "To divide the work between faculty and students in such a way that there shall be antithesis between the two groups, the faculty courses being, or being regarded as, intellectual and non-religious, the student courses as religious but not on a high plane intellectually, creates a highly undesirable situation," Burton contended. "The natural inference is that the students will lose their religion as they acquire knowledge." Burton's fear was that an institutional separation between the academic and personal, essentially a separation of facts and values, might actually foster habits of thought that would render collegians unable to live as both students and "lovers" of the text. In addition, this division all but guaranteed flaws in both domains. College courses would move toward dry intellectualism, while the association studies would find themselves "moralizing" texts apart from a thorough study of biblical literary contexts and genres. In either case, students would be getting a thin perspective on the potency of the text for learning and life transformation.[41]

Despite such protestations, during the first decade of the twentieth century this two-pronged system became the accepted norm in higher education. While Burton's warning went largely unheeded, more representative was the perspective of O.E. Brown, professor at Vanderbilt University. On the relationship between association Bible study and curricular Bible study in the colleges, Brown noted that both academic and association studies were necessary and complimentary. Curriculum study, he noted, dealt with the instructional life of the institution while association Bible study engaged the college "home life" of individuals. Curriculum Bible study attended to "factual information" and could be assessed with "intellectual tests" while association Bible studies attended to the religious "values" of individuals and could only be evaluated through a "character test." Since one was interested in abstract truth and the other in "concrete" character and service, both were absolutely necessary for a full-orbed religious experience.[42] This distinction between facts and values was a clear call to an institutional division of labor when it came to the study of the Bible.

While the rise of academic Bible study surely prompted this specialized approach, it was the association's changing understanding of the Bible that sustained this dualistic vision. Many within the association by the first decade of the twentieth century had accepted the scholarly understanding of historical criticism. At one level, such work seemed quite benign. Advocates of critical methods simply argued for a more thorough investigation of the text in its historical setting, looking carefully at the cultural dimensions of biblical times to correctly interpret textual themes. Some did fear that factual inaccuracies unearthed through such techniques would undermine confidence in biblical infallibility. However, the combination of such ideals with Darwinian notions threatened a far more corrosive outcome. Of particular importance was the growing tendency to view the Scriptures, not as a body of timeless truths but rather as the progressive religious experience of a particular people, emerging from less adequate to more adequate understandings of religious life. Within such an idealist and historicist framework, the Bible was understood less as a divinely revealed sacred text and more as a cultural artifact, suggestive of remarkable spiritual truths but not deemed authoritative in any absolute sense.[43]

While leaders within the college YMCA rarely entered into such hoary controversies in self-conscious ways, there is little doubt that the movement was deeply influenced by these larger debates. In an important address to the collected body of collegiate general secretaries in 1909, Northwestern University YMCA Secretary Herbert Gates stated firmly that "We have (in the Bible) a faithful record of the naïve conceptions of a primitive people at different stages of religious progress. To follow some of these conceptions would lead one astray, as it often has done." Admitting that the "ethical note" of the Hebrew prophets was advanced compared to "anything that had gone before," it was still true, he noted, that these were early stages of the historical evolution of religious progress. The time had come, Gates noted, to reject the position that the Bible was "authoritative" and an "infallible instrument of divine revelation, to be accepted literally and obeyed implicitly." Instead, this text was to find its authority in the "appeal which the Bible itself makes to the conscience and moral judgment of man." In fact, Gates suggested that the only reason that the sixty-six books of the Bible were selected as canonical was because they contained principles that pragmatically enhanced human character and service. Gates recommended that secretaries help students form their own "Canon of Scripture," consisting not of the entire Bible but rather of those writings that were most helpful to their own lives. Such a procedure, he reasoned, would be more in keeping with the laboratory sensitivities of collegiate academic culture. Men in the colleges, Gates noted, were "being constantly urged not to accept the statements of any authority, however great, without personal investigation and confirmation." Like all other subjects placed under the scrutiny of the scientific method, the "hypotheses" of the Bible would be tested in life experience in order to confirm their validity.[44]

Stemming from such a philosophy, it was the Bible's practical usefulness rather than its divinely inspired nature that now commended its continued presence in the YMCA character building curriculum. In 1905, C.H. Sutherland, president of the YMCA at Harvard, noted that "The voluntary Bible study work is primarily useful; it is frankly utilitarian . . . it looks to social utility and the transformation of the individual life."[45] Such a mindset also found a vigorous champion in Clayton Cooper, who served as national college YMCA Bible study secretary in 1902 and remained in this post until 1909. Author of *College Men and the Bible* (1911) and *The Bible and Modern Life* (1911), Cooper suggested that the Bible was one of the most "practical uplifting forces in the college community" chiefly because of its "utilitarian value."[46] He frequently cited practical results of Bible study, sharing examples of "tough" fraternity men who were taught to sing hymns, of men delivered from impure thoughts, hard drinking, and gambling, and of students inspired to participate in campus service. Suggesting that the Bible was now being interpreted not as a "storehouse of mysterious and sacred information" but as a "means leading to successful and normal human service," Cooper noted wittily that "A Bible class today should be for Acts rather than for Romans—it should be attached to deeds rather than to theology." "The Bible," he concluded, "is taking its place among the serviceable books of the world."[47]

The trajectory of YMCA Bible study curricula in these years clearly documented these emerging trends. Under the leadership of Henry Sharman (1896–1902), the first paid national college YMCA Bible study secretary, two potential four-year Bible study "cycles" were developed to provide a standardized curriculum from freshman to senior year. Under one model, students began with Sharman's own *Studies in the Life of Christ* (1896), moving to *Studies in the Acts and Epistles* (1898) by Oberlin's Edward Bosworth, W.W. White's *Studies in Old Testament Characters* (1900), and Bosworth's *Studies in the Teachings of Jesus and His Apostles* (1901) in the remaining three years.[48] A second cycle of studies, developed around this same time by former Yale YMCA President and General Secretary William Sallmon, consisted of *Studies in the Life of Paul* (1896), *Studies in the Life of Jesus* (1897), *Studies in the Miracles of Jesus* (1903), and *Studies in the Parables of Jesus* (1906).[49] While still roughly inductive in method, Sharman noted that these materials were to be "practical and devotional" rather than intellectual in orientation. While some of the questions were indeed factual in nature, most pressed students to consider personal character and social implications, eschewing Harper's more academic inductive approach. "All historical and critical study not directly connected with the living truths in question is avoided," Sallmon suggested, "in order to place emphasis upon the more important lessons for immediate practical application." In describing his rationale, he suggested that, since Bible classes tended to "present bare cold facts for mental disquietude," the association could "do much by providing attractive methods of Bible study, constructive in character and calculated to build faith, increase hope, sweeten life, and scatter joy."[50] The separation of facts and values could hardly have been more evident.

By the early twentieth century, the college YMCA had become even more committed to the new character-driven approach, exemplified in many ways by an altered understanding of the life of Christ. While former emphases on Jesus had typically highlighted both his atoning sacrificial death on the cross and his personal evangelistic methodology, the focus was now more closely related to the way in which he lived his life. Jesus was portrayed, in other words, less as a suffering Savior and more as an exemplar of muscular character. In fact, drawing upon the fourfold description of Christ in Luke 2:52 ("Jesus grew in wisdom and stature and in favor with God and man"), leaders throughout the movement articulated the Christian character building goal as an attempt to provide for a Christ-like symmetrical growth in the intellectual, physical, spiritual, and social realms. "The college association until very recently has felt only responsible for the religious development," one popular secretary suggested, "but it is beginning to see that the Christ-like man is the spiritually, socially, physically, intellectually developed man."[51] Rejecting the emphasis on the passive, peaceful, and otherworldly Jesus of Sunday school lore, YMCA leaders pointed out that the Jesus of the Bible was a muscular carpenter with a strong physique, honed through his rugged and nomadic lifestyle. He was popular, intelligent, and possessed an organizational acumen that revealed a clear picture of manhood as it was meant to be expressed. A perfect embodiment of the Rooseveltian "strenuous life," Jesus was a religious model young men could emulate without fear of feminization.[52]

Such an emphasis became "curricular" through the widespread use of Harry E. Fosdick's *The Manhood of the Master* (1913). Far from a "solemn, pale man," Fosdick suggested, Jesus had a "radiant nature," loved "company, food, and drink," and possessed a dominating intellect. He was a warrior who did battle against evil both in his own life and in the broader culture. Denigrating pictures of Jesus that elevated his peaceful agenda, Fosdick told collegians that

> The stern qualities of the Master, such as his fearlessness, have often been lost sight of in the emphasis placed upon the tenderer aspects of his character. . . . Against this too saccharine interpretation of the Master's spirit we have done well to guard ourselves by considering such austere qualities in him as his indignation, his loyalty, his self-restraint, his fearlessness. . . . Struggle, built into the fiber of manhood; temptation used as an opportunity for conquest and growth—this is the ideal which the Master presents. . . . Consider the ways in which the Master appeals to all that is strongest and most military in you.[53]

While he did not address women explicitly in the text, Fosdick did in fact note the importance of a gendered division of labor within the society.

While both men and women were to serve and emulate Christ in their lives, the paths set out for doing so were markedly different:

> The two sexes represent two spheres of character . . . the glory of their life together has never been in identity of function but in balanced harmony. Yet we take it for granted that both should find their ideal in Christ. Wherever men live by themselves apart from the civilizing influence of woman, their strength runs to roughness, their independent will to rudeness and vulgarity; wherever women live by themselves apart from the tonic, military attitude of man, their idealism runs to sentiment and their powers of loyalty lose their temper. The two need each other for completion.[54]

Fosdick's interpretation here was important because it reflected a desire to maintain the rhetoric of Victorian gender balance at the very time when economic and other factors were complicating these clearly defined roles. Fosdick desired men to reclaim their correct moral posture, seeking not an internal balance of masculine and feminine ideals as much as a mutual interdependence with women who would correct the potential abuses of the male code.

Many of these changes were consolidated between 1911 and 1915 under the leadership of Union Theological Seminary Professor Harrison Elliott. Working together with YWCA Bible study leader Ethel Cutler, the Sunday School Council of Evangelical Denominations, and Charles Foster Kent's committee on Teachers' Training in Colleges and Universities, Elliott suggested a broad-based evaluation of voluntary collegiate Bible study.[55] Through a series of surveys of students, professors, and general secretaries, he noted that many were discontent with earlier and more directive studies, describing them as "too theological," unrelated to the real problems of men, and "too pre-digested," thus limiting the ability of students to think about and discuss the texts in light of their own problems. Because the Bible was "not an end in itself but exists solely as one of the factors in the general plan of the Association in developing Christian character and inspiring men to service," he suggested that such study should be "frankly functional," addressing such questions as, "What does the Bible mean to me? What does it mean to me in relation to the life of my home, my community, my college, my country? What am I going to do about its challenge to Christian social service?"[56]

Under Elliott's leadership, the college YMCA Bible study program changed in important ways to reflect the muscular paradigm. While formerly Bible studies provided for the systematic analysis of a text and then moved to life application, newer studies started with life and social questions and then consulted Scripture passages for assistance in finding suitable answers. In fact, Elliott set the committee to the task of exploring the "dominant moral and religious problems of the college students in the various periods of their development" so that curricula could be produced to address these issues.[57] Furthermore, because the fundamental aim was the solution to these student problems, the new Bible study vision both opened the door to extrabiblical materials and provided a context for viewing some parts of the Bible as more "useful" than others. If the purpose of such study was not a knowledge of the text but rather the practical improvement of character and service, then it made sense to move to a position where only the most helpful sections of the Bible would be used and where other sources of greater usefulness would supplement biblical sources.[58] Reference to the Old Testament was increasingly diminished as leaders contended that its character and service ideals were antiquated compared with those of Jesus and Paul. They also noted that the Bible would no longer retain its privileged position as the source of all "solutions" to life problems.[59] Explicit Bible study was relegated to the fall semester, while missionary studies and the analysis of North American social problems now occupied the spring agenda. As one committee member put it bluntly: "Young people of 18 don't care about the Old Testament and the New as much as they care about new solutions of old difficulties."[60]

For exemplars of the new paradigm, one had to look no further than the original studies published for freshmen students, *Student Standards of Action* (1914) and *Christian Standards of Life* (1915). The first, written by Elliott and Cutler as the fall Bible study, was an analysis of Christ, but not from the standpoint of a "chronological study of the facts of his life." Instead, this was a study of the problems of college freshmen, looking "directly to the life and words of Jesus for the solution of these student problems." Chapters dealt with issues of adjustment, time management, the use of money, friendship, achievement, honesty in academics and athletics, chivalry, and loyalty, essentially covering all of the dimensions of the fourfold life. Significant in this study was the introduction of extrabiblical material. Because the goal was the solution of life problems, Elliott noted, *Student Standards* included not only snapshots of Jesus' life, but also information from scientific bulletins, popular authors and philosophers, and magazine articles to substantiate proper values of life.[61] The spring semester curriculum of mission and social study, *Christian Standards of Life*, analyzed topics related to such issues as enthusiasm for service, the aggressiveness of Christian faith, and training for social efficiency. Still containing short Scripture passages as daily readings, the bulk of the study took up the biographical examination of historical and contemporary exemplars such as Jacob Riis, Mary Lyon, Adoniram Judson, and Phillips Brooks. Extrabiblical in essence, such materials reinforced the activist, service-oriented muscular philosophy of the college YMCA.[62]

As the aims and methods of such studies were transformed, those attending these groups grew more diverse. During the 1909–1910 academic year, of the 28,562 men who were attending these studies for more than 2 months, an estimated 6,156 were not professing Christians. At Yale, Cooper mentioned that one Bible study group he observed included Christians, pantheists, agnostics, Jews, and "a vegetarian" studying the Scriptures together.[63] When curricular aims were devoted more self-consciously to ideals of moral uplift and social service rather than doctrinal or evangelistic themes, Bible studies grew more appealing to members of other faiths. W.D. Weatherford noted in 1909 that 50 percent of the male students at the University of Louisiana were enrolled in voluntary Bible classes despite the fact that more than half of the students were Catholics and Jews. When a local priest asked twenty-five of the Catholic students to quit the studies because of the Protestant affiliation of the YMCA, these individuals "flatly refused," he noted, "because they had gotten from it such help for moral victory that they felt they could not afford to miss its stimulus."[64] The students informed the priest, Weatherford commented, that the studies did not address issues of a theological nature, focusing instead on practical living and social service. Here was specific evidence of Bible study's new muscular Christian orientation.

Concomitant with the transformation of the content of the Bible study program was a shift in pedagogy. In the early years of the movement, Bible studies were often facilitated by professors, local ministers, and even college presidents. By 1910, however, most voluntary Bible study on American campuses was taking place in small groups led by the students themselves. At a campus like the University of Illinois, for example, faculty served as leaders of Bible study classes until 1897, but within ten years there were only a few such classes offered. Alternatively, just over a thousand students were enrolled in small groups led by their peers. By 1910, a national survey revealed that 73 percent of all groups possessed student leadership. The need for large numbers of leaders provides a straightforward explanation for the rise of student control, but there were also deeper ideological reasons for this change. Students, the leaders suggested, would be more likely to facilitate the character and service goals of the new Bible study, resisting the faculty's tendency to focus on cultural, literary, and scientific studies. Students were obviously in a better position to understand and articulate their own needs and problems, forging solutions that would be more in keeping with the practical exigencies of campus life. For muscular Bible study, student leadership seemed a more efficient approach.[65]

Thus, beginning with Harper's close inductive studies in the early 1890s, the association moved next to specific textual studies, such as those by Sharman and Sallmon, which provided more emphasis on practical application. From here, it was just a short move to thematic studies which started, not with the biblical text but with topics of student importance, utilizing the biblical material to provide insight on these issues. Finally, by 1915, the college YMCA was prepared to start specifically with student needs, bringing the Bible to bear on these issues when needed but also leaving room for extrabiblical materials when they were more appropriate.[66] Far from a simple debate over semantics, this change was indeed revolutionary. If the Bible was simply one text among many that could contribute to life problems, there was very little sense in arguing for its exclusive place in spiritual guidance. While movement leaders unequivocally asserted that the Bible was indeed pragmatically powerful and the best text for producing changed lives in this era, the consequentialist ethic implied in such a stance provided little assistance when others argued that alternative resources were more helpful. When the Protestant Christian hegemony of the colleges eroded after World War I, it became more difficult to sustain confidence in this seemingly antiquated text.[67] Despite the enormous success of association Bible study in this era, therefore, the victory was perhaps pyrrhic in nature. These popular revisions actually set the context for the Bible to assume an increasingly minimal role.

Muscular Membership

With such a clear programmatic vision for a Christianity marked by character and service, it soon became clear that traditional membership criteria were no longer appropriate. At the 1907 YMCA International Convention in Washington, D.C., leaders took up a resolution for the modification of the membership criteria for the college associations, drawn up by a number of church and higher education leaders. The proposal specified that, for collegiate chapters, active membership should be granted to students who were "either members of evangelical churches or accept Jesus Christ as He is offered in the Holy Scriptures as their God and Saviour and approve of the objects of the Association, which are as follows: to lead students to become disciples of Jesus Christ as their divine Lord and Saviour, to lead them to join the church, to promote growth in Christian faith and character, and to enlist them in Christian service." While this did not at first blush appear to be a dramatic departure from long-standing commitments, it did represent the first time any branch of the YMCA had requested membership status based on a criterion linked to Christian character and service and divorced from church membership. To cushion the negative reaction, the resolution provided that these changes would be permissive rather than mandatory, allowing but not requiring chapters to change their policies. In addition, while nonchurch members would maintain active voting membership, they would be prohibited from holding leadership positions in the local chapters.[68]

The rationale for changing the membership criteria for students was closely linked to the exigencies of college life. Students rarely joined churches in their college towns, and many had not yet taken up membership in their home churches before leaving for school. In some cases, churches of chosen denominations did not exist in the college community, leading students to resist forming memberships until after graduation. In addition, many students had not yet come to full conviction on a particular denomination, making it difficult for them to assume church membership until fairly late in their college career. Despite Mott's assertion that students and "practically all of the State and International student secretaries" approved the idea, the debate surrounding this alteration was vigorous. Famed fundamentalist advocate William Jennings Bryan, for example, saw it necessary to draw a line in the sand, suggesting that "a man who is not ready to say that he will some time connect himself with a church is not in a good position to join a work that has for its expressed object the bringing of young men into the church." Others suggested that this provision, designed for the specifics of the student situation, would serve as an "entering wedge" to universalize this pattern in the larger YMCA. If college students were given special membership privileges devoid of church membership requirements, in other words, there would be little ground to stop other branches of the association from claiming similar hardship so as to gain membership advantages.[69]

Though Mott resisted this "slippery slope" argument, critics of the revised platform were actually quite astute in predicting larger changes. The questioning of the church membership requirement was soon expanded to include a broader reconsideration of the evangelical basis as a whole. While many schools maintained church and doctrinal membership requirements up until World War I, a growing number of chapters in the East and Midwest desired a shift away from a belief-driven membership appeal to one rooted in character and service.[70] At Harvard, YMCA General Secretary Gurry Huggins was disenchanted with the organization's membership restrictions from the start. Interested in cooperative work, he spearheaded the formation of a Phillips Brooks House Association that formalized a working relationship between the YMCA, the Episcopal St. Paul's Society, and the liberal Harvard Religious Union in matters of social service. Upset that influential students, including Episcopalian Franklin D. Roosevelt, were unable to meet the YMCA criteria, Huggins attempted to drop the Portland evangelical test from the constitution and form a purpose statement linked to the formation of character through life and service. As the YMCA president noted, "I can't see why men who are actuated by the spirit of Christ and ready and willing to do His service should be kept from a union of fellowship because of questions of belief, which, to me at least, have little if anything to do with the effectiveness of service."[71] Their new statement noted that the Christian association "was open to all members of Harvard University who desire to be disciples of Jesus Christ in life and service, and to associate their efforts in the extension of His Kingdom among young men." While they hoped this would not generate a breach with the YMCA, they held firmly that they required a statement that would admit men willing to serve "regardless of their theological views."[72]

Many in leadership felt that creedal and church membership requirements had grown both difficult to secure and less important to the aims of the organization. A committee formed to investigate this issue suggested, "Both assent to creedal statements and membership in a church can become static; the assent to a purpose, a definite program of life implies not an obligation already filled, but an obligation still to be met, an inspiration and an ideal for the future."[73] At Brown University, students resisted membership statements that would require church membership or "willingness to subscribe to a statement of belief in a certain metaphysical conception of the nature of Jesus Christ." Jesus himself, chapter leaders noted, would not "insist upon the assent to a particular metaphysical conception of his nature, or upon membership in any specific Christian churches as a necessary initiation into the privileges and responsibilities of his service and discipleship." The University of Chicago chapter put it simply in 1911: "Membership in the Association is not conditioned upon creed or cash, but simply upon service which shall be helpful in varying degrees to one's self or to one's neighbor."[74]

Some were deeply opposed to such changes, claiming that a revision of the evangelical basis would destroy the purpose of the organization and move it quickly toward complete secularization. One Amherst alumnus responded that "If the YMCA is to stand on grounds with Jews and Mohammedans, the friends of religion will be deeply grieved," while another, writing in *The Congregationalist*, recommended that the campus YMCA drop the name "Christian" because of its vague commitment to the creeds and institutions of formal Christianity.[75] Such voices however, were increasingly marginalized within the public discourse of the college movement. The majority of faculty members at these institutions were invariably supportive of the broadened membership policies and in some cases were the chief stimuli for altered requirements. In addition, a growing number of YMCA leaders supported the shift to a purpose basis rooted in character and service. David R. Porter, the secretary for high school and preparatory work who would eventually replace Mott at the helm of the college movement, noted that membership should be broad enough to include Catholics and Unitarians. Many of the members of these groups, he told an assembly of national leaders, were "just as Christian as most of the men in this room" and should not be excluded from the ability to join others in activities of Christian service. The key was to focus not on a "certain sort of theology" but rather on "Christian character" and "definite kinds of service."[76]

The debate over the membership basis, initiated at the 1907 convention and continued in the years following this initial furor, represented a critical turning point in the collegiate association's history. Commensurate with the larger changes in the movement to themes of muscular character and service, the shift from a creedal to a purpose basis signified not only a separation from churches and an opening to nonevangelical religious groups but also a more generalized transformation in the very meaning of the "Christian" focus of the organization. Increasingly, the purpose of shaping Christian students had far less to do with forging personal salvation and beliefs and far more to do with a generalized commitment to symmetrical character development and service to others. One perceptive observer, writing in 1908, commented that the revised aims of the college YMCA were equivalent to those of the newly developed Rhodes scholarship: intellect, a fondness for outdoor sports, qualities of manhood, and the "moral force of character."[77] Such an approach obviously had democratic appeal in inviting many groups to participate in common efforts. At the same time, it provided a very direct breach from the faith-oriented intercollegiate organization that had emerged in 1877.

The new vision of college YMCA muscular Christianity was immortalized in 1913 by Daniel Chester French's famous bronze statue, "The Christian Student," constructed on the Princeton University campus to highlight the role of the institution in the origins of both the intercollegiate YMCA and the SVM. This life-sized statue, which was placed directly across from the YMCA's Murray-Dodge Hall, depicted a student garbed in a football uniform but also sporting an academic gown draped over the shoulder and a number of books under his arm. Commissioned by Cleveland Dodge, this work was designed as a memorial representation of Class of 1879 graduate Earl Dodge, who had unexpectedly died at the age of twenty-five. Serving as captain of Princeton's championship football team, Earl Dodge graduated near the top of his class and served as president of the YMCA's Philadelphian Society.[78] In combining the chief elements of Christian character—spiritual, intellectual, physical, and social excellence—with service to alma mater and to the YMCA, Dodge epitomized the desired posture of a manly Christian who was devoted to the strenuous life of commitment to God and institution. "The Christian Student" was a fitting symbol of the muscular Christian vision of this era.

In assimilating this vision, the intercollegiate YMCA was undoubtedly imbibing both the language and the tone of turn-of-the-century Progressivism. In fact, higher education historian Frederick Rudolph, commenting on the spirit of collegiate institutions in this era, suggested that the elements for which the Progressive temper stood included "honor, character, a certain wholesomeness bordering on utter innocence, a tendency toward activity rather than reflection, an outlook that one day would make a good Boy Scout or Girl Scout and, at the time, a good member of the campus Christian Association."[79] In this light, it is perhaps not surprising that the college associations received unabashed commendation from Theodore Roosevelt himself. Writing in 1903, Roosevelt asserted that "The work of the YMCA has grown so much among college students because it has tried not to dwarf any of the impulses of the young, vigorous man, but to guide him aright. It has not sought to make his development one-sided, not to prevent his being a man, but see that he is in the fullest sense a man, a good man."[80] The seamless blending of moral righteousness and active service under the larger banner of "character building" demonstrated a resonance with Progressive-era themes within the culture and on the American campuses.[81]

It was also true, however, that muscular Christianity possessed the capacity to diminish the explicit place of religion in the college YMCA. Movement leaders were correct in recognizing that the loss of male interest in Christianity could foster a declining Protestant hegemony in American culture. At the same time, the "solution" of muscular Christianity sustained similar risks. From a centralized position that informed every aspect of organizational programming, the spiritual life was now just one component of the broader fourfold character. Of course, the move to such a vision was always described as an expansion of the Christian ideal—faith now encompassed every aspect of human existence. However, the widening perspective of "character" did create a context within which religion was part rather than whole. In the end, as leaders explored implications for student service and outreach, it become clear that the college YMCA possessed, not only a new program, but also a new gospel.

Chapter Six
A Conservative Social Gospel

The new character and service emphases of the intercollegiate YMCA were indicative of the larger muscular Christian transformation of the student movement in the years leading up to World War I. The gradual decline in studies focused upon the apologetic defense of the faith revealed that the singular evangelistic focus of the Wishard era has been compromised to a significant degree. In the midst of the changes, however, the goal of campus evangelism was not discarded. In fact, from the time Mott assumed his leadership position until World War I, evangelism remained an important, though less dominant, component of association work. The goals and methods associated with such efforts, however, were altered to reflect the new emphases of the movement. The ideology of conversion-oriented personal work, so prominent in the organization's early years, was reconceptualized to reflect more adequately the new character and service ideals of the college movement. While the need to make a "decision" for Christ remained entrenched in association rhetoric, the nature of that decision was changing.[1]

"Moral Evangelism"

Overturning a former certainty about these issues, the college YMCA "Commission on Evangelism," formed in 1910, noted that there was "by no means uniformity in opinion as to what constitutes evangelism." What was clear was that older definitions were no longer compelling to many in the movement. Suggesting that some still viewed evangelism in terms of "winning men" to Jesus, they noted that many now envisioned it as an "appeal to personal character" that was related to "everything that true rational living means, that it is basal to manhood, loyalty, courage, and genuine success." The aims of evangelism, in their estimation, were not only to "urge men in a strong, plain, forceful way to accept Jesus Christ as their Savior and Lord" but also to "present in an attractive way high ideals of Christian life and service" and "to create and maintain a high moral and religious sentiment in the college community." E.T. Colton, a national collegiate secretary, noted that evangelism should proceed along "character" rather than intellectual lines, avoiding apologetics and instead focusing on "those aspects of temptation which prevail among students."[2]

This new evangelistic focus was perhaps best exemplified in the tours of S.M. Sayford, which included more than 150 colleges between 1888 and 1894. A graduate of the University of Pennsylvania, Sayford's messages regularly implored students to resist the allure of sinful practices, guarding their minds and hearts against the wiles of Satan while dropping habits that were retarding personal spiritual growth.[3] Utilizing military metaphors with great frequency, he spoke of waging war against the dark forces that battled against body and soul, telling students to "Reckon this life a warfare, this world a great battle-field, that we must fight if we would win."[4] Quite direct in his words, Sayford was not shy about naming student moral foibles that might elicit personal decay. Frequent appeals to drinking, smoking, profanity, and gambling laced his messages, and on a number of campuses he also held "confidential meetings" with male-only audiences to discuss matters of sexual sin. Meetings ended with an appeal for students to sign cards committing themselves to the "higher ground," a phrase meant to designate moral cleanliness and self-control in the mental, physical, and social domains of life. By 1900, over 20,000 students had covenanted to the "higher ground" through Sayford's preaching, some committing to Christianity for the first time and many offering pledges of recommitment.[5] At Illinois, students relinquished "the tobacco habit, profanity, intemperance, and other practices," while several YMCA chapters initiated temperance unions complete with signed pledges to give up all "strong drink" for a year.[6] President David Starr Jordan of Stanford recalled that "His talk before the young men of the University was in all respects the best confidential talk to young men, on the dangers that beset college life that I have ever heard from any source whatever," later informing Mott that "It is not putting it too strongly to state that Mr. Sayford has been more largely used in leading students into the Kingdom of Christ and purifying the life of the colleges than any other living man."[7]

Sayford's success throughout the 1890s in many ways heralded a new style of campus evangelism. While personal work was still occasionally recommended, by the turn of the century the primary mode of evangelistic effort was the "campaign." Modeled in many ways after Sayford's campus visits, large annual evangelistic crusades, typically lasting three or four days, were led by a host of college association leaders. Mott would typically hold three or four large-scale campaigns per year, but many other national college secretaries, most notably men such as E.T. Colton, A.J. Elliott, A.B. Williams, and C.C. Michener, also enlisted to address campuses in their own regions. In terms of attendance, such campaigns were conducted on a grand scale. Between 1899 and 1901, for example, individual meetings attracted 250 students at Wesleyan (out of 280 men enrolled), 650 at Stanford, and 700 at the University of Michigan. In 1907, each evening, 940 students came to hear A.J. Elliott present his evangelistic talks at the University of Illinois, while 800–900 attended nightly evangelistic meetings at the University of Wisconsin in 1910. In many cases, campus leaders stated that these were the largest religious meetings ever held at their respective institutions.[8]

Addresses focused little on issues of personal belief or the power of an instantaneous religious conversion experience, instead emphasizing student decisions to live a life of fourfold character. Building a "clean and pure" life of victory over temptation was an evangelical passion, to be sure. It was, however, distinct from Wishard's desire to bolster student evangelism with biblical arguments on the nature and atoning work of Christ in salvation. In Mott's estimation, this approach reflected a need to demonstrate that character and manliness were compatible, rejecting the common idea that self-control was a feminine ideal and that masculinity was tied irrevocably to immoral practices. "When we go 'off on a bum' with the crowd just to show them we are not 'sissified,' " he suggested to students at the University of Virginia, "we are exhibiting to the world that we are moral cowards, that we are afraid to stand up for the right."[9] Likewise, Stanford President Jordan's enthusiasm for Sayford reflected his own sense that this new evangelism was the ideal means of championing a muscular Christianity on the campus. "I believe that virtue belongs to the young and the strong; not exclusively to the prig, the milksop, or the invalid," he suggested. "The man we dream of as the Stanford man will stand up against vulgarity, rowdyism, and mean temptation as he stands up against the bold dash of his opponent's rush line."[10] The exertion of an indomitable will against temptation was a struggle that could be undertaken only by a true man, following the balanced character of Jesus.

It was the engagement of the will, a popular muscular Christian theme, which became the central lynchpin of evangelistic efforts.[11] Henry B. Wright, YMCA secretary at Yale, claimed that in distinguishing successful and unsuccessful campaigns, those winning men to allegiance to Christ were the ones that mobilized student willpower. "Such a campaign, conducted on common-sense lines, wholly devoid of sensational or professional methods, and appealing to the will rather than to the emotions," he stated, "could not fail to bring about a definite cutting with sin by many men."[12] Because students were habituated to live in the world of ideas, Mott felt that their wills would atrophy to the extent that resistance against sin would decline. The work of YMCA evangelists, then, was in part to fortify the will through appeals to the better path of Christ. The linguistic turn from "conversions" to "decisions for the Christian life" was thus more than a simple semantic shift.[13] Resisting emotional appeals, an evangelism directed at college men would result in "calm deliberation" and "forceful decision," utilizing a logical appeal to Christian living to prompt a "free action of the will."[14]

Such a philosophy revealed an important shift in association thinking at this time. Before 1890, efforts were directed toward a religious conversion experience characterized by the instant regeneration of the heart through the agency of the Holy Spirit. The new vision focused more distinctly upon the gradual development of moral power through human effort and active Christian work. Unlike conversion, which was a function of received grace and divinely initiated revelation, character and service were self-initiated, forged by personal effort and willpower. While grace was a gift, in other words, character was to be conquered.[15] Anchored by a Kantian emphasis on the individual as an active creator rather than a passive recipient of reality, the new evangelism placed more of the burden upon students themselves for such growth to take place. In a culture where ardent willpower and efficacy were diminished by religious and vocational changes, here was a place where the self-made Christian man could thrive.

"Social Evangelism" and the Conservative Social Gospel

While the shift to a character-oriented evangelism clearly represented muscular Christianity and its desire for "clean living," this temper was perhaps even better represented by an equally powerful and increasingly dominant emphasis on "social evangelism." Spearheaded by political and social reformer Raymond Robins, this approach, which assumed prominence between 1905 and 1920, centered the Christian appeal on the need for active service and moral reform work in the broader community. A founder of the Progressive Party, Robins had worked with civic reformer Graham Taylor in Chicago, laboring in a university settlement and fighting for a variety of urban improvement projects. One of the main forces contributing to a service orientation in the Men and Religion Forward Movement, he played a similar role in fostering this mindset in intercollegiate YMCA evangelistic work. Robins was deeply interested in personal religious experience, but he also brought a significant emphasis on American social problems to the evangelistic task. While Mott and others had worked hard to attack prevailing student sins, Robins sought to demonstrate the need to look beyond such flaws to enhance the civic good.[16]

Together with John L. Childs, who would later serve as a leading progressive educational theorist at Columbia University's Teachers College,

Robins delivered addresses in 48 colleges and universities, traveling 45,766 miles during the 1915–1916 academic year alone. Speaking to large audiences, a typical four-day campaign on these campuses included addresses to students and faculty on "The Challenge of the Changing Social Order," "College Men and Community Leadership," "Fundamentals in the Industrial Conflict," and "The Redemptive Principle in Education." Dealing with issues such as immigration, urban poverty, and the plight of black Americans, Robins spoke of how students possessed a responsibility to attend to the social needs of local communities. Importantly, he claimed that such work, properly conceived, would help the dispossessed to develop a life of character, marked by intellectual, social, physical, and spiritual excellence. In other words, social service had as its chief purpose the development of character within individuals and communities that lacked such ideals. Moral reform through active service was in many ways a crystallization of the evangelistic tenor of muscular Christianity.[17]

The tight connection between character and service in the college YMCA linked the movement irrevocably to a conservative social gospel theological stance. Because service was always tethered to the muscular association aim of character building in this era, social problems were frequently attributed to individual moral failings and personal inadequacies rather than larger structural issues. In other words, many espoused the perspective that the development of fourfold character was the key to eventual social improvement, secured essentially one person at a time.[18] Two theorists, Shailer Mathews and Francis Peabody, provided the foundations for such a position in the college work. In *The Social Teaching of Jesus*, quite popular among local YMCA chapters, Mathews noted that "It cannot be too often emphasized that social regeneration according to the conception of Christ cannot proceed on any other line than that of the replacing of bad men by good men." Peabody, in the equally popular texts *Jesus and the Social Question* and *Jesus Christ and the Christian Character*, wrote that "The secret of national welfare is in personal morality. . . . Economic prosperity is the social consequence of personal righteousness; political prosperity is the corollary of individual holiness. . . . To plant in the soil of the world the strong seed of Christian character was to be certain of an abundant harvest of social consequences."[19] This is why individualistic moral causes, such as temperance, could be envisioned as plausible solutions for far-reaching social degradation. If character was the chief source of social disease, it could also be posited as the engine of social renewal.

New missionary and American social study texts within the spring college YMCA curriculum clearly demonstrated this conservative social gospel perspective.[20] While early SVM publications focused exclusively on the need for conversionist salvation and apologetic argumentation, books in the early 1900s located more of the rationale for missionary service in the need to promote social change through fourfold character reform in foreign settings. Many of the texts specified the backward character qualities of individuals in nonwestern locales. In *The Cross in the Land of the Trident* (1895), for example, Volunteer Education Secretary Harlan Beach spoke of India as a land populated with people of weak characters, marked by low intellect, physical "feebleness" due to poor diets and intermarriage, and a diminished willpower to avoid sinful temptations. In his *Dawn on the Hills of T'ang* (1898), he similarly critiqued Chinese culture and informed his readers that, despite the desire to spread the gospel to this land, "China's greatest need is character."[21] In Robert Speer's depiction of Latin America, this region was viewed as educationally bereft of a literate base, decimating the intellectual side of life. In addition, the country possessed physical maladies linked to alcoholism, sanitation and hygiene failures, and rampant disease. Morally, the citizenry was profligate, the lustful Spanish culture evidenced by the large numbers of illegitimate children sired in this setting. The source of these flaws was to be found, not in economic or structural injustice, but rather in the weak characters of individuals. Speer suggested that "the tone, the vigor, the moral bottom, the hard veracity, the indomitable purpose, the energy, the directness, the integrity of the Teutonic people are lacking in them. . . . The deepest need in South America is the moral need. The continent wants character."[22]

By the early 1900s, local association chapters began to add studies of North American social problems—often referred to as "home missions"—to the broader collegiate program. While Bible study lent a potent theological rationale for social service within the college YMCA during this era, the more direct analysis of social conditions provided relevant data for this activist task. Mott hired Richard Henry Edwards and A.W. Trawick to serve as national secretaries for "social study and service" and both crafted influential texts to lead students in social thought. Edwards, formerly a general secretary of the Yale YMCA and Congregationalist pastor at the University of Wisconsin, authored influential pamphlets on *Volunteer Social Service by College Men* (1914), *The Challenge of American Social Problems to College Men and Women* (1914), and *How to Work out a Service Program* (1914). In addition to his larger study on *The City Church and its Social Mission* (1913), Trawick published a series of pamphlets on *College Men and Community Service* (1912), *Social Investigation, with Special Reference to the Race Question in the South* (1912), and *Service Visits to Families* (1914). With the inclusion of such works as Josiah Strong's *The Challenge of the City* (1907), G.W. Fiske's *The Challenge of the Country* (1912), W.P. Shriver's *Immigrant Forces* (1913), Norman E. Richardson's *The Liquor Problem* (1915), Harry F. Ward's *Poverty and Wealth* (1915), Jeremiah Jenks's *Social*

Significance of the Teachings of Jesus (1906), Mott's *The Future Leadership of the Church* (1908), and W.D. Weatherford's *Negro Life in the South* (1910), the college movement generated a significant social service library that became a collective theoretical seedbed for new program approaches. As an Amherst College YMCA broadside announced for its service program, "A bunch of angels on a pinpoint used to be a hot topic for discussion but now we want—FACTS."[23]

The key themes of the conservative social gospel were clearly evident in these works. In Mott's text, which called students to enhance social gospel activism within the broader church, he suggested that, amidst all of the social unrest caused by immigration and industrialization, the nation would ultimately be "saved and conserved only by Christian character." "Follow far enough any one of the grave national problems," he noted, "and you will come to the one point—the need of better men." Largely discounting efforts to bring about social reform through political channels, he suggested that "Only in bringing to bear upon the hearts of individual men" the work of character transformation "is there any hope of effecting any thoroughgoing and permanent changes in their social conditions and relationships."[24] In addition, Jenks, a Cornell professor of political economy, posited that Jesus had not attempted to reconstruct the social order through political upheaval or the alteration of government policies. Instead, he attended largely to the character needs of individuals. Jesus, he noted, was interested not only in relieving the distress of the poor, but also in teaching them that they could rise above circumstances through education, hard work, and moral improvement. The key for students was not to seek economic reform, but rather to ensure that the poor would remain industrious, avoiding the pauper spirit and growing in a sense of "responsible manhood." Since a "right heart was more to be desired than riches," the best service was to be found in character building, helping the poor develop qualities of life that would enable them to make their own way.[25]

Such a conservative social gospel perspective was seen most dramatically, perhaps, in the movement's perspective on the plight of immigrants and black Americans. Like so much of the immigrant service work of this era, YMCA labor along these lines included heavy doses of Americanization, the attempt to enhance the character of foreign laborers to promote worthy citizenship. In a conference on YMCA college work with immigrant populations held in Boston in 1911, speakers continually referred to immigrants as morally and socially backward in every dimension of the fourfold life. Speaking of how the American college student could "save" the immigrant, one speaker suggested that "To let alone a class of people who have come from a barbarian country, and who are nothing less than grown-up infants, would not be a wise policy. . . . We have got to do something to make out of these grown-up children equal men as we are." As one author stated, the goal of immigrant service was not so much to change conditions as to "save them as men and women" through education, health training, and moral improvement. The social problems facing immigrants, it was reasoned, could be solved if students would "amalgamate" immigrants and "inspire them with noble ideals" of character.[26]

African American issues were handled in similar fashion. In 1910, Southern Secretary W.D. Weatherford published *Negro Life in the South*, the fruit of two years of research and personal interviews.[27] Aiming to provide white students with ideas of how they might serve this population, Weatherford's depiction was quite dire. On an educational level, illiteracy was rampant. On a moral level, cheap whiskey and cocaine were "doing there deadly work" and profanity, gambling, and debauchery were "everywhere prevalent." On the report of a black physician, he noted that 98 percent was a "low estimate" for the number of black men who had been "socially impure."[28] Despite some strides in Christian understanding, Weatherford stated that blacks possessed a "tropical imagination" that "revels in the strange, mysterious or supernatural." Sermons were characterized by "wild chants" rather than reasoned expositions, prayers were "too loud and repetitive," and ministers were uneducated, lazy, and immoral, devoid of the education necessary to exercise informed leadership.[29] Weatherford's "fear" was that "the negro race in the South will remain so backward, that it will remain so ignorant, that it will remain so far in the rear of civilization, that we of the South will forever be held down by the weight of our helpless neighbors, and allow the people of other sections of our country to march on and leave us hopelessly behind in our wealth, in our civilization, and in our culture." Because of his sense that blacks and whites would "stand or fall together," he stated that it was in the interest of "self-defence" that "every intelligent white man must study this question, and be prepared to take his part in the physical, social, intellectual, and moral regeneration of this neglected race."[30]

Not surprisingly, in the contest over the proper solution to the "race problem," Weatherford placed himself squarely in the camp of accomodationists such as Booker T. Washington over and against the more radical egalitarian ideals of W.E.B. DuBois. DuBois, Weatherford suggested, "stirs up race hatred in the radical negro," pressing blacks to "demand and take certain so-called rights" rather than "helping the colored race to be worthy of position."[31] Because the problems of the black community related to issues of personal character rather than institutional or structural matters, Weatherford's goal was to promote education among black Americans in order to teach punctuality, self-discipline, silence, and a strong work ethic. He argued vehemently for the humane treatment of African Americans, even using Kantian philosophical principles to demonstrate that they should be treated as ends rather than means and people rather than "hewers of wood and drawers of water." Yet though the vision of interdependence was organic in nature, the vision for black advance was decidedly individualistic. No "sane white man," he noted, believed in "promiscuous mingling between blacks and whites" on a level of complete equality.[32] Equality would be achieved, not through immediate measures of rights-based social mixing but through the gradual and separate promulgation of black education in the values characterizing the more advanced white culture. "However difficult the task," he concluded, "we must bring the negro to believe in himself. We must make him feel that he is capable of being a true man."[33]

On the cusp of World War I, such a mentality was still largely intact. In 1914, Mott officiated at a Negro Christian Student Conference in Atlanta designed to "give the present generation of Negro students in the United States a strong spiritual and moral impulse," to help students analyze potential life callings, to raise consciousness about the needs of Africa, and, finally, "to consider what light Christian thought may throw on present and future cooperation between the races."[34] The conference condemned the "habitual injustice and unkindness" characterizing white interactions with black Americans, stating that such practices negated the values of brotherhood inherent within the Christian faith. Yet while Mott posited that the conference was a critical first step in healing racial antagonisms, it was assumed that progress among African American students would take place chiefly through moral and religious development rather than through changes in social policy. Distinguishing between forced "segregation" and voluntary "separation," conference speakers did not attack Jim Crow legislation but rather sought to enhance the equality and dignity of African American students within the "separate but equal" framework, justified just eighteen years before in the Supreme Court's *Plessy vs. Ferguson* decision. As Trawick suggested, "No word was said at Atlanta to cause the Negro to feel that he was less of a man and less entitled to respect because of the fact that he was a Negro. On the other hand, much was said to cause him to be proud of his racial identity and to persuade him to accept his place in the divine ordering of things and to strive hopefully for the accomplishment of the evident purpose of his creation." Despite the goal of analyzing "cooperation between the races," the gathering did not address issues of an interracial nature. Instead, it focused upon means of enhancing personal morality, social service, missionary efforts in Africa, church commitment, and family life in the black community. Washington was present at the conference, and, as of 1914, his accomodationist philosophy was indeed alive and well in the college movement.[35]

By all accounts, blacks in leadership of the "colored" college YMCA work were also conservative on race policy. William A. Hunton regularly described the YMCA in terms of its potential to elevate the black race as a whole through moral uplift, providing them with the religious and character-building tools necessary to achieve self-reliance, a strong work ethic, and eventual equal footing with the "enlightened" white community. He described the social and religious condition of blacks as having "improved but little since the abolition of slavery." Ignorant, superstitious, and "groping in darkness," black Americans, he claimed, lacked the "enlightened self-control" and "Christian culture" of their white counterparts and were thus in need of the Protestant values that would "be highly creditable to any race."[36] Despite the repeated indignities suffered in his significant rail travel throughout the south, black college YMCA leader George E. Haynes held a similar perspective. Deeply influenced by conservative white teachers who came to Fisk through church missionary channels, he claimed that his people should have "faith in themselves, faith in the white man, and faith in democracy" as they worked for self-improvement and an eventual place alongside whites as American citizens.[37] Such a philosophy perhaps explains Washington's enormous praise for the intercollegiate YMCA. In an address at the YMCA Jubilee convention in 1901, Washington proclaimed that Hunton and Moorland, through their work among the colleges, were "making that young black man the most useful, the most reliable, Christian man in his community."[38]

In all of these works detailing both global and local social problems, the college YMCA clearly demonstrated that it was still a representation of what Martin Marty called "private Protestantism," a religious perspective consumed with the development of personal faith and moral character and with personal service to the poor and needy.[39] Public Protestants, on the other hand, advocated appeals for social and economic justice—implying a focus upon structural considerations. For the college YMCA in this era, while there were certainly public dimensions to this task, the fundamental focus was upon symmetrical physical, intellectual, spiritual, and social growth. This mentality resonated quite well with the tenets of muscular Christianity in this era. For muscular Christians, the development of character was a matter of individual effort and willpower. A focus on structural arrangements, on the other hand, appeared to highlight environmental rather than personal influences in shaping life and behavior, thereby sapping human agency and leaving men reliant upon outside forces. Such a perspective was obviously unappealing to those seeking to restore the efficacy of the individual in the decision-making processes of life. Peabody suggested, in fact, that Jesus would have "no part in the limp fatalism which regards character as the creature of circumstances. He makes a masculine appeal to a man's own will. A Good Society comes from good character."[40] For muscular Christians, the elevation of individual human potency was a cherished aim, not to be pulled asunder by claims of environmental determinism. A conservative social gospel framework was, it seemed, the ideal concomitant to the formation of a masculine Christian faith.[41]

Building the University Service Station

The appeal to service highlighted by the conservative social gospel was actually quite generalized on the campuses of the early twentieth century. Practical service for social improvement had become a core rallying cry of the academic mission of the university.[42] Schools established closer links to public life through the creation of state historical societies, agricultural services, and strategic committees providing expertise in politics, environmental science, and public health. In many institutions—most notably the University of Wisconsin—technical and nontechnical extension courses were offered to the public on issues related to topics as diverse as railroad regulation and forestry. In addition, extracurricular expressions of these ideals were also emerging across the nation. The growth of student government was of course one expression of the greater sense of public conscience, as was the nascent growth of collegiate "honor systems," particularly in the southern states. Even more characteristic of this impulse was the university settlement movement, an attempt to provide material and social assistance to the casualties of urban industrial squalor.[43] Lyman Abbott's sense that the American university could be distinguished from the English ideal of the "cultured gentlemen" and the German ideal of the "scholar" by its signal goal of "service" highlighted a critical dimension of emerging collegiate purpose.[44]

Yet even in the midst of this flurry of activity, it was the college YMCA that historian Frederick Rudolph characterized as "the undergraduate expression of the mood of Progressivism and of the developing service rationale of the colleges and universities."[45] Service, of course, had not been absent from the early college YMCA program. The "neighborhood work" that marked the movement from its origins, however, reflected a more specifically evangelistic purpose in providing for preaching, teaching, and personal work in area jails, hospitals, and churches. While such work did not completely disappear, a new service movement was clearly becoming dominant in the first and second decades of the twentieth century, one that viewed outreach as a Christian form of social betterment. Paralleling many other Protestant agencies in this era, the desire was for the YMCA to combine the passion of Christian servanthood with the benefits of academic training to generate a powerful engine of communal moral improvement.[46]

Of all the manifold forms of social service, boys' work was the most common and comprehensive. While only five of fifty-five surveyed chapters were engaged in Sunday school teaching in 1910, 40 percent was actively involved in work among boys in the local college community.[47] Such work took many forms. Wisconsin students focused upon underprivileged and delinquent boys, working in some cases together with probation officers in order to serve as "Big Brothers" to these youth. Students at the University of Virginia devoted their attention to local "news boys," while Brown University undergraduates conducted clubs for "street arabs." At times, students continued the long-standing tradition of engaging in "deputations," though these events differed quite significantly from previous models. At Yale, YMCA leaders spoke of deputation work to preparatory schools as a chance to "give coming Yale men a good square talk, right out from the shoulder, on the importance of clean morals in college."[48] University of Virginia students sent speakers, "men in whom manliness as well as genuine interest in the work is unmistakably present," to bring the message of clean athletics and academic honesty to surrounding schools.[49] While cards asking individuals to document conversions had been a staple of earlier work, the cards used by this time revealed a changing emphasis, documenting the boys' willingness to "make a strong stand for clean living," "contribute to Christian work," and "give up the habit of (blank)."[50]

At times, YMCA chapters introduced long-term programs that reflected the desire to facilitate development in every fourfold dimension of a boy's life. Oberlin students started a manual training school for town boys in 1908–1909 in order to facilitate free opportunities for job training. They, together with many other chapters, also held track meets for the boys in order to encourage physical activity. At Cornell, boys clubs instituted in 1903 brought youth together once a week to take part in athletic activities, learn basic skills of reading and chemistry, and hear an "instructive talk" on subjects such as self-restraint and profanity. Princeton students in 1900 founded a "town club" for boys and young men of the area, leasing a house complete with a reading room, gym, and recreation center. Maintaining a strict nightly schedule from 7:30 to 9:00 p.m., boys would spend thirty minutes in each area so as to provide for balanced intellectual, social, and physical development.[51] The purposes of the work, Porter remarked, were to guide boys into successful careers, to provide recreation, to secure physical health, to elevate education and discipline, and lastly, to train them in "religious life and service."[52] No longer the solitary focus, religious work was now simply one dimension of the broader purpose of developing boys' fourfold life into a balanced and holistic character.

While boys' work was the most common form of social service, work among the sick, the poor, and the destitute was perhaps the most representative facet of the new social vision. At Yale, about 175 students were engaged in outreach projects to the poor, working in areas as diverse as Yale Hall, the Oak Street Boys club, the Bethany rescue mission, and the local African American YMCA (which included a carpentry shop for practical work, classes in American history, and manual labor). Harvard students established a reading room on T wharf, patronized daily by 160 fishermen, and also gathered clothing for the poor, distributing them through the Salvation Army. At the University of Tennessee, students started a night school in the poorest district of Knoxville and afternoon schools featuring sewing for girls and a model garden for the boys. While many of these efforts were focused on urban poverty, the college YMCA did not completely ignore the rural context. In light of the "country problem" and its attending despair, many campuses provided clubs and social entertainment for children and their parents using country schoolhouses on midweek nights. Working in institutions ranging from hospitals, jails, and almshouses to reformatories and reform schools, the scope of such projects was as diverse as the institutions themselves.[53]

In addition to these efforts, one of the most expansive components of the student YMCA service program was an attempt to reach out to industrial laborers in urban settings. Recognizing that churches were largely ignoring the plight of these workers, Fred Rindge, secretary of the Industrial Department of the YMCA, established in 1910 a program for college industrial service. Educational programs were core feature of all service work among laborers, with classes offered in such topics as mathematics, mechanics, and "plan reading." Lectures on health, alcoholism, and personal efficiency were also quite common at this time, geared to provide workers with healthy habits of living to facilitate symmetrical growth. In addition, YMCA groups provided outlets for physical recreation among these groups, organizing sporting events, holiday dinners, playground trips, and weekend camps. For immigrant laborers, Americanization work was quite common, including lessons in civics, American history, and the English language. One report, in fact, noted that such men were "taken out of their ditches" at the noon hour so that students could help them "swallow an English lesson between their sandwiches and beer."[54] In 1913 alone, over 30,000 industrial laborers, representing 45 nationalities, studied English in classes taught by college YMCA students.[55]

Such service work, Rindge suggested, would be as helpful for employers as it was for those in the laboring classes. At a monetary level, he was certain that the improvement of the workingman's character would serve to enhance productivity. "Engineers are realizing more than ever that improved working and living conditions, sanitary measures . . . recreative and social features . . . and the treating of all men as men is not only a self-satisfying but a paying investment of time and money," he stated. "This work which improves the character of men is bound to increase their efficiency; therefore it pays, both from the human and economic standpoints."[56] With near-utopian enthusiasm, Rindge viewed industrial service as the solution to two mutually related social problems. On the one hand, college-educated engineers were typically trained for expertise in the theory and practice of industrial work but relatively unschooled in human relations. There was a distinct need, therefore, to cultivate human sympathy, "born of personal contact," for working men and their conditions. On the other hand, thousands of immigrants, "ill suited by nature and training to fit into the economic life of a new country," were "peculiarly receptive" to outside influences. For Rindge, the purpose of the industrial service movement was "to give these two 'opportunities' a chance to re-act on each other."[57]

Such a plan was implemented quite successfully at the Sheffield Scientific School at Yale University. During the 1907–1908 academic year, YMCA leaders, recognizing that engineering students were in a "position to influence the working out of the new relationships of modern life," met at Northfield to discuss a program that would "awaken in students of the engineering courses the realization of this great opportunity and to inspire the devotion of their lives to it."[58] English classes were organized in a rented house on Chapel Street for seventy Poles, eighty-nine Italians, and ten French Canadians. In addition, lectures were given over the course of the year related to industrial labor, boasting an average of sixty-four students in attendance at each session. By 1912–1913, eighty-seven Sheffield YMCA students were involved in industrial service among over four hundred workingmen.[59] For leaders of the Sheffield association, the industrial service movement was incredibly important, in part because it validated the Christian career motive of those studying in the engineering field. As one leader suggested, careers in ministry, missions, and the YMCA had previously been the only occupations deemed worthy of the title "religious work." Now engineering was a viable alternative, equally "religious" because of its link to human service and character development. By 1913, 160 colleges and technical schools were engaged in YMCA industrial service.[60]

Other attempts were also made to provide environmental settings that would enhance character and citizenship among the urban poor. At the University of Pennsylvania, for example, a social settlement was erected in 1900 by the YMCA and other campus organizations housing boys clubs, girls clubs, a rescue mission, and various classes for both children and adults. Designed to serve as an alternative to the "clubs" in the city that lured boys and girls with card playing, gambling, and drinking, the house initially maintained a staff of about sixty-five college students and faculty.

By 1906, the settlement had grown so significantly that a new building, costing $60,000, was constructed to continue the work. Entitled "University House," the four-story structure boasted a roof garden, a large gym, bowling alleys, showers, an auditorium, club rooms, a kitchen, and a savings bank to encourage the development of proper spending habits.[61] Akin to these attempts to provide Christian environmental settings within the urban context, many associations also organized and maintained camps in secluded areas designed to remove urban youth from their social contexts. At the University of Michigan, Princeton, and the University of Pennsylvania, students took groups of underprivileged boys to wilderness camps for two-week summer sessions.[62] "After a year's work in the hot glass factory or dusty woolen mills an outing is a physical necessity," one Princeton leader noted. "More than this, the boys and girls get a definite conception of how a vacation may be spent profitably by unconscious imitation."[63] Befitting the conservative social gospel mindset of the movement, such work was to "raise the ideal of life for those in all social classes, such that they will rise out of their present conditions and make good for themselves."[64]

With all of these provisions for social service, YMCA leaders hoped that philanthropic habits would be developed in students' lives that would continue beyond the college years. By this time, in fact, leaders had determined that the association was preparing students, not only for church and missionary careers but also for vocations in "social work." Richard Henry Edwards, the primary agent of many of the trends, suggested that "Alongside our campaign for able and trained leadership in the home and foreign service of the churches, as such, we propose to set, without apology to anyone, the immediate and great importance of strong college men going into professional social work. . . . We shall do this alongside our recruiting for leadership in the Church, and on the basis that such organizations are playing an essential part in the world for which the Church exists."[65] Edwards noted that those in YMCA leadership had begun to recognize that social service was as "ministerial" as professional church and missionary work, providing students the opportunity to engage in labor that brought character and service together in seamless ways. Ironically, while the call of muscular Christianity was to get masculine men into the pulpit, the logic of the program in this era moved in the opposite direction, highlighting alternative venues for service outside of the churches.[66]

Service and Secularization

In the eyes of many educators and theorists, "social evangelism" was reflective of a larger sea change in the nature of student Christian faith. From a previous focus upon creeds and the inner status of the soul, the new emphases on character and service promoted a perspective highlighting active and lived religion. Though "disinclined to much profession of piety," Francis Peabody suggested, the student was yet "extraordinarily responsive to the new call for human service." "The first serious question which the college student asks," he claimed, is not "Can I be saved?" "Do I believe?" but "What can I do for others? What can I do for those less fortunate than I?"[67] Much as the "laboratory principle" converted students "from bloodless bookworms to dynamic human beings," YMCA advocate Henry Cope suggested, a new laboratory conception shifted attention away from metaphysical abstractions to the practices that constituted the religious life. Cope affirmed this posture, noting that "They may be little at prayer meeting—a normal college man prefers to pray with his feet—but you will find them in settlements; conducting boys' clubs; groups of them going out to the high schools, guiding athletics, stimulating the younger lads to finer toned living . . . seeking their brothers they find the Father."[68]

These ideals suggested that this shifting tenor was not only a move from intellectualism to activism but also from subjective to objective dimensions of Christianity. Charles F. Thwing, president of Adelbert College and Western Reserve University, noted that among students there had been a "decline of conversation upon religious topics" because piety was "becoming less self-conscious, less self-centered, and Christian work at the same time is becoming more active and more vigorous." "The rise of the Young Men's Christian Association in the colleges," he suggested, "is a very great embodiment of the decline of self-centered piety showing itself in talk, and of the rise of piety that shows itself in Christian service."[69] Arguing that religion was moving from the subjective to the objective "phase" of its growth, Charles Gilkey told a college YMCA audience in 1905 that "We do not insist upon the necessity of a deep personal religious experience as did our fathers, nor do we give ourselves to the analysis of our own spiritual state as they did." "The sincere Christian student," he claimed, "ready to throw himself with enthusiasm into concrete Christian work, can state this purpose only very unsatisfactorily in the vocabulary of self-conscious introspection—the only 'prayer meeting' language he knows."[70]

The move from an evangelistic perspective anchored in personal salvation to a practical religion of ethical deeds made perfect sense within a university climate that was increasingly inhospitable to declarations of a priori "truth." As both George Marsden and Julie Reuben propose, this distinction between dogmatic theology and practical religion served, for university reformers, as one of the most powerful antidotes to sectarianism and doctrinal authoritarianism in these institutions.[71] Andrew Dickson White's perspective, articulated in his 1876 *The Warfare of Science* and greatly expanded twenty years later in *A History of the Warfare of Science and Religion* (1896), was becoming increasingly representative of the broader tenor of higher education. Science was at war, White professed, not with "religion" but with "theology." While creedal theology threatened to decimate scientific processes of inquiry because of its unchanging cast and resistance to the method of trial and error, religion simply implied that which the New Testament book of James called "pure religion and undefiled," the acts of service and righteousness highlighting "living" religiosity.[72] To claim that religion was defined by deeds rather than creeds, therefore, was to render the Christian spirit compatible with the climate of higher education. Colleges could be religious even if they could not be theological.

While college YMCA leaders rarely reflected on the philosophical foundations of their programmatic positions, it is clear that this new vision was at least partially shaped by pragmatist ideals. "Unlike the German or Indian," Clayton Cooper suggested, "his seriousness is not associated with metaphysical or theological discussion or expression. He asks not so much 'What?' as 'What for?' His aims belong to 'a kingdom of ends.' "[73] Not only was this brand of Christianity more self-consciously ethical, but it was also overtly consequentialist in its ethic, evaluating all Christian principles in terms of their results in the world rather than through any means of independent validation. At the University of Michigan in the 1890s, both John Dewey (who served as a YMCA board member) and university President James Angell regularly implored YMCA members to embrace a religious truth anchored to the scientific ideal of hypothesis testing, stating that the actual effects of religious ideals would be the best means of proving their legitimacy. Angell, in fact, proposed that Christianity could only be deemed helpful because it promoted truthfulness, a higher position for women, constructive charity work, and democratic government.[74]

One of the results of this shift was the elimination of hard boundary lines between religious groups. In the industrial service program in New York, for example, educational classes among immigrant workers were formed in collaborative ventures between YMCA students and three Catholic priests, Russian Jews on the lower east side, and the proprietor of an Italian saloon. Work among immigrants all but assured that students would be coming into contact with different religious persuasions, and the elevation of character and service over conversion meant that doctrinal nuances were often left in the background. Furthermore, many chapters began encouraging students of other faiths to join them in these activities. Because of his work with the YMCA in the industrial service movement in New York, one Jewish student remarked that he believed "the day is advancing when all will be brothers, and creed, race, and prejudice will be forgotten in common service."[75] Concurring, Rindge stated plainly that "If life is the test of Christianity, we must utilize and develop every right impulse, whether a man's motive is avowedly Christian or earnestly altruistic." Service was the ultimate common denominator for an organization seeking to enhance its muscular vision, rendering particularistic theological positions less substantial as Christian boundary markers.[76]

Service therefore diminished the evangelistic impulse in quite direct ways. While some continued to see social service as an evangelistic inroads, paving the way for a more powerful gospel presentation, many others began to argue that service alone was enough to commend such work to the associations.[77] Students were regularly reminded that their service among boys, immigrants, and workingmen was not to include direct appeals for religious conversion. Recognizing that some employers or laborers would not desire explicit religious prosyletization, Rindge urged collegians to tell foremen "you haven't any axe to grind, but just want to be of service."[78] While local practice surely varied, the official literature on college YMCA social service required that students not call into question the religious beliefs of those they were attempting to serve. Thus David Porter recommended the elimination of New Testament stories when working with boys' clubs enrolling Jewish members. When working with foreigners, Rindge suggested that students not critique immigrant religious traditions but rather assist in "helping men to be loyal to the best there is in their own religious faith."[79] The goal of such work was rarely framed in terms of Christian conversion. Instead it focused on growth in character, evidenced by developing discipline, health, education, and commitment to American values. Jews and Catholics, as well as Protestants, could be "saved to character" through social service.

Some recognized as well a tendency for social service work to compete with and even obfuscate the religious work of the college associations. At Sheffield, one alert critic noted that, while "experience everywhere" revealed that service could attract students "never touched by the regulation student prayer meeting," associations were in danger of allowing religion to be "timidly obscured" in order to "court their support." The object of service, he claimed, was not "to substitute concrete service for Christianity but to vitalize Christianity by bringing it concretely to bear on the lives of those about us."[80] Ironically, a similar critique was directed at the college YMCAs by Shailer Mathews, who was by all accounts one of the theorists responsible for generating social interest among students. He reminded YMCA students that social service was best described as "an expression of religious motives" rather than a "substitute for religious faith." Such work was essential to Christianity, he confirmed, "But to say that the tree is known by its fruit is not to say that the fruit can take the place of the tree." For Mathews, service was a result of religious commitment, a natural outgrowth of a life lived in a close personal relationship with God. To believe that a religious life could grow only through self-expression, he suggested, was to neglect the fact that "Spiritual devotions should precede spiritual devotion."[81] To proceed with service untethered to a vital religious life was to invite the despair that often accompanied humanitarian efforts divorced from Christian truth.

Surely part of the new emphasis on action and service was related to the fact that YMCA members desired an integrity of belief and action. The deep-seated antagonism toward "pious affectation" and "hypocrisy," in fact, demonstrated this need to link doctrinal creeds to their practical manifestations in the world. However, there was a closer proximity than many thought between this perspective and the idea that the practical implementation of Christian ethics was the only important measure of Christian faith. When beliefs were just means to an end of servanthood, it hardly seemed helpful to maintain those beliefs or place them in a central position within the movement, provided that service was proceeding apace. From such a vantage point, some began to contend that beliefs were actually superfluous to the aims and goals of the Christian life. In reality, many within the YMCA saw themselves as uniquely positioned to facilitate a religious transition for students from the belief-oriented faiths of their families of origin to the new activist faith championed by the association. As one secretary remarked, "The Association, with no quibble about creeds and all its energy set upon the realization of active Christian manhood, should be able to point out to them the ultimate importance, not of belief, but of life."[82]

Many failed to recognize that in detaching active character building and philanthropy from their source in religious truth, they might have been severing an important resource for such work. Both the leaders championing such changes and a number of the early students coming under their leadership were themselves recipients of the very doctrinal and theological training they now downplayed. Thus, their own activism stemmed from beliefs and biblical education that anchored their practical muscularity. However, in time those who followed often lacked such a background and began to embrace the practical faith of the YMCA untethered to the religious beliefs that had once sustained them. Mott and many of his colleagues wanted to demonstrate the practical nature of the Christian faith, but many of those following their message began to suggest that Christianity was defined by these practices alone. While the call to service could be an interpretation and application of Christian principles, in other words, it could also be a catch-phrase to replace them. In the end, it was not such a long stretch from a service-oriented faith to a stance in which service was the final faith.

Section 3

Building the Kingdom, 1915–1934

Chapter Seven
A Shrinking Sphere

The College YMCA in Wartime

In the late nineteenth and early twentieth centuries, the college YMCA assumed a centralized position in American higher education. The expansion and institutionalization of the organization between 1890 and 1915 had placed local chapters at the very center of campus life. YMCA buildings were common landmarks on campuses nationwide. YMCA service activities were tightly linked to student life and culture. One critic even observed that the YMCA handshake was "known in every school in America."[1] Valued by students and administrators, the YMCA had established a near monopoly on student religious life and had developed a version of practical Christianity that was integrally connected to the progressive mood in higher education. By the beginning of World War I, it had become what the Wesleyan University student newspaper, *The Argus*, called "the largest and most important student fraternity in the world."[2]

It was within this context of ebullient optimism that John R. Mott announced that he was leaving the college work in 1915 to assume a position as general secretary of the larger international YMCA. His successor, David Richard Porter, was in many ways the perfect embodiment of the fourfold muscular Christian vision of the prewar era. An academic standout at Bowdoin College, he was chosen as a member of the first group of Rhodes scholars from the United States, studying at the Honors School in Modern History at Trinity College, Oxford in 1904. While most took three years of preparation to pass the final history exams, he completed his within six months, later earning a master's degree from this institution in 1910. He achieved some prominence on the Bowdoin football team, scoring a memorable touchdown against Harvard on a hundred-yard fumble recovery. He also distinguished himself in athletics at Oxford, excelling in rugby, cricket, and tennis. Upon graduating from Bowdoin, Porter assumed leadership over the newly formed high school YMCA, later officially termed the Hi-Y. With his academic experience, athletic prowess, and involvement in

YMCA service and boys' work, Porter was an ideal candidate to continue Mott's legacy at the helm of this powerful movement.[3]

It was not long before Porter realized, however, that he would be serving within a context quite distinct from that of his predecessor. World War I was a watershed moment for many colleges and universities. Despite the best efforts of faculty and administrators to speak of the long-term national benefits of a college education, institutions across the country experienced enormous enrollment declines, estimates revealing a 40 percent drop in male student attendance nationwide.[4] Faced with the imminent prospect of further declines initiated by the reduction of the minimum draft age, many leaders in higher education feared that the war would break the backs of key institutions. Such declines decimated many YMCA chapters and left enormous leadership gaps at both the national and local levels. Many of the collegiate secretaries were claimed for YMCA war work leadership. In addition, since older students were more likely to join the war effort overseas, student leaders, and in some cases whole cabinets, were removed for war service. At the University of Wisconsin, the *Daily Cardinal* noted in April of 1918 that the heads of nine YMCA departments had enlisted in the war effort and that the entire committee organization had been crippled. Many of the summer conferences were also decimated, and in several cases regions combined to sustain adequate numbers. Overall, the decline in male student enrollments, combined with the loss of student and full-time leaders, quickly hobbled the organization and its leadership infrastructure.[5]

Yet in addition to these obvious losses, the broader effects of World War I on the campuses also served to reorient the work of the student YMCA. Administrators and faculty sought government assistance to aid the colleges in stemming the flow of male students to the armed forces. Stated in terms of preserving the mental strength of the nation, this plea was well received by Woodrow Wilson, himself a former college president with a sense of public service in higher education. Just two years after the 1916 creation of the Reserve Officers Training Corps (ROTC), therefore, he convinced a hesitant Congress of the need to enhance college attendance through the development of a Student Army Training Corps (SATC).[6] Those colleges adopting the SATC program were to be organized as army training facilities, collegians enrolled as soldiers in training. By October 1, 1918, 140,000 students in 525 colleges were enlisted in the United States Army. Porter noted that nearly all schools with more than a hundred students maintained SATC units and that 90 percent of the students in many of these institutions were part of the organized military presence. Students experienced a combination of drill (no less than eleven hours per week) and study during the day, many of the classes directed purposefully to military application. On many campuses, fraternity houses and other college buildings were transformed into makeshift barracks and the typical extracurricular activities and student traditions were greatly diminished. Academics were also downplayed, leading the Princeton *Alumni Magazine* to cite that "Every day Princeton becomes less of an academic college and more a school of war."[7]

While the SATC program lasted only a short time—it was disbanded in December of 1918—the YMCA played an extremely significant role in this period of campus transformation. The War Work department of the larger YMCA, in fact, essentially took over the intercollegiate movement during the fall of 1918, organizing a vast program and contributing significant financial and human resources to the effort. A headquarters for this work was set up in New York, with Porter in charge, and any school with an SATC unit numbering over 250 was provided with an additional full-time YMCA secretary. A host of new secretaries were therefore sent to local campuses (some secured their first full-time secretaries at this time), increasing threefold the number of employed officers in the college movement. On every campus where a student YMCA chapter existed, the organization took a role as the de facto social and religious support center for the student "troops," directing the character and service ideals of muscular Christianity to this new context. At the University of Illinois, for example, where 3,100 SATC students were enrolled, several YMCA War Work secretaries teamed up with the campus YMCA leaders to deliver lectures on religion and sex hygiene, develop Bible studies, provide postal services, and organize movies and dances for the men. At the University of Wisconsin, where 4,000 student-soldiers were posted, 7 additional secretaries were sent by the YMCA War Work department to promote a vigorous program of lectures, Bible studies, personal interviews, and campus entertainment. At these universities, hospital visitation also became important as the campuses dealt with large-scale influenza epidemics. Particularly because fraternity and athletic life was largely discontinued during this time, it could rightly be said that the YMCAs organized the social and service life of the campus. Noting that most student organizations had "gone out of business," the Vanderbilt *Almunus Magazine* reported that "only one organization has taken on new life, namely the YMCA."[8]

Such centrality, however, was short lived. In the years following the war and throughout the 1920s, YMCA leaders began to recognize that the supportive environment of campus religion, service, and activism that had nourished local chapters was rapidly being eclipsed within a fragmented context far less conducive to a dominant YMCA presence.[9] From a numerical "high point" immediately following World War I, the organization experienced a precipitous decline throughout the 1920s and early 1930s. The total number of students involved in local college associations declined from 94,000 in 1921 to 51,000 in 1940. Between 1920 and 1940, a period during which student enrollments roughly doubled, the number of chapters declined from 764 to 489. Of course, the fact that this decline came at a time of massive expansion only served to accentuate the sense of marginalization. While approximately 33 percent of the male student population was involved in the organization in 1920, only 8 percent could claim the same in 1934.[10] By 1939, the National Council of YMCAs stated with unequivocal candor, "The YMCA is not the important factor in college life that it was twenty-five years ago."[11] The following chapters are devoted to explaining the changing postwar context, the evolving program of the YMCA in light of these changes, and the means by which the organization was marginalized within the life of American higher education.

The Fragmentation of the Campus

Higher education institutions were in many ways refashioned in the decade immediately following World War I. During the 1920s, colleges and universities experienced an unprecedented enrollment explosion that stretched campuses to the limits of their resources. The forces instigating and sustaining such growth were complex. The expansion of the high school surely provided a significant feeder system. While in 1890 fewer than 300,000 Americans were attending high schools, by 1930 that number had escalated to 4,800,000, representing a sixteenfold increase. As historian Paula Fass has suggested, rising incomes meant that parents could afford to allow their children a more leisurely transition to adulthood. The declining size of middle-class families also meant that parents were able to afford higher education and plan specific educational goals for each child according to individual needs and interests.[12] Economically, while perhaps only five professions had viewed college as necessary for training in 1870, fifty or more, highlighted by the expansion of business and engineering, viewed higher education as an essential prerequisite by the 1920s. As corporations began to hire white-collar professional accountants, engineers, and managers to handle escalating bureaucratic functions, they increasingly called upon college men to fill these nascent positions.[13] As David Levine has suggested, World War I was a higher education watershed in part because it linked the colleges to the society in such a way as to demonstrate the value of a college education, transforming the student "from a frivolous young person to prospective leader of society."[14]

Such trends supported both the growth and diversification of the student body. Far exceeding prognosticator's expectations, the undergraduate population tripled between 1910 and 1930, the largest absolute increases coming during the decade of the 1920s. Between 1919 and 1930, the total student population rose from 597,857 to 1,100,737. Such gains were often quite rapid. For example, at the University of Illinois, enrollments doubled from 3,000 to 6,000 between 1919 and 1922. Ohio State University saw similar increases from 4,000 to 8,000 in this same time frame. While in 1900 only 45,000 students (about 4 percent of the college-age population) were enrolled in state-controlled colleges and universities, by 1930 this number was approaching 500,000, representing nearly 20 percent of the college-age population. Emerging from the elite periphery of American culture, higher education was, for the first time, becoming popularized within the general public. This expansion of institutional size was accompanied by a parallel expansion in collegiate religious diversity. In 1890, a study of state universities revealed that Catholics represented about 3 percent and Jews about 1 percent of the student populations in these settings. By 1926, however, Catholics represented about 10 percent of all students in such institutions. Despite anti-Semitic tendencies and some admissions criteria limiting attendance, by 1919 Jews also constituted nearly 10 percent of all students in higher education institutions. In some eastern schools, Jews made up one-third to two-thirds of the student bodies.[15] While Protestants were still a majority, they were no longer monopolizing the campus.

In the midst of such growth and diversification, it was clear that no single club could serve as an all-student organization representative of universal student interest. While the YMCA had always been viewed as a central and unifying agency in campus life, leaders recognized that the movement was rapidly losing its place as a centripetal force. "YMCA leaders," according to Building Secretary Neil McMillan, "saw that the majority of the state and other large universities were growing exceedingly complex . . . too complex, in fact, for it to be possible for any one organization to dominate more than one phase of student life as had been possible in early days or in smaller institutions."[16] The growth of institutions necessitated a degree of specialization not mandated in small, homogenous settings where single organizations could serve multiple purposes. Because of this, the YMCA was increasingly viewed as an organization appropriate for those with religious interests but not for the whole student body. As one stated,

> Previously the Association . . . could be called an all-student organization. The organization included the leaders of student sentiment in all aspects of student life and it . . . was looked upon as one of the leading, if not the leading, student organization. Now the Student Association is usually one among a number of student organizations, gathering into its membership and work only a section of the student body. It represents those particularly interested in religion or who happen to have made this their principle student activity. As a result the Association is sometimes looked upon as made up of the

"pious" group. It cannot be said to be a representative all-student organization, as it previously was.[17]

The diversification of the student body had a similar result. As the Protestant evangelical consensus of the nineteenth century began to crumble following the expanded enrollments of members of other faiths and increasingly strident interpretations of church-state separation, many leaders could no longer philosophically commit to the YMCA's privileged status as a common center of student activities and service. "The student movement," one leader commented, "developed its philosophy and program in compact colleges where the student religious background both in the church and state college was largely Protestant. Assumptions which then could be made with regard to the common social and religious background of students are not now tenable."[18] While the YMCA had previously appealed to a generalized audience for whom Protestant religious interest was a kind of assumed backdrop, it now appeared to be relevant only to a segmented group—the so-called pious group—for whom such interests were paramount.[19]

Furthermore, administrators were growing increasingly interested in providing supervisory oversight in the extracurricular realm. In the late nineteenth and early twentieth centuries, many of the larger colleges and universities nationwide had adopted rather laissez-faire policies toward the extracurriculum, allowing student-led activities and organizations to spring up naturally and haphazardly with little administrative intervention. As John S. Brubacher and Willis Rudy have suggested, the curricular and extracurricular integration characteristic of the antebellum college had, in the period between the Civil War and World War I, become a "bifurcated" world marked by an independent student society working at cross-purposes to the academic aims of the institutions.[20] For a number of reasons, however, postwar administrators became more attentive to "student life." Some were influenced by new progressive educational models that placed emphasis on the mental, social, physical, and emotional aspects of student experience.[21] A more organismic psychology, which placed emphasis on the education of the "whole person" and spurned dualisms between the intellectual and the social, was taking its place as the educational "common sense" of the era. With such an acknowledgment of the potency of the YMCA "fourfold" life, college bifurcation was increasingly viewed as an untenable context for student growth.

Administrators also recognized that the extracurriculum, both with regard to student activities and student services, marked a critical entry point for student moral influence and for a restoration of the integration lost through electives and curricular specialization. Since student life consumed so much attention within the broader framework of a college education and since both faculty and curriculum seemed incapable of sustaining a comprehensive ethical resource for students, the extracurriculum appealed to many administrators as the best opportunity to shape the moral lives of students.[22] Such centralization, of course, had more than logistical benefits for campus leaders. By consolidating the various strands of voluntary activity under a university-sanctioned individual or program, administrators gained a degree of control over the potent (and sometimes troublesome) world of the college extracurriculum. Diversity was deemed healthy to the undergraduate life of the campus, but diversity contained within a larger boundary of administrative oversight was a far safer endeavor. While such changes were still far removed from the in loco parentis mindset of the nineteenth century, they did provide a mechanism for the preservation of administrative order on the American campus.

Administrative Inroads

The result of these forces was a greatly enlarged administrative reach into the domains of student activity and care, a process that had significant ramifications for YMCA chapters invested in similar activities. For example, due to expanding enrollments and a desire for enhanced "parental" moral influence, campus leaders in the 1920s developed a renewed interest in student housing. In the late nineteenth and early twentieth centuries, the construction of campus dormitories had not kept pace with burgeoning student enrollments, forcing students to look to fraternities and sororities, clubs, and private boarding houses for their lodging. Many saw academic facilities such as libraries and laboratories as far more critical to institutional success than residence halls, and this bias generated a rather laissez-faire approach to student living conditions. In the postwar era, however, administrators again took up the issue of dormitory construction, seeking to provide closer oversight for students and more inclusive contexts for positive peer moral influence under adult supervision. Administrators desired these structures to become homelike in atmosphere, possessing a resident director/proctor and, optimally, several resident faculty members as well to serve as parent figures. Even when students were forced to take up residence off campus, college leaders now required close inspection and supervision of these facilities. The YMCA's work in this area, it appeared, was no longer needed.

New facilities also compromised the centrality of YMCA buildings across the country. By the postwar period, the YMCA building movement had fallen upon hard times. While such structures often did serve for a time as social centers on the college and university campuses, financing and maintaining these buildings proved to be a costly endeavor in terms of money and time. Because funds had been raised largely through private donations and few had endowment funds to meet continued expenses, economic shortfalls were commonplace. At the University of Illinois, for example, deficits grew from $6,000 in 1914 to $40,000 in 1916, an amount nearly twice the annual budget. In addition, many were beginning to recognize that buildings could prevent secretaries from attending to the spiritual purposes of the associations.[23] At the University of Virginia, secretaries complained that they spent much of their time with roommate conflicts, curfew violations, and vandalism, confirming the opinion of the student who wrote that there were "too many secretaries tied so thoroughly to the inexorable task of running a small hotel and amusement center that they have consciously to give up much of the vastly more important work for which they were called."[24] In the eyes of YMCA Building Secretary Neil McMillan, the big problem with college buildings was the failure to recognize the distinctions between urban and collegiate contexts. While proliferation of staff was often possible in urban settings with wealthy donors, the limited financial resources of college students prevented such expansion on the campus. Because of this, either the religious work on the campus suffered or the buildings themselves suffered through lack of sufficient supervision.[25] As university of Illinois historian Scott Peters suggests, "By (1916), the new building was looking less like something that would help to usher in a new epoch and more and more like an albatross that threatened to bring the Association down."[26]

While such factors generated tremendous challenges, it was competition from the institutions themselves that struck the final blow to many YMCA buildings.[27] New student unions, gymnasiums, and dining facilities constructed in the 1920s featured many attractions that had previously served as the defining components of YMCA buildings, thus instituting a competitive marketplace for these services. The dynamics of such competition worked out differently on different campuses. In some locales, institutions assimilated YMCA facilities for their own purposes, converting these buildings either to academic facilities or to student unions. At the University of Illinois, for example, the university purchased the YMCA building to use for the School of Military Aeronautics and maintained possession once the war ended. Desirous of a more steady income, some YMCA chapters rented out space within their buildings to the institutions, an arrangement that typically furthered the marginalization of the movement. At the University of Wisconsin, economic necessity forced the YMCA to allow the incipient student union to occupy and manage the first floor of the building. Although the advisory board noted that this decision was rooted in a desire to allow association hall to be "enjoyed by all the students of the university," the actual result was a complete university takeover.

President Van Hise supported the opening of a cigar stand in the building, and regular "Union smokers" with cigars "in abundance" were held from the day of opening, dispelling "the odor of sanctity which was popularly believed to hover over the place." The Union president proclaimed triumphantly that the "rooms belong to the student body. They will have no connection whatever with the YMCA." Despite the fact that the Union controlled only the first floor, YMCA leaders found that the entire facility was equated in the student mind with the Union, one secretary remarking that he had to remind students using the second-floor rooms that they were "guests of the YMCA."[28]

When YMCA chapters attempted to compete with campus-sponsored facilities, the results were equally revealing. At the University of Michigan, tensions emerged when, during a YMCA building campaign, a number of students and faculty members registered opposition to the YMCA's centralized presence on the campus. At issue was the assumed competition between the proposed new YMCA building and the Michigan Union. Many wrote in to the *Michigan Daily* to protest the construction of a building that would detract from the center of campus life and splinter the unity of the campus. One noted that, because of its attempt to "centralize and magnify that ineffable sentimentality banally known as love for the Alma Mater," support for the Union as the democratic center of campus life was "morally obligatory." The *Daily* reported that, in a canvass of "representative men" on the campus, students felt the YMCA had "overstepped its bounds" and should focus on religious issues alone, leaving social activities to the Union. Eventually, the YMCA group at Michigan did construct a building, but this 1917 edifice, Lane Hall, revealed that the YMCA acknowledged its diminished role. While the former Newberry Hall had been designed for student service and entertainment, Lane Hall was primarily constructed for religious purposes, devoid of the recreational space afforded to the previous structure. The Michigan Union, rather than the YMCA, became the democratic center of student life.[29] As the case on this campus suggests, students often desired university sponsorship of campus life, centering events within the administrative structure rather than in this "religious" organization.

College YMCA leaders by the 1920s were forced to admit that the building movement was a grand miscalculation, fueled by expansionist fervor rather than focused philosophical vision. Those buildings that were not sold or rented to the universities were often scaled down, like at Michigan, to eliminate duplication with new facilities. When new buildings were constructed, their scope and design reflected the minimalist and specialized attempt to focus on religion alone. At the University of Washington, for example, leaders noted that their new 1923 facility "profits by the best experience in other universities" in that "it is primarily a headquarters for religious and other group work rather than a 'union' building or an imitation of a city Y.M.C.A." In 1928, Dean Babcock at the University of Illinois suggested that a new building should not serve as another Union—they rejected proposals for a cafeteria and residence hall—but rather as a "spiritual powerhouse" on the campus. As Henry Wilson noted at its dedication, "The University provides facilities for physical education and recreation, therefore we do not have locker rooms, gymnasium, or swimming pool. We do not have dormitories, nor restaurant, nor other commercial enterprises. We have set ourselves to the tasks which fulfill our purpose, the meeting of moral and spiritual need and opportunity." The expansive buildings constructed prior to the war had, by the 1920s, become colossal monuments of a bygone era, physical evidence of a central position that was no longer possible or necessary.[30]

In addition to these "bricks and mortar" projects, new personnel with expertise in student care also complicated the role of the YMCA on the campus. Bolstered by war-time mental testing and counseling, the postwar personnel movement was an attempt to provide a degree of specialized scientific and psychological expertise to student development in the areas of morality, health, vocation, social relationships, and even religion. Even before the war, some schools had hired Deans of Men and Deans of Women to oversee the general public morality of the campus and to serve as personal counselors for students, clearly assuming roles that had at one time been the province of YMCA secretaries. From this more generalized position, a series of other counselors, including those supervising admissions, placement, health services, employment, and housing, emerged to expand burgeoning student development staffs. These innovations reflected a clear sense that such tasks required professional—even scientific—expertise, conducted in specialized fashion by those with credentials in these domains.[31]

Provision for student employment assistance marked one concrete example of this larger trend. While the YMCA had supervised such work on many campuses in the prewar era, by the mid- to late-1920s much of this work was taken over by the universities themselves. Dartmouth named an associate dean for employment in 1919, and by 1925, 44 percent of all large universities, 12 percent of medium sized universities, and 11 percent of small colleges had hired paid staff to handle such work. While at times YMCA employment bureaus were maintained, nearly all were terminated by the 1930s in favor of university-directed options.[32]

A similar pattern emerged in freshman work. Obviously, the YMCA had long assisted freshmen through their work in greeting them at trains, facilitating student housing selection, and assisting in registration. In the early 1920s, chapters around the country also developed "freshman camps" and "freshman fellowship groups" to facilitate the transition to college. The camps, three- or four-day events held just before the beginning of the academic year, took place at remote campsites either rented or owned by the YMCA. In many cases, chiefly because of the capacities of the camps themselves, these programs were available only to about 10 percent of the male students in the entering class, chosen through recommendations from high school principals, parents, and local Hi-Y secretaries. Led by campus, student, and YMCA leaders, the idea behind these events was to educate students on topics related to registration, majors and course selection, study habits, extracurricular activities, relationships, and campus traditions. In addition, while not explicitly religious, many of these orientation programs provided students with patently didactic moral directives, giving practical advice on how to live lives of moral purity amidst the temptations characterizing collegiate culture. Freshman fellowship groups, led by upperclassmen and faculty, continued a similar curriculum throughout the year while also establishing informal mentoring relationships for new students. Heralded by campus deans, these gatherings also proved to be quite popular among freshmen. By 1923, at the University of Illinois, for example, there were already eighty such groups on the campus.[33]

By the late 1920s and early 1930s, however, less than ten years after their initiation in most cases, most of the freshman camps and fellowship groups were squeezed out by encroaching administrative programs. Deans and other administrators recognized that, as the university culture increased in size and complexity, new students needed some form of introduction to the academic and extracurricular domains of student life. No longer content with voluntary approaches to such work, therefore, many colleges and universities themselves began assuming responsibility for orientation activities, typically in the form of "freshman week" programs held prior to the beginning of the academic year. According to the Association of American Colleges, 60 percent of the 281 colleges surveyed in 1928 maintained some program of formal, administratively organized orientation for new students, most initiated in the 1920s. Many campuses also began instituting their own discussion groups for freshmen, making these available as regular provisions of the generalized orientation.[34]

For a time, freshman camps did coexist with university-sponsored programs, the camps being offered immediately prior to or following the campus orientation. Because the campus version dealt with so many of the same issues, *however, this was a somewhat awkward combination*. When camps were retained when camps were retained, they became far less an expression of general collegiate preparation than a specialized event for those interested in the explicit moral and religious fare of the YMCA. More commonly, however, such camps folded and the YMCA was offered a slot within the freshman week program, advertised as an extracurricular option for religiously minded students. As with YMCA buildings, the options were simple: either narrow the scope of the program, pass it off to the schools, or admit that there was little need for such work in a diverse campus community.[35]

Ironically, the very success of the prewar college YMCA movement thus proved to be its undoing in the years following the war. Many chapters, for either philosophical or practical reasons, had positioned themselves as key agents in campus service and recreation. However, it was these very fields of labor that were rapidly removed from association auspices as the institutions themselves became enamored with the "four-fold" life. As historian G. Grey Austin noted of the situation at the University of Michigan, "Paradoxically, no sooner had (campus service) become the backbone of the SCA when the University administration and other agencies usurped these very functions, leaving the association in a greatly debilitated state."[36] In other words, the muscular Christian vision of the prewar YMCA set the organization on an ambitious project of character building and campus social service, only to see these very elements enveloped by the institutions themselves. It appeared that the movement might be required to move back to Luther Wishard's original vision of a club dedicated strictly to religious purposes.

Leaders of the student movement, however, were beginning to recognize that even the religious domain was no longer the exclusive province of the association. While the institutional assumption of student housing, health, entertainment, employment, and orientation responsibilities might have been predictable in an era of expansion, less predictable was administrators' growing concern for the religious climate of the campus. There were likely a number of reasons for this expanding religious interest. In some cases, student moral and religious decline, typically linked to popular accounts of "flaming youth," was a stated impetus for greater efforts along these lines. Furthermore, sensitive to accusations of secularization, some administrators responded to outside pressures demanding greater attentiveness to religious issues. In other cases, administrators recognized that the increasing size and diversity of the campus mandated formal rather than voluntary efforts to cultivate student religion. As students of varying religious backgrounds began attending colleges at both private and state universities, the need for an inclusive work representing all students was seen as a distinct need, particularly in the wake of renewed administrative devotion to church-state regulations.[37] Whatever the motivating factor, however, few campuses were immune to this national fervor. A 1928 YMCA report documenting new administrative initiatives along these lines claimed that "This tendency has continued until it may be said that there are few colleges which are not either operating under some recently formed committee or are planning to form such a committee with a view to assuming a larger responsibility for their students' development in religion."[38]

The new activist role of college administrators in student religious life was manifested chiefly by new physical structures, new personnel, and new academic courses on religious topics. A number of universities, including Duke, Syracuse, Princeton, and the University of Chicago, constructed new chapel facilities in the 1920s. Institutions also hired new personnel to assume responsibility for the religious life of the campus. In some locales, it was still common practice for the university to subsidize the YMCA secretarial salary. Increasingly, however, institutions sought to coordinate such efforts themselves through a centralized administrative position. It was during this era, for example, that influential religious leaders such as Charles W. Gilkey, dean of the University of Chicago Chapel, James Yard, dean of Religion at Northwestern University, Dr. Robert Wicks, dean of the Chapel at Princeton, and University of Michigan Counselor in Religious Education Edward Blakeman were hired to supervise the religious work on their respective campuses. Particularly common in independent colleges and universities, the resurrection of such personnel in charge of student religious life was indeed a tacit admission that the era of voluntary religious leadership in the colleges was over.[39]

Furthermore, as a response to outside critics and occasional student requests, a number of institutions added coursework in Bible and/or religion in the postwar era. In 1925, in fact, David Porter commented that "There can be no doubt but that within a decade or two there will be some type of school of religion with some sort of academic affiliation located in every public-controlled university community."[40] While Porter's comment was exaggerated, by 1940 most private colleges and universities and approximately 30 percent of state universities possessed departments or schools of religion.[41] At one level, the provision of such courses should not have threatened the college YMCA. Association leaders had been arguing since early in the century that institutions should provide academic religious training to complement the voluntary offerings of the student chapters. However, many of the courses offered by the institutions in this era clearly trespassed on the "turf" of YMCA studies. Traditional academic courses in Old and New Testament, Church History, and Biblical Archeology were offered in many locales, but they were often accompanied by classes in Religious Education, practical religious work, and issues related to the relationship of religion to modern social problems. In reality, therefore, the conflict over academic coursework was a function of the coalescing of association work with academic religious studies. As the university approach to religious studies also shifted from traditional biblical themes to the sociological, psychological, and educational domains of religious thought, there developed a considerable overlap of content, if not method, in these domains. To the extent that courses duplicated the issues increasingly characteristic of YMCA study groups, they again represented an extension of administrative reach into the domain occupied by the association.[42]

On a number of levels, YMCA chapters were able to commend administrative initiatives in expanding religious work on the campus. "These experiments," one general secretary predicted, "may make available to the voluntary religious groups in a new way the resources in good will and effective cooperation of the whole university situation."[43] Yet while YMCA leaders uniformly applauded the efforts of administrators to enhance the religious climate of the campus, they were also understandably threatened by the professionalization of collegiate religious work. In 1928, the YMCA student division began complaining that university presidents seemed to doubt the capacity of the association to organize student religion, one noting that some administrators "are questioning the place of an autonomous student directed fellowship in the complex life of the modern large university."[44] Another suggested that college leaders evidenced a pervasive "lack of trust in the mind of youth" to exercise proper discernment outside of administrative control.[45]

The expansion of institutional religious work was envisioned as a process of consolidation rather than hostile takeover. Consolidation under a bureaucratic head, however, proved to be injurious to student initiative and ownership. At the University of Michigan, new President Alexander Ruthven hired Dr. Edward W. Blakeman as counselor in Religious Education and charged him to direct university religious affairs while also serving as a spiritual guide to students. In this role, Blakeman initiated a Spring Parley with faculty and students to foster discussion of religious issues, sponsored lectures on religious topics, and took charge of the freshman camp. Perhaps most importantly, however, he decided to form a Student Religious Association to consolidate all of the student religious activities on an interfaith basis under administrative auspices. Already in the throes of decline, the campus YMCA recognized that it would no longer be able to maintain its central place on the campus. The organization transferred its properties to the institution and was absorbed into the Student Religious Association, thus effectively terminating its seventy-six-year reign on the campus.[46] This pattern was largely replicated at Princeton in the 1920s when President Hibben developed a new position, dean of the chapel, as a means of centralizing the religious work of the campus under a single head and providing "general supervision of the religious interests of the students." The first occupant of the new post, Robert R. Wicks, was charged by Princeton President Hibben to consolidate the religious work of the campus, "centering the religious life of the University in the Chapel, rather than in a separate organization." By 1930, Wicks had succeeded in replacing the 105-year-old Philadelphian Society with a new Student Faculty Association, comprising the dean of the Chapel, interested faculty and students, and the Episcopalian and Presbyterian campus ministries.[47] Interestingly, such policies overturned the "free inquiry" models of the previous era. Whereas a previous generation of administrators had allowed students to generate their own religious subcultures in the name of freedom, they now saw that a degree of control was necessary to expand scope, to create an environment hospitable to all competing groups, and to secure a modicum of professionalism in religious affairs.

The Church Follows Its Students

While the expanding social and religious program of the university threatened to encroach upon YMCA activities from within, an equally potent challenge came from the outside in the form of expanding church and denominational movements. Starting gradually in the second decade of the twentieth century and then quite rapidly through the postwar period, demands for a heightened presence on college campuses emerged through a variety of overlapping forces. Church representatives began to recognize, first of all, that increasing numbers of their youth were enrolling in public rather than denominational institutions. In 1918, Presbyterians estimated that one-seventh of all state university students nationwide came from Presbyterian homes. They also determined, much to their surprise, that nine-tenths of Presbyterian collegians were studying outside of Presbyterian institutions. In 1925, 33 percent of the students at Ohio State were declared Methodists, giving a "parish" of approximately 5,000 students, while the 2,500 Methodist students at the University of Illinois constituted a greater total than the students enrolled in the Methodist colleges of the entire state. Baptists also recognized this growing disparity, noting in 1913 that there were more than twice as many Baptist students in state institutions as denominational institutions in the western states.[48] It was not unusual in the early twentieth century for national denominational leaders to label public university students as marginal in religious commitment. However, these statistics soon forced them to recognize that loyal churchmen might indeed attend such institutions. In a report to the Religious Education Association in 1907, the distinguished Committee of Six reported that "It is a mistake to suppose that these students are what are left after the churches through their own schools have claimed their own." Instead, the committee reported, many of these youth were "the flower of the land and the flower of the church," making church and state "not so much different things as activities of the same people in different fields."[49]

Church leaders, like association leaders, recognized the enormous potential of collegians. Joseph Cochran, corresponding secretary for the Board of Education of the Presbyterian Church, U.S.A., calculated that college students constituted 73.41 percent of the "prominent men" listed in "Who's Who" publications. "Making all due allowance for differences of opinion as to what constitutes prominence," he stated, "the conclusion is forced upon us that the college man's potential is somewhere between one and two hundred times that of the man who has not gone to college."[50] Such a perspective was indeed a striking reversal of traditional opinions among church leaders. Many had resisted vigorous work among collegians because of students' temporary residence in college communities, failure to invest personally and financially in the life of the church, and imprecise fit with the family-oriented nature of congregations. One denominational leader, however, suggested that such a perspective failed to recognize the critical nature of these youth in the larger work of the Kingdom. While church representatives had been known "to complain that students wear out the carpets and hymn books, and jostle the regular pew holders from their accustomed places without making adequate financial contribution to the support of the church," they now recognized a larger purpose.[51]

Of course, church workers might have hesitated to devote significant resources to this effort if they had full confidence in the program of the YMCA. As it was, however, denominational representatives were increasingly doubtful whether the association could effectively develop vital Christian faith and strong church loyalty among collegians. These leaders asserted, first of all, that the YMCA chapters were not growing quickly enough to adequately serve the burgeoning population of religious students. Cochran, speaking for the Presbyterian churches, acknowledged that there was "little danger of overdoing religion with a half dozen university pastors" when it required several hundred faculty "to take care of students scholastically."[52] Dr. Wallace Stearns concurred in 1908 that "The Association does not cover the field, does not increase in a degree commensurate with the growth of the universities and colleges, nor does it need to feel any jealousy of any agency seeking similar ends."[53] Like administrators who argued that the YMCA was no longer capable of providing services for campuses, denominational leaders were convinced that the expansion of higher education had opened the door for a multiplicity of student religious options.

While practical factors of rapid growth were of significant importance, however, there were deeper philosophical issues at play as well. Church leaders were convinced that YMCA chapters were unable to provide students with the biblical and theological foundations necessary for mature Christian understanding and growth. A survey conducted in 1923 by YMCA leader Paul Micou found that church leaders as a whole felt that the

YMCA did not do enough to teach Christian truths to religious students, replacing concrete teaching with activist service. Rev. Thomas White, Presbyterian University pastor at Indiana University and president of the Conference of Church Workers in State Universities, spoke for his body in stating that the YMCA was giving "too large a proportion of its time to things that are social rather than religious." The move from doctrine to "practical religion," the legacy of the Mott era, in many cases stirred denominational leaders to document biblical and theological gaps within student culture. Few doubted the biblical principle that faith without works was "dead." Many were convinced, however, that works without an attending system of belief would tear students away from their communities of faith and leave them vulnerable to heterodox movements that would seek to utilize their activism for unworthy causes. Cochran put the matter bluntly when he suggested that the YMCA had left students "unpastored and unchurched." The associations, he suggested, "entered the universities, not to father or mother, but to brother him." The churches alone, he reasoned, could provide the necessary "parental" influence.[54]

Perhaps of even greater import, church leaders claimed that the YMCA had lost its earlier commitment to promoting church membership and loyalty among students. While the larger YMCA reaffirmed its commitment of "absolute loyalty to the Church" in 1913, student pastors and the larger denominational boards were less certain that the student movement embraced this call in practice. Cochran, reporting in 1911 to the Religious Education Association, noted that the associations "do not assist the churches in securing attendance upon public worship; that they fail to encourage students to unite with the church of their choice; that they are largely occupied with building up their own membership."[55] For church leaders, lofty association rhetoric of commitment to the universal church often foundered when put to the test in local settings. If it was true that associations were loyal to the churches, one Ann Arbor pastor questioned, why was it that "there has not been one single instance where the Association has brought down a man to me to join our church on confession of faith?"[56] For pastors and denominational leaders, it appeared that the YMCA had cultivated a stance of loyalty to the church universal while withholding enthusiasm from local churches in the college neighborhoods. The fact that many secretaries were accused of neglecting church attendance certainly did not ease these concerns.[57]

In this sense, church officials felt that the YMCA created within students a "sense of superiority" in their own form of religious work over and against that of the church. At the root of such an accusation was a growing opinion among church leaders that the YMCA was portraying itself, not as a support to the church, but rather as an evolutionary advance over purportedly antiquated church forms. There was indeed much evidence to validate such a claim. YMCA leaders often critiqued denominational divisions and spoke of the development of their own interdenominational organization as a sign of clear spiritual progress.[58] In addition, they often depicted their socially oriented faith as a more Christ-like alternative to the static concern for biblical accuracy that characterized local churches. While YMCA leaders denied that they were antagonistic toward the church, they were quite willing to admit that they embraced the role of pushing religion beyond typical church limits. "The youth of a majority of our members and the comparative freedom of our Movement," one national leader noted, "make it natural that we should have more to contribute to the transforming or revolutionary function of the Church than to its conserving function. We owe it to the Church to be true to this pioneering genius."[59] This brand of religious hubris served only to forestall attempts at cooperation and to incite church leaders to fortify their efforts. As one suggested poignantly, "The spirit of Christ does not dominate when a graduate will refuse to have anything to do with the local religious institution in his community because of the sense of superiority he may have as a result of his experience in Christian work at the university."[60]

For all of these reasons, church groups after World War I began investing significant resources in a determined movement to occupy the campus. Sporadic church-affiliated groups had actually been organized in colleges and universities long before World War I. As early as 1887, for example, Episcopalians at the University of Michigan built Harris Hall for their "Hobart Guild," while Presbyterians formed the Tappan Presbyterian Association to reside in Macmillan Hall, dedicated in 1891. It was not until 1908, however, that student pastors began meeting collectively in national conferences. At a 1910 conference, a nascent body of local university pastors formed a permanent organization entitled the Association of Church Workers in State Universities (ACWSU), renamed in 1923 as the Conference of Church Workers in Universities and Colleges in the United States (CCWSU). At the same time, there also emerged a series of conferences for national denominational secretaries responsible for higher education, resulting in the 1912 formation of the Council of Church Boards of Education (CCBE).[61]

Successful prewar experiments in this field, coupled with growing church funding through national denominational subsidies after the war, made for a rapidly growing movement. While in 1912, 32 men were employed as denominational student pastors, by 1925 this number had risen to 150, most employed in slightly less than 50 large institutions. While there was some attempt to standardize this position nationally, in reality a great variety of forms existed for this role, often depending upon the size of institution and community. In large institutions, a student pastor might become an associate minister of the local denominational parish church, giving full time to work with students. In other cases, typically in larger cities with multiple churches of the same denomination, student pastors might serve the denomination "at large," representing a number of local churches in the work with students but officially connected to none of them. Alternatively, a student pastor might direct a social center placed near the campus to serve as a point of ministry to students from his denomination. Particularly in communities where churches of one denomination were located far from campus, this provided a means of attracting students to a denominational center or "guild house" for mutual connection. In smaller communities, student pastorates were sometimes formed on an interchurch basis. Where none of the churches had a student population large enough to warrant a pastor, individual congregations might join with other churches and employ a pastor to serve students collaboratively. Such a position could also be combined with the YMCA secretaryship, funded equally by the denominations and by contributions raised by the associations. These "cooperative pastorates" were popular in New England immediately after the war.[62]

While association leaders valued the option of a student pastor tied to a local church and found the cooperative pastor concept a plausible solution for some small communities, the other options were roundly condemned. Both the "at large" denominational pastor and the interchurch pastorate, they maintained, possessed the danger of linking ministry too closely to a single personality stripped of church ties.[63] Because of the obvious parallels between independent university pastors and YMCA secretaries and between guild houses and YMCA buildings, these attempts also served as direct challenges to the association presence on campus. Since one Baptist publication noted that these pastors would meet students arriving on campus, help them find rooms and employment, enroll them in Bible and mission classes, provide for their social life, advise them with regard to vocations, and cultivate the spirit of service, the charge of programmatic "plagiarism" by one association general secretary likely had much basis in fact.[64] Because denominational leaders concentrated on the largest schools with the greatest concentration of students, they inevitably came into conflict with YMCAs that were the most highly organized with general secretaries and buildings in place. While the YMCA had always envisioned itself as a supporting arm of the church, it now found itself in competition with church groups for the allegiance of religiously minded students.

One of the most direct sources of competition came in the area of religious education. In some cases, individual denominational foundations or their student pastors offered courses that were approved by the institutions and counted for college credit. At the University of Illinois, the Wesley Foundation of the Methodist Church, the Catholic Columbus Foundation, and the Illinois Disciples Foundation offered courses that were accredited by the University in the 1920s. While all three were teaching courses simultaneously, there was no coordination between these groups, and courses were often duplicated from one denomination to another. In other cases, teachers representing different denominations and faiths joined together to provide for a collaborative school of religion, cooperating in a shared curriculum. Formed in 1925 on an interfaith basis, the Iowa School of Religion allowed Protestant, Catholic, and Jewish groups to each recommend and support a professorship in the school. With salaries paid by the religious groups themselves, the school had an enrollment over two hundred by its third year, offering courses in Old and New Testament, Comparative Religion, Rural Church Work, the Expansion of Christianity, and Materials and Methods of Christian Education. At the University of Oklahoma, a mutidenominational school of religion was entitled the Department of Religious Education and linked administratively to the institution's School of Education. Since these courses addressed both academic and practical topics, and since they were offered for academic credit in many locales, they served as a direct threat to the YMCA Bible study curriculum.[65]

With these multiple sources of potential overlap, church and association leaders spent a great deal of time debating whether the church or the YMCA would serve as the center of campus religious work. Mott's opinion, repeated frequently in these years, was that churches and associations should devise an appropriate division of labor based upon the particular strengths of both groups. "In everything which can be best done interdenominationally the initiative, leadership and responsibility should be with the Christian Associations," he posited. "In all that can be best done denominationally the initiative, leadership and responsibility should be done with the regular parish churches."[66] While such a statement was often embraced in theory, however, the practical definition of what could be done "best" by each group often generated tensions on local campuses. At the 1915 and 1916 "Cleveland Conferences," designed to explore these issues, association leaders stated that cooperative labor was essential, even going so far as to suggest that members of the various denominations should be represented on YMCA advisory boards. They still contended, however, that the YMCA, with its "distinctive and providential" mission, was the primary interdenominational agency through which the churches should act.[67] Church leaders were less sanguine regarding this approach. Rev. H.L. Rote, Methodist pastor at the University of Michigan, noted that the YMCA had started because churches were unwilling to take their responsibility seriously. Noting that he had "nothing but the greatest appreciation" for the YMCA, he also noted that, in the organization's process of "going around the church," the church "naturally woke up" to its responsibility. For the association to resist the primacy of the church would be, he suggested, a fundamental misrepresentation of its original purpose.[68] In the end, the conference committee encouraged continued "right of initiative" for church and YMCA groups but also cooperation in areas of mutual interest, specifying that "while some functions may be more particularly those of the churches and others of the Associations, each should feel its responsibility for co-operation in the work of the other."[69]

It was the precise nature of such cooperation that became the vexing issue over the next decade. Attempting compliance, Porter recommended that secretaries provide books for students on denominational creeds and church history, secure schedules of local religious services, provide church leaders with lists of students organized by denomination, promote joint evangelistic and "Day of Prayer" campaigns, encourage church and Sunday school attendance, and avoid scheduling activities during church service time slots. He also encouraged each chapter to form a "committee on church connections" and to meet regularly with church workers to exchange plans, coordinate activity schedules, and share resources. By 1916, Mott and Porter could both contend that significant progress had been made.[70] At the University of Chicago, a YMCA committee was formed to help foster connections with local churches, promoting a "church week" in January and a "go-to-church day" to encourage church attendance. At Iowa State University, the association sponsored an "All-Out-To-Church" Sunday at the beginning of the year where cabinet and student members went to student residences and invited them to church services. In 1918, the summer conferences had two sessions where students met with other members of their denominations, partaking in daily celebrations of communion.[71]

Some local YMCA chapters attempted more radical cooperative efforts. At the University of Pennsylvania, innovative general secretary Thomas St. Clair Evans established the YMCA as an umbrella organization for the work of the churches on the campus. The denominations were encouraged to place their university pastors on the secretarial staff of the YMCA to do work both for their own denominations and collectively for the association. The Presbyterians were the first to enter this alliance in 1914, and they were joined eventually by the Baptists, Methodists, and Lutherans (1917) and later by the Reformed Church (1922). Reflecting his belief that the YMCA was a movement "of the church" rather than "on behalf of the church," the association at Pennsylvania was "simply the Clearing-House through which the Church leaders of Philadelphia and the National Church organizations are bringing their forces to bear upon the University of Pennsylvania."[72] YMCA work at Cornell followed a similar path in the 1920s. Under the leadership of Richard Henry Edwards, former national secretary of social study and service, the Cornell United Christian Association adopted a model in which denominational student pastors would work independently with their flocks while working cooperatively as YMCA leaders on issues related to foreign students, religious education, devotional service, or vocational counsel.[73] Demonstrating his openness to respond to university pluralism, both Jewish and Catholic representatives were added to the secretarial force in 1929. For obvious reasons, the addition of Rabbi Isadore B. Hoffman required a change in the name of the association from the Cornell University Christian Association to the Cornell United Religious Work. Cornell University President Livingston Farrand felt that this change was "the most interesting and valuable thing in recent religious history at Cornell." "That is what the University is now concerned with," he suggested. "Officially it does not care whether you are Jew or Christian if you are sincere and seeking the true values of life."[74] By 1933, Friends and Unitarians were included as well.

Despite their desire for cooperation, YMCA leaders were hesitant to embrace these approaches. Mott's Executive Secretary Charles Hurrey noted that such plans did not account for students unaffiliated with a particular denomination. More importantly, leaders felt that the so-called Pennsylvania plan would diminish a sense of membership in the broader YMCA, SVM, and WSCF, a significant blow to the worldwide student movement. Finally, while they found at Pennsylvania a plan preferable to "six free lance student pastors," it was clear that they also sensed a loss of student initiative. The general secretary at the University of Illinois registered his own resistance to the plans at Pennsylvania and Cornell by suggesting that these two campuses had become "oversecretarialized and pastorized," diminishing student program ownership.[75] Evans and Edwards did not deny such accusations. In fact, both stated that the YMCA had placed too much emphasis on student initiative, denying collegians the wisdom of church leaders in their program and policy planning. Evans noted that "Students are neither children nor fully matured men and women. . . . They love to 'start something.' They need what all wise young people desire: experienced advisors, the historic sense, and careful adjustment." Edwards was no less candid in describing the need for greater adult leadership, stating that the history of student religious movements had "revealed a large number of lively beginnings and many poor finishings." "It has been the purpose of the United Work at Cornell," he suggested, "to provide both for student initiative and also for what we have called 'finishiative.' "[76]

With such cooperative attempts set in motion, denominational leaders soon suggested a more radical plan of establishing a denominational federation as the interdenominational organization on the campus, independent of the YMCA. Predictably, this drew Porter's immediate condemnation. In a book-length study entitled *The Church in the Universities*, he suggested that there was "always a blight upon the Christian community where religious experience is judged coincident with denominational attachment" since the "true" Church was "superdenominational." A federation, he reasoned, would be interdenominational in name only, bringing together denominational units but lacking the common purpose and loyalty to Christ promoted in a true interdenominational organization. Since the college years were the ideal time to demonstrate the importance of the universal Church and to help students see the power of united action beyond creed, he was adamant that "the power of the Christian gospel is inhibited by nothing so much as by surrender to lesser and relative loyalties." To church leaders, it appeared that Porter had become antidenominational in his convictions, but for him it was a matter of the future of the church. The presence of the YMCA was critical, he reasoned, "when one examines the present dismembered state of Christendom and realizes that impotence will continue to hound the footsteps of the Christian community until some truer conception of the Church Universal is born in men's hearts."[77]

Although such tensions were not resolved uniformly nationwide, cooperation did occur in ever more inclusive ways. As mentioned, when university leaders assumed supervisory authority through religious functionaries, denominational groups and the YMCA were often brought together under a common head. Also common were voluntary "student religious councils" consisting of the YMCA, YWCA, and denominational groups. In some cases, these cooperative efforts expanded to the regional and national levels. By 1935, in fact, the National Intercollegiate Christian Council was formed in order to provide a national context for regional leaders of these voluntary groups to discuss matters of collective importance.[78] While some spoke of the formation of such councils in terms of a philosophical commitment to cooperation, it was clear that this move was also rooted in pragmatic concerns. As the expansion of the student population began to far outstrip growth in religious agencies, feelings of marginal presence could be assuaged by a larger and more diverse movement. As one university pastor at Ohio State put it: "Perhaps this change has come about because religion is much more on the defensive today, due to a complex of forces. We feel that we must work together if religion is to 'rate' at all upon the modern campus."[79] While such cooperative ventures did not eliminate connections to the broader YMCA, they were indicators of the fact that the college YMCA was no longer an independent or dominant entity in student religious life.

As one national leader put it, "The Christian Association had at one time practically a prescriptive right to the ordering of the voluntary religious activity of the students on the American campus. . . . It has that right no longer."[80]

Neither Fish Nor Fowl

In the end, YMCA chapters were neither religious enough nor secular enough for many critics. The movement lost its close identification with the churches because of its more secular service bent. Yet to many students and secular agencies on an increasingly heterogenous campus, it was now considered to be an exclusively Protestant religious organization. Unable to compete with the more explicit and doctrinal religiosity of the denominational church groups and equally unable to match the inclusiveness and scale of administrative movements for service, recreation, and even religion, the organization was placed in a challenging position. The YMCA was set up as an "all-campus" organization, attending to multiple domains in an attempt to provide holistic care. The growth and fragmentation of the campus, however, meant that organizations had to choose niche areas of interest rather than providing coverage of student life. Within an increasingly specialized campus culture, in other words, the fact that the YMCA could embrace all aspects of the fourfold life was now a liability rather than an asset. For many, the college YMCA had become "neither fish nor fowl," unable to appeal to a strictly religious clientele and unable to satisfy the diverse and expansive needs of a broader secularizing campus.[81]

There was obviously some concern that, with the removal of such activities from association auspices, the organization might be viewed as an unnecessary vestige of a former era, an important voluntary organization in the period preceding the professionalization of student services and church work but now largely anachronistic in a complex university culture. While the organization had been indispensable to institutions that were at once homogenously Protestant and also in need of student care, association programs now appeared duplicative and amateurish when compared to the provisions of the schools and churches. Some found solace in the fact that the YMCA had indeed been a pioneer in student life. Others argued that the YMCA still possessed a unique ability to reach certain types of students. Despite the work of the churches, many contended that there was still a need for an organization that would work on behalf of students who were either nondenominational, affiliated with underrepresented congregations, or desiring to express their religious impulses outside of church contexts. University of Illinois Dean Thomas Arkle Clark asked administrators to remember that "three-fourths of all the men who enter the University have no affiliation with any religious institution, and few of these get any after coming here. For these men who are unorganized spiritually, the YMCA is a home to which they can come for guidance, for spiritual and for material help."[82]

Porter, however, was a bit uneasy with the strategy of claiming for the YMCA those aspects of student religious culture that were unclaimed by others, suggesting that this served as a tacit concession to a marginalized role for the movement in the future. Recognizing the fragility of the organization's continued existence within higher education, Porter and his colleagues saw a need to organize a public relations campaign bent on convincing students of the continued indispensable nature of the organization. That task would both be complicated and clarified, he soon discovered, by the changing nature of the student body.

Chapter Eight
In Search of a Youthful Religion

While the internal and external attempts to provide new leadership for student life clearly marginalized the YMCA's contributions, organization leaders soon recognized that this was not the only challenge to the association's dominance in postwar higher education. Throughout the 1920s, the YMCA found itself at odds with the interests of the student body, generating a deep rift between movement aims and undergraduate sentiment. This tension was substantial enough for a major national YMCA report to document in 1928 that, even if the encroaching presence of administrators and churches had not occurred, "factors at work within the life of the colleges" would have made a "redefinition" of the movement "imperative."[1] The 1920s youth culture provided a markedly different social context within which to promote the muscular Christian themes of the early twentieth century. Morally lax and resistant to the piety and service themes that characterized prewar emphases, students demonstrated both active antagonism and passive apathy toward the movement. Previously a strong representation of student opinion, the organization soon lost its favored position and emerged as the target of what historian Paula Fass has described as "barely veiled contempt."[2] It did not take long for leaders to recognize that the new dynamics of student culture would require a new approach to campus ministry.

The Struggle at Princeton

On the fiftieth anniversary of the founding of the intercollegiate YMCA movement in 1927, Princeton's flagship chapter of the YMCA was in turmoil. Membership numbers had been gradually declining since the beginning of the decade. Several of its traditional service programs had been taken over by the administration, leaving the chapter bereft of a unique purpose. The opinion of the organization in the eyes of both the student body and the administration had plummeted, raising questions about its long-term viability on the campus. Attempting to explain this decline, General Secretary Samuel

Shoemaker pointed to a radical shift in campus climate. "There is a great deal in the life of the Princeton undergraduate that is not altogether congenial to Christianity," Shoemaker contended. "He feels the freedom of University life, and he feels the uncertainty as to the reality of religion which he finds in the minds of not a few of the Faculty, whose mental leadership he is set to follow." Shoemaker noted that students had become more relaxed morally and spiritually, free from the dictates of conscience, less anxious to conform to external regulations, and less eager to strive for deeper religious purposes. "It is a pretty loose generation," he noted, "and we have more money in Princeton than is good for us. . . . These men are modern types of the rich young ruler, in large part." For this secretary, the changes in student culture placed the association in a difficult position. "That kind of ground is a thorny patch to grow a revival in," he concluded, "and it is no good saying it is anything else."[3]

Shoemaker's comments appeared to be on the mark. In the years immediately following the war, students and faculty seemed less willing to bear up under the older codes of character and service that had marked the prewar era. Removed from the evangelical moral earnestness of previous decades, students were reticent to seek wisdom from authority figures, including the secretaries of the Philadelphian Society. In fact, Shoemaker became so frustrated and perplexed by students' disregard for the advice of YMCA leaders that he was compelled to remark cynically in 1924 that "I think that if undergraduates will not listen to these, neither will they be persuaded though one rose from the dead."[4] He noted that, while some of this resistance was attributable to the natural independent spirit of youth, there was also a deeper resistance to the basic outlines of evangelical faith and morality beginning to infect not only students but also faculty and administrators. F. Scott Fitzgerald, whose book *This Side of Paradise* eloquently documented this shift, actively resisted the YMCA and sought a decrease in financial appropriations to the organization during his time as a Princeton student. President Hibben had recently denied evangelist Billy Sunday access to campus buildings while encouraging a series of lectures by modernist Albert Parker Fitch. Shoemaker's predecessor Thomas St. Clair Evans noted that "Princeton cannot be looked upon as the religious center that it was a generation ago. In the midst of the beauty of its surroundings there is a conspicuous absence of religious fervor, and an ethical rather than an evangelical Christian atmosphere."[5]

Tensions came to a head beginning in the fall of 1923 when students began attacking the YMCA for its links to controversial campus evangelist Frank Buchman, a Lutheran minister who had formerly served as the general secretary of the Penn State YMCA (1909–1915). Even before establishing his connection with Princeton, Buchman had earned a significant reputation for his ardent moral zeal and uncompromising attitude toward student sin.

Labeled as "pure John" by those at Penn State, Buchman had even been parodied in a vaudeville show by a student who wore a placard announcing himself to be "99–99/100 percent Pure."[6] After working for a time in a Lutheran settlement house in Philadelphia and teaching at Hartford Theological Seminary, Buchman served as a missionary in China before returning to college work as an independent evangelist. Having established strong followings at Harvard, Yale, and Oxford, Shoemaker invited him to address the Philadelphian Society on several occasions in the early 1920s.[7]

The core technique of Buchman's methodology was the weekend "house party," a gathering of students for hikes, swimming, tennis, and personal confession in a corporate setting. While practices varied among his many disciples, Buchman's "soul surgery," as it was called, required students to confess hidden sins to groups of peers, thus "washing out" the soul and purifying the will.[8] This search for victorious Christian living was viewed cynically by many in the media. An article on Buchman in *The New Yorker* suggested that his house party represented a potent mix of "mysticism, mesmerism, spiritualism, eroticism, psychoanalysis, and high-power salesmanship," putting "the emotion of the camp meeting into the country home" and making "sexy conversation the key to paradise."[9] At the same time, Buchman cultivated impressive external support, receiving contributions from John D. Rockefeller and staying at one of the philanthropist's homes on a regular basis. The attraction of educated students did not surprise Shoemaker, who noted that "Buchmanism does for educated people what the Salvation Army does for down and outs."[10] Soul surgery, he remarked, was precisely what affluent and self-satisfied collegians needed.

At Princeton, the battle between those who favored and those who opposed these techniques soon generated a significant campus uproar. Students began accusing both Buchman and his student disciples of forcing confessions through undignified emotional appeals and public peer pressure. Several noted that the evangelist greatly overemphasized sexual sin over and above other maladies, enhancing rather than removing students' prurient interests and encouraging the doctrine of "the more sin the more forgiveness."[11] They also accused YMCA students of prying into highly personal domains of life (*The New Yorker* called them "sin-ferrets"), lacking the qualification to deal with such issues in an intelligent fashion. President Hibben, recognizing the peril of broader campus unrest, prohibited Buchman from returning to Princeton, noting that, for the sake of general peace, "As long as I am president of the university, and I think I speak for the whole administration, there is no place for Buchmanism in Princeton."[12] As a result of this ruling, Shoemaker resigned as YMCA general secretary.

The matter, however, was not settled. Shoemaker's successor, Ray Foote Purdy, was an equally impassioned devotee of Buchman's methods, and, despite Hibben's direct orders, invited students to meet with the evangelist at Murray-Dodge Hall in the spring of 1926. That fall, Purdy also secured several members of the Philadelphian Society to join students from Yale and Harvard in a series of Buchman-inspired "house party" meetings in Waterbury, Connecticut. After an indicting *Time* magazine article appeared, accusing Princeton YMCA students of "delving with dangerous ignorance into delicate problems," over four hundred undergraduates came to a YMCA fund-raising event to debate the relative merits of Buchmanism.[13] In the wake of this meeting, Hibben appointed a committee to investigate the Philadelphian Society and its links to the evangelist's methods.[14] The Commission spoke with thirty-two "witnesses," including both undergraduates with first-hand knowledge of Buchman and members of the association's leadership cadre. YMCA leaders attempted to deflect criticism by pointing to the helpful work of the association, but others spoke of forced conversions, moral invasiveness, and a required adherence to particular methods of prayer, conversion, and the "morning watch." Overall, they expressed dissatisfaction that the association was concerned with moral transformation rather than advanced religious thinking. One particularly venomous rebuke suggested that YMCA was "exceedingly distasteful" to most "thinking members of the university." "It is a sublime insult for Princeton students to be considered material for missionary work," the student noted. "We are not Aborigines."[15]

The pronouncements of the special commission came quickly, accusing the Philadelphian Society of "aggressive evangelism" and moral reform but largely exonerating them from serious accusation. Despite the acquittal, however, the Philadelphian Society was left in a deflated state.[16] With word that their contracts would not be renewed in 1927, Purdy and five associate graduate secretaries resigned from their posts. This was followed by the resignation of the majority of the cabinet, including student president Charles Howard, shortly afterwards. Although he was reelected, Howard noted that the student attitudes toward the Philadelphian Society and toward religion in general had been soured. While the organization was now battle tested, he predicted that after the controversy had passed, the challenge on the Princeton campus would be "amused indifference" rather than hostility.[17]

Though undoubtedly true, there were some "hostile" acts. Students, in reaction to the conflict, formed the Society for the Propagation of Moral Turpitude, a group devoted to celebrating the impurity condemned by their pious YMCA peers. Even more symbolic was the treatment of "The Christian Student," the statue outside of Murray-Dodge Hall that had served as the emblem of YMCA muscular Christianity since its dedication in 1913. In the wake of the Buchmanism controversy, it became customary for students to paint the statue, hang obscene slogans and drawings on it, and stack empty gin bottles around its base. Alva Johnston reported in *The New Yorker* that "one morning early risers found him wearing a pink brassiere." The statue was torn down by seniors on the eve of Commencement in 1929. Then, during a pep rally in 1930, it was removed from its pedestal and abandoned two hundred yards away, significantly damaged by the student prank. Fearing further vandalism, university authorities stowed the statue in a nearby garage. Muscular Christianity at Princeton had succumbed to a rather undignified end.[18]

The Collegiate Religious Malaise

The events at Princeton in the 1920s provided clear evidence that student culture was changing in the postwar era. While Buchman's methods differed little from those of S.M. Sayford, John R. Mott, and other campus evangelists a quarter of a century earlier, many students were now antagonistic toward such perceived religious invasiveness. As one outside critic, writing for the popular *Christian Century*, observed,

> Not so long ago, the campus Christian associations would put on an annual "religious week" when, after months of prayer-meetings and special discussion groups, the imported speakers and personal workers made a frontal attack on student wickedness. Similar attempts since the war have been increasingly futile. Resenting such pressure, students keep on saying "Religion is my own business." The technique of other days, mass meetings, signing life-purpose cards, membership drives, prayer groups, even bible study classes—distasteful to most undergraduates, is probably gone never to return.[19]

Both college and YMCA leaders were quick to identify a generalized religious malaise among students of the Jazz age. Robert Cooley Angell's assessment of college students in 1928 was that "About a third are seriously interested in religious matters and the remainder show various degrees of indifference and dislike." Ernest Wilkens, president of Oberlin, presented in that same year an even more troubling picture, estimating that, in a typical college of a thousand men, one hundred would be "religiously minded," one hundred would have "dispensed of religion," and eight hundred would be largely unconcerned about such issues. In 1924, a memorandum written to University of Chicago President Ernest Burton noted that "those with any vital interest in religious affairs relegates one to the category of the queer, the 'back numbers,' to a hopeless and insignificant minority, marginalized within the campus social hierarchy."[20]

Such anecdotal evidence of student religious apathy was largely confirmed through empirical data. The number of students at the

University of Michigan and Columbia University in this era declaring any religious preference on registration forms declined rapidly, while at the University of Washington in 1924–1925, 1,895 students cited "no church preference" in the appropriate section of the form, the highest tally for any category on the list. In a longitudinal study conducted from 1920 to 1928 by the YMCA, students nationally were asked to describe the place of religion in their lives from among the following choices: (1) I have no interest in religion and a man can live a good life without it, (2) I find religion an interesting subject for discussion but do not consider it to have any impact on my life, or (3) I feel that religion is a vital part of my life and that it is essential for good living. While in 1920, only 6 percent of students chose the first response, the number had escalated to 20 percent by 1928. The percentage that answered that religion was essential dropped from 34.5 percent to 20 percent over the same period. Most students (about 60 percent) lived in the middle range where discussion was acceptable but the pursuit of an ardent religious lifestyle was discouraged.[21]

As might be expected in light of these trends, traditional marks of religious devotion were also on the decline. YMCA leaders acknowledged, for example, a growing biblical illiteracy within the postwar student population. Many commentators found that students were unable to locate biblical metaphors in works of literature, confirming University of Michigan sociologist Robert Cooley Angell's assessment that, "they have a faint recollection of a few Old Testament stories, or at least they recognize names like Jonah, Samson, Goliath, and Daniel; they can repeat six or seven of the Ten Commandments; they have a general idea of the teachings of Christ, though little accurate knowledge concerning his life; and they can recall the names of a few disciples and St. Paul. Beyond this all is a haze." While statistics are rare regarding student habits of Bible reading, two surveys confirmed these reports. Among students at the University of Michigan in 1924, only 5 percent of both sexes read the Bible regularly, while 67 percent read occasionally, and 28 percent never read at all. Similarly, the landmark study of undergraduates conducted by Richard Henry Edwards, Joseph Artman, and Galen Fisher in 1928 found that 7 percent of collegians read the Bible regularly, 39 percent of men and 47 percent of women read it occasionally, while "most" did not read it at all. Such realities lent credence to Williams College Professor James Bissell Pratt's widely quoted description of 1920s students: "Their grandfathers believed the Creed; their fathers a little doubted the Creed; they have never read it."[22]

Similar trends could be found with regard to church attendance, although statistical data was often contradictory. A group of church workers and association leaders in 1916 noted that less than one-fourth of the student body at large universities attended churches on any given Sunday.

While this statistic was in no way substantiated, it did seem to hold quite true for large institutions in this era. At the University of Michigan, about 25 percent of students attended church services in 1928, but only 10 percent were characterized as "regular attenders." Only one in twenty in this setting stated that they attended church more frequently after leaving for college, while one in three admitted to attending far less after leaving home. While statistics could vary, the general sense among YMCA leaders and other critics was that churches were declining in influence among collegians when compared with more optimistic prewar tallies.[23]

Empirical surveys also revealed a general decline in orthodox evangelical belief among American collegians. Many more students by the 1920s, for example, denied literal interpretations of the biblical text. Edwards, Artman, and Fisher found that part of the reason for student apathy with regard to Bible reading was a loss of the sense of its revelatory origin and divine power. One student, when asked about the nature of the Bible, stated that it was "Formerly a very holy though not infallible book of strong inspiration which should be constantly studied" but "Now a book of most uneven merit containing much beautiful literature and a great very many primitive ideas about the spiritual life."[24] Many, they noted, no longer believed in the miracles cited in the biblical text, and preferred to attribute apparent miracles to natural causes. Beliefs about Jesus, similarly, reflected a decline in confidence regarding his divine nature, most students now claiming him to be a "great thinker and teacher" rather than the incarnation of God. In other studies, data revealed that students on the whole no longer believed strongly in the power of prayer, the exclusivity of Christianity among world religions, and the reality of hell, which Pratt referred to as an "extinct volcano."[25]

This decline, of course, paralleled what historian Robert Handy has labeled a larger national religious "depression" in the two decades after World War I, fueled by the Scopes trial, declining missionary interest, and the rise of religious pluralism.[26] For YMCA leaders, however, the growing antipathy toward evangelical religion in general and the YMCA in particular was in many ways the function of a new collegiate culture. Some blamed the academic environment of the universities for this religious declension. Whether the stance of the intellectual life was deemed dismissive or hostile to religious claims, some suggested that higher education highlighted naturalistic themes to the exclusion of spiritual interests, relegating religion to the periphery of the curriculum and heralding "scientific agnosticism" as a normative mode of thinking.[27] Some students found themselves in a state of religious confusion, perplexed by the seeming incompatibility between Christianity and modern thought. "To students of former decades," one YMCA leader noted, "religion was perhaps more meaningful than it is to the students of 1928. Formerly students knew just what to do and just what to believe in order to be saved. The church creeds, dogmas, rituals, certain beliefs and the Bible literally interpreted were infallible guides and the criterion and basis of authority were found in these. The modern scientific method has made untenable many of the former practices which to our parents and grandparents were very vital and real."[28] YMCA proponent Bruce Curry suggested that many students were receiving from their professors the implicit notion that "You cannot be educated and a Christian at the same time." Both the "belittling of organized religion" and the "deprecation of the Christian point of view in the classroom," he urged, "must not be ignored."[29]

While seeds of doubt were surely planted through academic channels, YMCA leaders were certain that the extracurricular world of "student life" was equally culpable in collegiate religious decline. Students of the "Roaring 20s," it appeared, were increasingly liberated from the moral strictures that had once anchored the Protestant conscience. The implicit rebellion against Victorian codes of conduct, demonstrated in student drinking, dancing, movie attendance, and sexual exploration, threatened to decimate the core themes of the YMCA's prewar character emphasis. At the University of Virginia, the YMCA general secretary commented in the 1920s that the campus had turned against the organization because of its moralistic opposition to campus excesses. "This is a feeling of opposition to the YMCA as an organization with definite religious and moral standards which cut across some of the practices that exist in the university," he suggested. "It is more directly an attitude of antagonism for attempting to do away with the use of liquor in the building, and for trying to stop the immorality which existed near the Secretary's home." Student leaders, he commented, could not redirect the opinion of the majority of students who were embracing the new "Virginia tradition" of "wine, women, and song."[30] As YMCA standards increasingly came into conflict with commonly accepted practices of the student subculture, the spiritual earnestness of the chapter increasingly took on an oppositional cast. "Association loyalists were not liked by the general run of campus opinion, especially by the fraternities," Paula Fass has suggested, because YMCA activists were considered "overly earnest, uncomfortably committed, and self-righteous."[31]

While the YMCA's character emphasis was rendered problematic by student moral rebellion, many leaders felt that the organization's service orientation was equally blunted by new student ideals. Whether attributed to declining admission standards, postwar disillusionment, or larger cultural forces of consumerist hedonism, many had a sense that this new generation of collegians came to campus with a pragmatist, selfish, and materialist bent reflective of the broader society. Rather than preparing for service to church and state, students were perceived as status conscious, crafting opportunities to secure personal advancement through what historian Lawrence Veysey called "stylized social ambition."[32] In addition, as Michigan's Robert Angell recognized, students were less willing to view service in the heroic terms of the prewar mentality. Instead, many now equated active benevolence with self-righteous posturing. Explaining the demise of the popularity of the YMCA on his campus, Angell explained,

> [The average student] regards the YMCA as an organization which has a narrow, stereotyped formula for goodness and does not feel it incumbent upon him to support it. This vague hostility is often increased by the dislike which most people have for anyone who starts out, however sympathetically, to improve his fellows. . . . The average student if he wishes to get in touch with a religious institution, therefore, commonly keeps away from an organization of, to him, self-conscious doers of good.

Interestingly, Angell noted that the YWCA was less affected by these currents since service within the culture was becoming known as a "peculiarly feminine function." The days of a service-driven muscular Christianity, it appeared, were over.[33]

While the new resistance to character and service surely compromised the centrality of the YMCA on the nation's campuses, the proliferating extracurriculum also played a significant role along these lines. YMCA leaders were quick to recognize that the burgeoning opportunities available to students in this era dwarfed prewar options and diversified potential areas of student involvement. The average undergraduate male spent thirty-two hours per week in campus activities of various kinds.[34] On a surface level, the most immediate ramification of the "overorganized" extracurriculum was the frenetic pace of campus life, social events crowding out opportunities to take part in religious work. Student President Alfred Henderson of the University of Rochester YMCA noted in 1930 that discussion group attendance was down because of "lack of time in the college man's calendar," which meant that "conflicts with other college activities is the rule."[35] As one secretary put it, "Unless some strong drive is made, a powerful wedge thrust into the midst of daily engagements, compelling the attention of busy students, the likelihood is that many will pass through college halls without ever having been brought face to face with the claims of religion upon their lives."[36]

Of course, the growth of fraternities nationwide served as a source of direct competition to the YMCA. The enrollment explosion of this era, combined with the spread of the elective system in collegiate coursework, had acted to erode class loyalties and fellowship to a significant degree, and fraternities now served the purpose of cohesive fellowship for many male students. In 1912, there were 1,560 national fraternity and sorority chapters. By 1930, there were a full 3,900. In the 1920s alone, the number of fraternity houses nationwide jumped from 774 in 1920 to 1,874 in 1929.

By this time, nearly half the number of the male students enrolled in higher educational institutions belonged to collegiate fraternities, and the property value of fraternity houses had reached ninety million dollars.[37] The large numbers of students in the Greek system militated against strong YMCA chapters simply because of the pervasive and all-encompassing nature of fraternity involvement, sapping time that could be given to other agencies. In addition, while there had been significant cross-fertilization between YMCA chapters and fraternities in the prewar period, such was not the case in the 1920s. As the fraternities themselves sought to maintain an elite coterie of students removed from the nonfraternity "independents" on campus, the YMCA was no longer viewed by many as a viable venue for joint participation. As one secretary from the University of Virginia recalled, "To belong to the YMCA was to hobnob with the independents and this was not considered proper in frat circles."[38] In the end, this division and the comprehensive social calendar of these houses made it all but inevitable that the YMCA would become a haven for students not involved in fraternities.

Increasingly, however, YMCA leaders began to assert that the cause of religious apathy was to be found, not only in academic culture or in the multiplicity of campus distractions, but also in the inadequacies of the religion offered by the college movement. Some mentioned that students, influenced by science and philosophy coursework, had begun to question Christianity's intellectual credibility. Fundamentalist attacks on science and on higher educational institutions during the 1920s served the unintended purpose of alienating many collegians from conservative religious perspectives. Because every "thoughtful man" was "completely loyal to the truth-seeking processes of science," one report argued, the "anti-revolution and anti-science activities and pronouncements of an influential minority of clerical and lay leaders" had resulted in a pervasive "shaking of student faith."[39] Dedicated to their institutions and appreciative of teachers who were attempting to synthesize academic content with religious expression, students viewed fundamentalist diatribes as demonstrations of a world-denying, ostrich-like rejection of cultural and academic trends.[40] The YMCA president at the University of Wisconsin, in fact, pointed out that famed fundamentalist William Jennings Bryan had "made agnostics of hundreds of us with a single speech" because he told students, "with arms and finger pointing straight at us," that "The next time you hear one of those college smart-alecks question the virgin birth, the immaculate conception and the divinity of Jesus, you ask him how it is that a red cow can eat green grass and give white milk."[41] Such appeals were unlikely to secure the enthusiasm of thoughtful collegians.

In addition, traditional religion was deemed unhelpful because of its static nature. Surely anachronistic in a society that had supposedly embraced the evolutionary nature of both species and ideas, the issue was in some ways broadly epistemological. President Hibben of Princeton noted that many students were resistant to religion because authoritarian presentations of religious truth differed so markedly from the inquiring spirit, candor, and open-mindedness of the scientist.[42] Science demanded "tests" of moral hypotheses, while Christianity seemed to require an unthinking adherence to tradition regardless of tested outcomes. For progressive students, the idea that behavioral prescriptions and religious morals could be defined and simply affixed to contemporary issues was nothing less than a bastardization of ethical religion.

At its root, the difficulty with traditional religion along these lines was its tacit individualism. Instructing individuals to focus on the status of their personal virtue and the need to secure personal salvation, traditional religion threatened to sacrifice the welfare of society on the altar of the individual soul. Such concern for personal religion, according to many, had created a kind of moral schizophrenia in which Christians could be both privately "good" and publicly either immoral or irrelevant.[43] Many looked around and saw a society plagued by war, racial hostility, religious division, and economic inequality and began to question why Christians had neglected such injustices.[44] The problem, it seemed, was that traditional religion was impotent in the face of such systemic flaws, preferring to focus on personal moral vices or simplistic social service and therefore failing to recognize the inconsistencies between private morality and public decadence. "His mind runs toward prodigious problems on which the weal of humanity depends—problems of international organizations and social justice," one leader commented, "and these petty questions of individual behavior awaken in him no ethical enthusiasm."[45] As Porter himself suggested, unless students could be convinced that Christianity offered an "adequate solution for the burning social and international questions of this day," they would "consider Christianity and social reform as alternatives and in many cases will choose the latter."[46]

In this sense, YMCA leaders were largely positive about the student dismissal of fundamentalist Christianity, praising students for their thoughtful rejection of positions that seemed hypocritical and inadequate for the modern world. They were concerned, however, that students would simply reject religion as a vestige of the past without considering the possibility of a more relevant version. To some, YMCA Bible study guru Bruce Curry noted, religion appeared "intellectually dishonest" and "impossible in the light of modern scientific knowledge." To others it appeared as "morally ineffective, weak, sold out to the existing order, a rather sorry stage play." And yet, he suggested, "the question persists: Is it religion that is open to these criticisms, or is it rather the shell of religion, which may be all we have seen? Is there below the outer layers a kernel of reality to which we might break through? . . . Could it transform life in line with our truest ideals?"[47] The movement as a whole responded in the affirmative to such questions. Still recognizing the importance of Christianity as a powerful change agent, the perception of a student reaction against traditional morality and religion drove many in the college YMCA to seek a means of demonstrating the continued relevance of Christianity to the modern world.

Searching for a New Faith

Among the most potent factors in this postwar religious reconstruction was the YMCA's growing affiliation with Protestant liberalism. While the movement had been firmly connected to evangelical Christian sentiment in the nineteenth century and had gradually embraced a more liberal evangelical and conservative social gospel flavor in the years leading up to the war, the college association in the 1920s unabashedly joined forces with the liberal wing of American Protestantism. Between 1900 and World War I, liberals and conservatives worked in cooperation toward social betterment, with conservatives emphasizing more of the personal and liberals more of the corporate dimensions of that task. An extensive middle ground existed for cooperative efforts. After the war, however, social issues became more exclusively the province of the liberals alone, many conservative Christians and fundamentalists eschewing such concerns and opting for a singular focus on personal salvation and piety. As these groups divided into separate spheres, YMCA leaders sided decisively with the liberals, denigrating individualistic appeals and moving to a social message rooted in the life and teachings of Jesus.[48]

Labeled the "new orthodoxy" by eminent psychologist of religion Edward Scribner Ames in 1918, the Protestant liberalism of the postwar period retained many of the principles of late nineteenth and early twentieth century liberalism while also incorporating the pervasive scientific empiricism of the 1920s. Deeply influenced by philosophical pragmatism and by new systematic theologies emanating from Germany, cutting-edge liberals had, by the time of international détente, largely dismissed all forms of "idealism" in favor of an unabashed confidence in scientific method and a justification for religious faith strictly connected to ethical consequences. Finding especially nourishing settings at the University of Chicago and Union Theological Seminary, liberals in the postwar period displayed a tight adherence to the core ideals of religious modernism: the adaptation of religious ideas to modern culture, the belief in an immanent God revealed in human cultural development, and a confident sense that society was moving, through the application of religious principles, toward a realization of a Kingdom of God on earth.[49] During the postwar era, the intercollegiate YMCA increasingly came under the spell of a number of liberal theorists, including left-leaning theologians Walter Rauschenbusch, Harry Ward, Reinhold Niebuhr, and Kirby Page, and liberal progressive and religious educators such as John Dewey, William H. Kilpatrick, George Albert Coe, Goodwin Watson, and Harrison Elliott. These individuals wrote for YMCA publications, spoke on campuses, and became staples at collegiate conferences, revealing not only a growing allegiance with liberal social causes but also a growing tendency to look outside the movement for theoretical grounding.

Influenced by multiple liberal theorists, YMCA leaders were probably most deeply indebted during and immediately following the war to Rauschenbusch. This theorist spoke as the invited guest of many local college YMCA chapters between 1908 and 1918, but his chief influence on the movement came through his written works and study materials. As early as 1908, student Bible study groups had been reading his 1907 *Christianity and the Social Crisis.* His 1912 text, *Christianizing the Social Order*, was also well received within association circles, serving as the basis for a popular study curriculum that was used on local campuses and in the summer conferences.[50] Perhaps his most powerful influence, however, came in 1916 when his work, *The Social Principles of Jesus*, was integrated into the College Voluntary Study Series.

This social gospel advocate was in many ways the perfect theorist to bridge past and present for the intercollegiate YMCA. While he pushed Christians to reject purely individualistic notions and to move forward in adding goals of social justice, he did so from a strongly biblical and theological grounding. Because of this, his works were ideally suited for those who were affiliated with conservative Christianity but dissatisfied with the seeming impotency of the faith in the light of modern social problems.[51] His deepest concern was for a faith that had lost its social message, evolving from a socially relevant force to an ascetic and other-worldly crutch for the perpetuation of the status quo. As a means of conforming people to ideals of worthy individual character, Rauschenbusch noted, Christianity was still unparalleled as a positive social force. At the same time, the radical tenor of the faith, epitomized by Jesus, had been blunted by conventionality. "The morality of the church," he noted, "is not much more than what prudence, respectability, and good breeding also demand."[52]

For Rauschenbusch, a move out of this malaise required a new biblical theology. He noted repeatedly that the vision for a vibrant social gospel was contained within the pages of Scripture, previously obscured by individualistic readings of passages brimming with social implications. If true believers were to read with a sense of both the personal and the corporate dimension of faith, he suggested, they would recognize deep patterns of sophisticated social thinking and action within beloved passages. Important to Rauschenbusch was the broad social vision of the prophets, marked by the fact that they confronted social sin against widows and orphans and "said less about the pure heart for the individual than of just institutions for the nation."[53] As the primary embodiment of the prophetic tradition, Jesus was for Rauschenbusch the clearest fulfillment of the Old Testament vision. His healing of the sick was described in terms of the importance of proper health and sanitation for lower-income residents. His triumphal entry into Jerusalem became an opportunity to show the working class backing of his political crusade. The Sermon on the Mount became a manifesto for social action, while prayer to the Father was his means of "straightening out the affairs of the world." Ultimately, his death was a clear indication that prophetic work would be accompanied by suffering and persecution. "Let us clear our minds forever," Rauschenbusch stated unequivocally, "of the idea that Jesus was a mild and innocuous person who parted his hair and beard in the middle, and turned his disciples into mollycoddles." Instead, Jesus was portrayed as an activist who fought to reconstruct social values in line with the ideals of biblical radicalism.[54]

The key for Rauschenbusch was Jesus' conception of the Kingdom of God, which he noted had been wrongly interpreted by many Christians as "being saved and going to heaven," "the organized church," or "the hidden life with God." As Jesus' teaching passed from the original Jewish context to the Greco-Roman world and to Hellenized Jews, he claimed, the radical political importance of such a phrase was abandoned in favor of one that placed this Kingdom in an other-worldly context, an apocalyptic future hope removed from the realities of life. For Rauschenbusch, Jesus was "not a Greek philosopher or Hindu pundit teaching the individual the way of emancipation from the world and its passions" but rather a "Hebrew prophet preparing men for the righteous social order" who was now universalizing this message for all people. Sin, rather than a personal affront against God in one's private dealings, was expanded to consider as well the issues of public immorality imbibed and passed on through social traditions and institutions. Salvation, therefore, had to reckon with these "present and active sources of evil," seeking redemption of social sins as well as personal immorality. Thus, such a Kingdom was "not a matter of saving human atoms, but of saving the social organism" and "not a matter of getting individuals to heaven, but of transforming life on earth into the harmony of heaven." Since Jesus was "too great to be the Saviour of a fractional part of human life," Rauschenbusch pressed Christians to be inclusive about the domains that would command attention, spreading not only "peace and charity" but also the "leaven of social unrest."[55]

In response to these challenges, Mott himself called together twenty-five leaders and professors of social work and fifty general secretaries of the college YMCA and YWCA at Garden City, New York for a "Conference on Social

Needs" in 1914. Characterized by a spirit of repentance both for inattentiveness to broader social issues and for personal complicity in many of the sins of the social order, the conference highlighted the fact that, in addition to the personal and global evangelistic task, students were "becoming alive to a third element in religious work and life, i.e., a 'corporate' gospel; the establishment of the Kingdom of God, here are now, over every department of social life." Assuming the obliteration of all social divisions resulting from prejudices of race, class, religion, and nationality, and the provision for equal opportunity and participation for all, this ideal society would generate a context of love and democratic human brotherhood organized under the common Fatherhood of God.[56]

Previously, the conference report noted, the associations were "so conservative" that their spiritual message was "entirely individual and not corporate." Now, however, they were prepared to present religion as a great regenerative social force, extending the reach of Christ not only to the individual heart but also to the "whole perturbed area of man's social existence." They rejoiced, for example, that some students were beginning to discover that the biblical Exodus was more than an "allegory of the soul" but was instead a narrative in which "walking delegates came down to Egypt to the Bricklayers Union" and where the ten plagues were examples of work-related "sabotage."[57] The "larger evangelism" spawned by this social sensibility recognized that interracial relationships would need to be strengthened, economic injustice analyzed, luxurious spending habits curtailed, and dishonesty halted before it worked itself out into municipal graft. With a growing awareness of the needs of the world, there was a powerful desire, Mott contended, for a "fuller gospel."[58]

YMCA leaders completely reconstructed notions such as sin, salvation, and the importance of Jesus to coordinate with this new vision. While sin in traditional terminology designated a behavioral rift between humans and a transcendent God, this term now represented the failure to engage in an active reconstruction of society in cooperation with an immanent God.[59] Movement leaders argued that sin in an industrial and global society had a greater reach than in previous eras. Market capitalism, combined with advances in communication and transportation, had raised awareness of human interconnectedness in ways that demonstrated the potentially expanding reach of sin. While nineteenth-century moral action in self-contained agricultural communities had been restricted to easily defined face-to-face situations, individual actions now had the potential to influence far greater numbers of people. Without doing anything personally immoral, individuals were frequently taking part in immoral structures that had negative implications for "unseen" people. More importantly, since the results of such "sin" were often invisible, few felt responsibility for complicity in structural immorality. With the effects of actions concealed within lengthy causal chains of interdependence and mediated by structures relying only indirectly on personal contributions, the potential for unwitting compliance in structural sin was enormous. Goodwin Watson, research executive for the YMCA, suggested that "The picture is of a civilization in which blind injuries are being inflicted upon people on a huge scale with no one personally responsible, no one doing it out of meanness or out of lack of affection for anybody else in the world."[60]

In fact, by the 1920s, it was clear that a growing number of leaders viewed such social sin as far more egregious than personal moral flaws. "As to moral scruples like Sabbath, dancing, smoking and drinking," one college YMCA report suggested, "we have come to feel . . . that they are pretty largely aside from the main business of life . . . and that a religion which keeps harping upon them as of primary moment both loses by being purely negative and by passing over some of the weightier matters of the law."[61] While such "scruples" had been a major focus of prewar evangelistic and moral campaigns, they were now deemed less critical than more pressing social concerns. World War I actually fostered this elevation of social over personal morality in concrete ways as leaders began to speak of the triteness and insufficiency of "clean living" for a troubled world. In their significant labors on the front, YMCA workers were given almost no choice but to forego traditional religious prohibitions against smoking, drinking, and gambling in their war-time canteen work. In addition, they viewed the morality of "religious" men in the war as colorless and narrow, assisting in maintaining personal goodness but ineffective in understanding the broader policy issues raised by the conflict.[62] In the midst of such tragedy, an inordinate concern for personal morality appeared as the worst form of ethical individualism.

The Buchmanism controversy certainly highlighted this rebuke on the Princeton campus. When Sherwood Eddy, a liberal YMCA theorist of this era, heard about Buchman's pressured confession approaches, he found one "fatal and persisting defect in the movement," namely the solitary emphasis on personal sin. He noted that, in the house parties, students would be chastised if they "marred the harmony of the meeting" with controversial social issues. While students would dramatically confess and cheer for those confessing sexual sins, there would be a "sickening silence" if a member mentioned "social justice in industry, the amassing of unjust wealth, the abolition of war, etc." Conversion and the confession of personal sin, Eddy noted, was not enough to secure dedication and commitment to the social values of the Kingdom of God.[63]

The Demise of Muscular Character and Service

With such a backdrop related to sin, salvation also took on a vastly different hue. College YMCA leaders were no longer content to sanction versions of redemption that were rooted either in the salvation of the human soul, the development of character, or the work of social service. Because of the social interdependence of the modern world, it increasingly appeared atavistic to speak of saving individuals apart from the larger society of which they were members. Because people were organically connected with one another and with the larger structures surrounding them, a proper goal of salvation was, as one leader suggested, to "turn all of the forces of society into saving forces so that structures as well as individuals could reflect the righteousness of God." One leader assured students that "today you cannot retain an effective belief in God as Father unless you aim your life in the direction of a social order so organized that human beings will be treated as though they actually were sons of God. Anybody in this crisis who tries to save his soul without at the same time trying to change society is playing a risky religious game." The goal was now to make Jesus' principles a reality in society and not "substitute an evangelical formula for an ethical fidelity."[64]

While the attack on the primacy of soul salvation had a prewar heritage, the urging to move beyond character and social service was a fairly new ideal. The prewar student YMCA certainly had a social vision and a deep—in many ways central—focus on service work among the downtrodden. However, it was assumed that social problems could be rectified through the continual regeneration of individual lives, working a form of cumulative benevolence to blunt the hard edges of industrial woe. After the war, however, leaders of the movement viewed such notions as both individualistic and naive. The war itself actually served as an important challenge to the idea that character was sufficient for world betterment. Here was an event in which individually "good" people engaged in prolonged conflict because of unjust and evil systems governing nations. The "seeds" of war were, in many cases, found not in personal immorality but in economic, class, and racial injustices that overflowed internationally. To many YMCA leaders, it was becoming increasingly clear that the system of manly fourfold character building, while helpful in the small-scale and close-knit relationships on the campus, was incapable of generating social righteousness because it neglected the "character" of larger social structures. A righteous world order required not only pockets of humanitarian concern but a wholesale restructuring of the political and social systems that governed and perpetuated injustice.[65]

Resting beneath this new vision was a sense that activist service would need to be coupled with incisive thinking for the purpose of social reconstruction. The prewar college YMCA, taken as it was with notions of practical, activist, and muscular Christianity, placed primary emphasis on the outward expressions of Christian service and moral cleanliness. However, after the war, leaders' attention shifted to the development of thoughtful and scientific engagement with vexing social concerns. Such a perspective made perfect sense in light of the changing notions of sin and salvation associated with liberal Protestant thought. If sin was found not only within the person but also in social structures, and if that sin was hidden within webs of structural interdependence that were often obscured from plain view, dealing with such sin would require intellectual resources far beyond those of previous eras.[66] The service mentality, the Advanced Program Commission noted, had generated a flurry of activity but had also resulted in "discounting and slowing up thought," producing students who were "motor-minded" rather than thoughtfully engaged in social issues. Speaking of the changing YMCA vision in a "new Christian world," leaders, by 1917, recognized that "While there has been no relaxation in getting individual students to throw themselves during free hours into various forms of local social service, it is not surprising that there has been a growth of conviction that the chief burden upon students is in the realm of thought, discussion, and the making of conviction. Such foundations are necessary if immediate social service and lives of service are to give highest results."[67] It was this perspective that guided one influential commission to state that the new purpose of the movement was to produce, not active servants, but "spiritual intellectuals."[68]

The confidence placed in the human intellect and in the capacity of the social sciences to reconstruct personal, national, and global problems was a clear indication of the optimistic view of human nature in the movement by this time. While notions of sin had perhaps expanded to include both personal and social domains, the belief in the potency of human potential had grown in parallel fashion. At the YMCA student conference on the nature of religion in 1926, Harrison Elliott noted that those with a pessimistic, hereditary view of human sinfulness were incapable of generating legitimate social impulses. Social ills, when considered within a more pessimistic religious anthropology, appeared more or less inevitable, logical, and innate expressions of human evil that could not be rectified through external means. In addition, deterministic hereditary theories threatened to constrain human responsibility completely within a framework of a biological imperative. By attributing predetermined capacities to individuals, such theories guaranteed a kind of ethical and social fatalism that would decimate motivation toward reconstruction.[69]

The key to effective work within the Kingdom of God, in Elliott's estimation, was to be found in a more optimistic view of human nature that possessed a firm belief in the "modifiability" of human beings.[70] For Elliott, humans had infinite potential to enact positive change in a world that was deemed "soft" to human intervention. Far from a perspective that saw humans as mired in sin or dependent upon divine intervention, most in the

YMCA now saw individuals as inherently capable of new and creative possibilities. "Human beings have within themselves," Elliott claimed, "the capacities for their own improvement." The rhetoric of theological anthropology had evolved from sinful dependence to potential-laden activism bathed in human possibility.[71]

For Jesus and the Kingdom of God

These shifts in the larger understanding of sin and salvation were accompanied by a changing perception of the place and importance of Jesus within this "Christian" movement. The loss of confidence in the Old Testament, coupled with new historical studies of the gospels, elevated Jesus' life to the place of "core curriculum." While the early movement had focused upon his divinity and saving power and the prewar movement had drawn insight primarily from his life of manly character, postwar YMCA leaders emphasized chiefly his announcement and demonstration of the Kingdom of God. According to YMCA leaders, it was Jesus' radicalism that most commended him to 1920s youth. Like youth, YMCA Bible study leader Bruce Curry suggested, Jesus lived "bravely, colorfully, freely," killed as a radical because he "challenged the old order of his day."[72] While not all were convinced by Curry's description of Jesus as a "communist," they were certain that his principles moved in the direction of a cooperative social and economic order that eschewed selfish competition.[73] In fact, they noted that it was Jesus who saved the Christian religion from its retrogressive social policies and ethical malaise. "In thinking about the Christian religion," leaders suggested, youth were "coming to distinguish between the religion of Jesus and our current Christianity. Jesus challenges them greatly, Christianity almost not at all."[74]

The fact that the student movement became more explicitly fixated on the life and ministry of Jesus, however, could not obscure the fact that leaders and students were increasingly comfortable with religious pluralism and increasingly resistant to Protestant and even Christian exclusivity.[75] Because the Kingdom of God was devoted to unity, democratic ideals, and the brotherhood of man, it became logically inconsistent to drive a doctrinal wedge between individuals desirous of social betterment. The 1922 membership basis maintained Christocentric overtones, but experiences throughout the country revealed that these references were more symbolic than instrumental in framing association policy. In most cases, chapters admitted anyone who aspired to work on behalf of what many called the "program of Jesus" in building a better world. Excluding any necessary belief in the divinity and/or exclusivity of Christ, they simply requested that individuals agree to work for his democratic purposes, seeking justice and brotherhood in every domain of human existence. Ironically, the desire for the college

YMCA to move back to a more religious grounding in the life and teachings of Jesus created a context in which the centrality of Jesus in the movement could hardly be maintained. As Curry put it,

> For the solution of our problems we desire naturally the keenest insights and deepest understandings of the spirit of man and God. We shall welcome these from any course—any book, any thinker, any prophet, any religion. By almost universal consent, Jesus stands paramount in this sphere. In matters of human relationships where most of our problems lie . . . no one claims to have out-thought, out-taught, and especially, to have out-lived Jesus. Therefore we cannot claim to be scientific in our quest for life as it should be lived, unless we give full attention to the contribution which Jesus may give. This is not to say that His way is to be accepted as an a priori solution, to the exclusion of the consideration of other solutions; it is merely to say that His mind, method, and ideal cannot be minimized or ignored.[76]

Despite his "blessing" on the program of Christian democracy, this Jesus proclaimed the message of his own undoing. A morality of activism and a concern for social consequences entailed little regard for the source of moral action, and the Protestant exclusivity stemming from a belief in Jesus' messianic claims was no longer viable within the new framework.[77] Muted tendencies along these lines existed before war, to be sure, but the minimal numbers of non-Protestants on American campuses at that time reduced the importance of such concerns. With the growing numbers of such students after the war, however, religious diversity was raised to the fore.

By the mid-1920s, college YMCA leaders felt the need for a cohesive statement of the religious changes then characterizing the movement in the "new world." To that end, they formed in 1927 a "Commission on the Message of the Movement" to delineate possibilities for a reconstructed religion capable of guiding students through the complexities of the postwar era. The twenty-two member commission, consisting of national and local YMCA secretaries and college and seminary professors, noted that the days of a homogenous student body, marked by conservative Protestant consensus, were over. Interpreting recent history, they suggested that what they termed the "Moody-Sankey" Protestant ideal of individual conversion, piety, and service had been "utterly changed" by the war and its aftermath, rendering American religious life "increasingly complex." Many thoughtful students were inclined to view religion as a "cast-off insufficiency" marked by "facile" moral codes that had little power to guide practice in the complexity of the modern world. What students failed to recognize, either dulled by pleasure or simple ignorance, was that following Jesus was a path to ongoing adventurous living "above the line of convention." A curriculum devoted to Jesus' teachings, the report suggested, was the direct means of developing students with "prophetic personalities" within the college context.[78]

Ministering to students who were opposed to traditional religious sentiment, YMCA leaders hoped that the vision of the Kingdom of God would generate the excitement necessary to prompt eager engagement with Christian truths suffused with the passion for a renewed society. To do that, however, would require a new aim: the creation of a Christian youth movement in the American colleges.

Chapter Nine
Building a Christian Youth Movement

While a college Christian Association should offer protection and assurance to the tempted and buffeted student, it has a special mission to discover potential prophets of tomorrow.

—David R. Porter, 1933[1]

In 1919, Carl Rogers left his Oak Park, Illinois home to study scientific agriculture at the University of Wisconsin. Rooming with his older brother Ross, at that time president of the campus YMCA, Rogers proved to be both an able student and a devout Christian. Eager for cultivating friendships, he quickly joined the organization and spent much of his time with the "Ag-Triangle," a subset of the larger YMCA for students interested in agricultural pursuits. Rogers embraced many service opportunities, including leadership of a boys' club in Madison and a camp for indigent Italian immigrants at Sturgeon Bay, and he described association work as the most formative feature of his time at Wisconsin. He noted in 1920, in fact, that he had "learned more from the hearts and lives of men" in the organization than from his studies, attributing much of his growth to the YMCA.[2]

It was the YMCA that first directed Rogers to thoughts of ministry, and several of the discussions sponsored by the organization revealed his emerging desire for full-time service. At the International Convention of the YMCA in 1919–1920, Rogers heard an address by Sherwood Eddy on the global need for Christian workers and committed himself to this task. He changed his major from agriculture to history and began leading various religious meetings on campus and attending YMCA conferences across the country. Most importantly, he was chosen as one of ten students nationwide to attend the conference of the World's Student Christian Federation (WSCF) in Peking, China in 1921. The day before he was to leave, he questioned in his journal, "how much the trip will change me and whether the Carl Rogers that comes back will be more than a speaking acquaintance of the Carl Rogers that is going out."[3]

Indeed, the trip and his subsequent experiences with the YMCA were utterly transformational. Rogers had come from a deeply religious and conservative midwestern family and entered Wisconsin steeped in his parents' emphases on personal sinfulness, Christ's atoning work, and the need for moral purity and separation from the world. While his religious fervor was maintained throughout his collegiate career, his experience with the YMCA and his trip to Peking began to reconfigure his spiritual trajectory. Much of the conference dealt with issues of internationalism and the Christian warrant for pacifism, and Rogers was convinced of the need for disarmament and a "warless world."[4] During the trip, he was confronted with various manifestations of economic and political injustice, visiting Chinese jails, hospitals, and factories. On one visit to a Chinese silk factory, where he witnessed young girls working long hours over tubs of steaming water, Rogers commented that "anyone who could see those little kids, and say that such things were all right, is not a Christian, by my definition. I don't care whether he believes the whole Bible from beginning to end or whether he believes every orthodox doctrine that ever was—I couldn't call him a Christian." As his biographer has suggested, the trip to Peking shattered his conservative religious sensibilities, replacing concern for moral purity with an appeal to Christian social responsibility.[5]

These changes had marked impact on Rogers's relationships with campus YMCA members and his family. At times, the traditional social activities of the YMCA frustrated him because of their triviality. Critiquing the vision of the president of the organization, Rogers remarked:

> his deepest idea of change in Y policy was that we should fix up the parlor for members only and arrange for other privileges. Oh Hell!!! . . . The churches are split wide open on the question of whether a man can think in modern terms and still be a Christian. . . . Europe is breeding hell for future generations because France thinks that the way to ensure peace is to throttle Germany until she pays her last cent. . . . And yet with the thousand and one problems of our day begging for a Christian solution, we sit around and twiddle our thumbs and wonder whether an old Haresfoot act would be entertaining for the frosh.[6]

Such views increasingly generated tensions with his family. Writing home to his parents from Peking, Rogers informed them of how he was beginning to deeply respect Jesus, not because of his deity or salvific power but rather because, as a unique visionary, he "came nearer to God than any other man in history." The new allegiance to the human Christ struck his family as "queer" and "unsafe" and demonstrated to them that his faith had been skewed through this increasingly liberal movement. His parents wrote to him on several occasions cautioning him against throwing out solid scriptural realities just because of his enthusiasm for these new ideals. Attempting to deflect such criticism, Rogers suggested simply that "I want to know what is true, regardless of whether that leaves me a Christian or no." Such words could hardly have provided much comfort.[7]

For postwar college YMCA leaders, however, Rogers was a helpful case study of emerging ideals. The liberal religious consciousness that took shape in the postwar college movement had enormous ramifications for the organization's perspective on the "flaming youth" of the 1920s. As historian Paula Fass has so insightfully noted, while interest in the youth population was universal in the Jazz age, interpretations of youthful rebellion were decidedly mixed. "Traditionalists," as Fass calls them, despairing over the larger social and moral changes within the culture, used youth as a symbol of national decadence, calling attention to collegians as harbingers of a larger social decline. Fass's "progressives," on the other hand, utilized the image of youth in far more optimistic terms. For these educators, religious leaders, and critics, youth of the postwar era were "charting the path toward the future," embracing new moral codes that would be more adequate for the modern world. While traditionalists interpreted rebellion as a sign of cultural dissolution and a breach in youthful moral order, progressives were more likely to see such trends as hopeful predictors of necessary change.[8]

Deeply influenced by the opinions of theorists such as Walter Rauschenbusch and George Albert Coe, the college YMCA program in the 1920s and early 1930s relied heavily upon a perspective that highlighted the potential rather than the problems of youth.[9] Youth, they surmised, had discovered the flawed nature of traditional moral and religious ideals, recognizing that adherence to such notions had failed to secure global brotherhood and democratic justice. Leaders thus saw within students' flouting of authority a nascent ethical idealism that was attempting to stretch beyond the simplistic moral norms of a previous generation. Student moral flaws were perhaps unfortunate side-effects of the youthful tendency to press beyond adult-designated boundaries, but it was precisely this spirit of rebellion that was necessary to transcend the contemporary status quo and expand the domain of moral righteousness in the world. One YMCA general secretary suggested of collegians in 1925 that "They drink. They are immoral. They do the things in the whole gamut of flagrant commandment breaking and exult in it," and yet also contended that "There is a certain commendableness in all this." Rather than "holding up to recalcitrant and derelict members of the younger generation the shameful way in which they are neglecting the religion of their fathers," he maintained, the YMCA should capitalize upon such moral restlessness by pointing students to social causes that would arrest their attention.[10]

Implicit within this philosophy was a reversal of generational hierarchies. Adults were increasingly described as mired in tradition and so steeped within extant structures that they were unable to think creatively about social need and reconstruction. In addition, adults had generated a social order that was marked by economic inequality, racial strife, denominational hostility, and international warfare. Youth, on the other hand, were idealists, always striving to achieve a better world rather than somberly looking forward to the next. Given the opportunity to bring their exuberance and social intelligence to the vexing problems of a "sick society," young people would utilize their naturally creative and experimental spirits to accelerate the pace of progress and usher in the embryonic equitable society that seemed ready to burst forth. As long as leaders could free winsome youth from the torpor of adult moral traditionalism, there was a certainty that the ideals of the democratic Kingdom of God would be realized. As Rauschenbusch suggested,

> This matter of raising the moral standards of society is preeminently an affair of the young. They must do it or it will never be done. The Sermon on the Mount was spoken by a young man, and it moves with the impetuous virility of youth. The old are water-logged physically. They are mentally bound up with the institutions inside of which they have spent a lifetime, and they want to enjoy in peace the wealth and position they have attained. . . . But while we are young is the time to make a forward run with the flag of Christ, the banner of justice and love, and plant it on the heights yonder.[11]

Youths' openness to change, their willingness to adopt an experimental spirit, and their tacit critique of adult morals all made them ripe agents of change. As Mott contended, "When my eyes rest on the older generation who are and have been in charge of the world's affairs, I have a sinking of heart. But when they rest on the coming generation growing up all over the map, my heart beats high with hope. . . . My forty years has taught me to trust youth."[12]

Of course, not all students were so inclined to move in this creative direction. YMCA leaders recognized that these perspectives on naturally idealistic and radical youth were not generalizeable to the entire campus. Befitting the new liberal orientation, the basis for YMCA critiques of student culture emerged not from moral failures but rather from the anti-intellectual and socially conservative bent of many collegians, particularly those popular within the campus social hierarchy. Culturally liberal in their flouting of adult moral conventions, many students were nonetheless politically conservative, either supportive of current national policies or, more commonly, apathetic toward social issues altogether. In a series of YMCA discussions on college students held in 1926, A. Herbert Gray suggested that students were not nearly rebellious enough. "Morally, of course many are rebels," he noted. "They grasp at what they call freedom and in so doing make many heartbreaking mistakes in the way of sexual folly and general dissipation. But intellectually they are docile. I found among them little or none of that burning passion to discover a new way for mankind, which is the real hope of the world today."[13] There were pockets of political interest, of course, but when posed with the question, "Are students in revolt?" Bruce Curry noted, "My offhand answer would be: Not as much as one might wish."[14]

In part, many noted that the class orientation of students served as a barrier to sincere social reconstruction. Robert Cooley Angell noted in his book on student life in this era that "(Students') contact with misery, vice, or indeed with any of the social problems facing this generation has been meager." When they did possess such contact, he noted, their fathers, who generally possessed the point of view of the employing classes, had "minimized the evils of the existing order."[15] Economically, students from the middle and upper classes were typically beneficiaries of the present system, thus blunting social critique and fostering passivity with regard to vexing social problems. YMCA leaders suggested, in fact, that students were infected with commercialism and a "plutocratic" outlook, deriving financial resources from parents "whose wealth depends upon the maintenance of the present industrial system." Even worse, according to these critics, the colleges themselves had reproduced this mentality. One YMCA committee noted that, because of a curriculum focused upon financial success and an extracurriculum according prestige to wealthy fraternity members, "the teaching, the system of living, and the general atmosphere of colleges supports the endowed and privileged idea."[16]

In this light, movement leaders contend that the new liberal consciousness of the college YMCA was incommensurate with the association's penchant for soliciting the involvement of "representative students" on campus. This critique was in part rooted in the sense that the popular students, particularly fraternity members and athletes, were the least likely candidates to uphold the radical values of the Kingdom of God. Many spoke of how fraternities selected students on the basis of artificial divisions of class, race, and popularity. Encouraging exclusiveness, questionable campus politics, distorted attitudes toward luxury, and artificial barriers of wealth and appearance, fraternities had elevated selfish and parochial aims in such a way as to "make spiritual democracy impossible."[17] Because of its "purely social" function, many argued that fraternities narrowed student focus, pulling attention away from the greater issues of national and international importance. In such a setting, the fraternity became the greatest threat to the hope that college students would rise up to challenge the social structure with all of the restive energy of youthful idealism. A.J. Elliott noted that during his travels around the country he had witnessed these institutions "killing the spiritual life of hundreds of students, pressuring individuals to discard their Christian idealism in favor of the undemocratic political and economic realities of campus culture."[18] As one student leader asked rhetorically, "Will the fraternity embarrass any American attempts at a youth movement and go down in history as the bulwark of obscurantism and conservatism in colleges?"[19] The answer was clear.

College athletics received a similar critique from YMCA leaders. Despite their previous enthusiasm for muscular Christianity, organization representatives spoke vehemently about the ways in which "big-time" sports decimated the values of the Kingdom of God.[20] Students, accorded great prestige by the national media, were drawn away from studies and instead immersed in a world of continual competition. YMCA leaders, in fact, critiqued the competitive nature of intercollegiate athletics as a potential template for growing economic and international "competition" later in life. As one committee described with regard to this issue, the "might makes right" athletic philosophy was a direct harbinger of the bellicose international politics of many nations. Other colleges were viewed as "opponents to be defeated" and the songs and yells of collegiate contests were "overwhelmingly militant." Through such agencies, one leader queried, "May it not be that we are thus building into their lives the very attitudes and conceptions which already have shaken civilization to its foundations and threaten to drag mankind back to barbarism?"[21]

This growing disenchantment with the mainstream of the collegiate extracurriculum created a difficult dilemma for YMCA leaders. To "seek first the Kingdom" meant to many that campus popularity was becoming a barrier to rather than an agent of organizational success. At Brown University, for example, one perceptive student leader suggested that the YMCA was continually cycling between recruiting "devoted men who limit popularity" and "representative men who lack devotion."[22] Similarly, a prominent national leader spoke of the "classic alternative" within the organization between "an overloaded, partially Christianized campus hero" and "a more earnest Christian" without "remarkable social prestige."[23] There were, of course, ways to compromise. The general secretary at Vanderbilt surmised that each local chapter should attract popular students to "bring large groups of students a short way," while also attracting less representative "prophetic voices" who because of their stands "isolate themselves from their fellow students" but foster the central revolutionary purpose of the association.[24] Armed with a growing sense that both fraternities and athletics were detracting from the realization of the Kingdom of God on local campuses, the YMCA could not help but develop an active critique against these core institutions. At the same time, such a stance almost assured a minority position within the expanding extracurriculum.

Yet it was precisely this minority consciousness that increasingly pervaded the student movement throughout the 1920s. In a significant shift from the early years of the twentieth century, local YMCA leaders found themselves unwilling to sacrifice mission for market by seeking to pad the organization's rolls with popular students who might raise the prestige of the movement. Leaders instead placed their confidence in the smaller number of students who possessed a desire for a vibrant political and social engagement, encouraged by those who resisted the "sheep-like conformity" of the majority. It was with this group of earnest students—a prophetic minority—that the college YMCA would invest its time and energy.[25]

Building a Youth Movement Consciousness

In his 1922 text, *What is the YMCA?*, YMCA training coordinator Paul Super described what he called a "crisis" of definition within the larger organization, one that left the association unclear as to whether it would be best described as an "institution" or a "movement." While certainly recognizing the possibility that the YMCA might always maintain aspects of each side of this continuum, Super argued that there had been a general transition from movement to institution in the years leading up to 1920. The professionalization of the secretary position, the growth of multifaceted building facilities, and the large labor and financial investment required for its service program all combined to elevate institutional forces. The effort involved in simply maintaining such an organization almost mandated that the concomitants of institutionalization—bureaucratization, standardization, professionalization, and the emergence of a hierarchical managerial structure—would assume heightened importance. The college YMCA in the years of Mott's leadership certainly epitomized this growing institutional identity. During his tenure, staffs had multiplied, buildings had been constructed, and boards of directors had arisen to manage the business affairs of the sprawling chapters. Even more to the point, the college YMCA itself had become an institution on the campus, evolving from a fledgling community of student evangelists to a full-fledged service institution.[26]

After the war, leaders of the college movement became disenchanted with growing institutionalism and desired to regain the characteristics of a religious "movement" that had suffused the early mission and rhetoric of the organization. Many began to argue, reflecting Mott's earlier concerns, that the bureaucratic machinery of the associations had obscured the spiritual purposes of the YMCA. Surely prompted in part by the fact that the college YMCA was losing its institutional identity through competition with campus programs, by 1923 the leaders of the organization could suggest that they had invested far too much energy in the "machinery, stunts, activities, programs, and campaigns" of the movement, blunting idealism through the dulling and homogenizing effects of institutionalization.[27] Such a mindset was quite acute on a campus like the University of Illinois, where the institutional successes of the YMCA in the prewar period rivaled any around the country. The president of the association in 1927 lamented that students could only see the YMCA as "a building where they may indulge in swimming or gymnastics or other pastimes" rather than as an active promotional agent for the work of the Kingdom of God.[28] "If the YMCA here forgot for one year what the C stands for in its name," he proclaimed, "it would lose ten years of its growth. It might provide many comforts and interesting activities for the men students, it might appear to be more prosperous, but the growth would be unsound and the retribution would surely come. . . . The YMCA worships Christ and not the false god called organization."[29]

The cumulative result of such shifts was the emerging definition of the postwar college YMCA as a "youth movement." Rather than think of themselves as "the one all-inclusive campus-wide Christian grouping," an influential report recommended a vision that would "influence a few to deep, personal commitment to Jesus' attitudes and ways of life, making no attempt to reach the whole campus." Based in a student-initiated crusade for a better world and bathed in youthful idealism, such a movement would harness the energy of student thought, vision, and impulse for the Christ-inspired construction of a future social order marked by justice and equitable relationships. For these leaders, the institutionalization of Christian work had rendered it too passive and comfortable, either cursed with the same "spectatoritis" that characterized athletics or given to simplistic service that failed to challenge the intellectual resources of students. What was needed was a clear call to active engagement with the social forces of evil marring the progress of the Kingdom of God on earth.[30] At the YMCA-sponsored Princeton conference on religion, in fact, Alfred Stearns suggested that religious leaders must enlist youth in "something hard," commenting that, "we have lost something which made the religion of the disciples next to Christ something gloriously adventurous, something that sent hot blood coursing through their veins, an appeal which carried them through all kinds of hardships and sorrow, something in which they found unspeakable joy."[31]

Muscular Christianity had also encouraged "something hard" of its recruits, but this postwar movement had a decidedly different bent. Whereas prewar activism was rooted in practical activity through institutional and community service, such rhetoric in the 1920s pointed to enlistment in a cause—the Kingdom of God. In this sense, it may have actually had more in common with earlier evangelistic and missionary work than it did with the social service and character movement of the early twentieth century. The cause of social salvation mirrored in many ways the earlier quest for individual regeneration, and the spiritual zeal anchored in the person of Jesus actuated both movements. Many felt that the program and activity-heavy quality of the movement, particularly between 1900 and 1915, had drained the spiritual enthusiasm from the associations. In the eyes of leaders, it was time to reclaim a lost spiritual legacy, though one newly chastened by urban blight and the atrocities of an international war. Rather than simply involving collegians in planning dances and activities, he suggested, campus chapters should give them difficult social and global tasks that would elicit legitimate spiritual enthusiasm.[32]

Such a task, however, required a structure and method commensurate with these larger aspirations. While the association never denied the right for administrators and church leaders to direct student life in their own way, YMCA leaders also contended that "There must, therefore be in our colleges a third type of religious organization which, on the one hand, preserves the prophetic character of religion over against the caution of the administration and which, on the other hand, makes it possible for many students to find religion apart from its ecclesiastical associations." "Such a group," one leader continued, "would be voluntary, democratic, and irrevocably responsive to student needs, a minority group representing a more adventurous religion and more radical social ideals than majorities commonly accept."[33] For a movement highlighting the contributions of youth, college YMCA leaders determined that they would need to provide greater opportunities for these young charges to exert their influence free from excessive adult compulsion and indoctrination. As it turned out, the new liberal religious vision required not only a renewed theological commitment to Christ-oriented social reconstruction, but also an organizational and educational method that would liberate youth to foster the democratic Kingdom of God.

The Democratization of the Movement

One critical means of enfranchising youth in this way was to develop an organizational structure that would enhance their participatory power. The highlighting of a "youth movement" consciousness in the college YMCA made many students and leaders acutely aware of the seemingly undemocratic authority structures within their own association. Difficulties emerged largely because of the growing strength of both international and state secretaries within the movement, but it was the state-level bureaucracy that was the most troubling to local students and secretaries.[34] Because state secretaries were in charge of a variety of activities, ranging from city organizations to rural relief work, the college movement was often tangential to their daily efforts. Yet despite their limited knowledge of the collegiate setting, state secretaries often encroached upon local campus governance, thus limiting student self-determination.[35] Foreign observers noticed and commented upon the vast bureaucratic structure created by American YMCA leaders, one particularly astute German critic noting how the American system was structured much like a business, where "personal interests and joys must step in to the background and where students are treated to the ideas of more experienced workers."[36] For an organization brimming with new democratic conviction, such pronouncements were quite painful. There was indeed no natural means for students to contribute to national movement policy. No doubt leaders such as Mott and Porter consulted students but, although collegiate initiative was upheld as a sacred principle, the students often felt "hedged about and inhibited" by their "absorption" in the general YMCA movement, unable to exercise legitimate democratic force because of structural constraints. Since collegians desired a "student movement," consistency demanded a new approach.[37]

In the student section of the Detroit International YMCA Convention of 1919, a "Committee of Nine" was appointed to explore the "best possible statement of democratic procedure adapted to the present order." This committee reported its work in 1920 with a series of resolutions determined to enhance the democratic participation of students in their own program. While they did not specify an end to state-level control, they did ask that committees elected by student associations of each state serve as consulting bodies to state or state student committees, providing contextual information about college work and having a share in the determination of state guidelines. More importantly, the committee recommended the formation of a series of regional field councils in each conference area, comprising two students from each state, the chairmen of state student committees, other state student committee members (ex-officio), members of the international student department, a local student general secretary representing each different kind of institution (ex-officio), and three graduates no more than ten years out of college. These field councils were to elect two voting members to participate in a national Student Department Committee, at least one of which had to be a student. Since five of the voting field council members were either current or recently graduated students, such a structure would send a clear message about the critical nature of student participatory power.[38]

The appeal of more democratic control at the local level was widespread in the entire YMCA in the early 1920s. After discussions at the 1922 International YMCA convention in Atlantic City, the constitutional convention of 1923 determined to create a National Council in place of the

International Committee.[39] The National Council consisted of representatives of local associations chosen by "districts" of four thousand active members, meeting yearly to approve budgets (including state budgets), policies, and programs.[40] With two-thirds of the Council comprising local laymen, students felt that they had been provided a structure within which they could exert programmatic influence. As it turned out, however, the actual control of local associations in the determination of policy was slight, and state secretaries, despite new safeguards on budgets, maintained and in some ways increased their independence and control.[41] Desiring a YMCA centered on local contributions to movement policy, students found a new structure based upon state-level geographical control.

Their hopes dampened, students went on the offensive. Student secretaries set forth a variety of resolutions requesting supervision from an independent national committee with a separate budget for the college work. Although this was denied at the national level, the National Council passed a resolution for the formation of a Commission on the Supervision of Student Work, consisting of students and local, state, and national secretaries. Charged with evaluating present supervisory methods and making proposals for future organization, the majority proposal recommended that a Student Council be constituted in every state, democratically elected by the local student associations within that state. Such councils would be responsible for program, policy, and budgetary planning within the state and would send representatives to serve on a National Student Council, which was to be "fully recognized as the national program—and policy-making body of the Student Associations."[42] While such provisions were surely appealing to students, the plan also recommended the formation of state student committees, appointed by state committees, to serve as the Ad-Interim Committee of the Student Council. These committees would also have a role in electing the representatives of the National Council.

A dissenting report, issued by a confirmed majority of the student members of the Commission, revealed just how dissatisfied they were with this "dual authority." They recommended instead the formation of Area Student Councils, comprising one representative from each local association, the regional executive secretary, state secretaries, supervisory student secretaries, at least one local student secretary, one lay member, and any members-at-large elected by the council, mandating a student majority. These councils were to combine the functions of the state and regional student councils, and from its members would be elected representatives for the National Council of Student Associations, the major administrative body of the movement.[43] "It is our conviction," the analysis stated clearly, "that enthusiastic devotion of students to the Association, which has resulted in our greatest progress in the past, is commensurate with the degree of responsibility which the students are made to feel for its control."[44]

On the other extreme, another dissenting proposal came from a group of state secretaries who noted from the outset that the student department should be given no special privileges beyond those accorded to other YMCA departments. They suggested that the state organization was to be the primary agency of supervision in its field and that if local associations chose to affiliate with state committees, they would be required to abide by state constitutional provisions. Their recommendation was to secure, wherever possible, a full-time state student secretary in each state having twenty colleges, and a part-time secretary in those states without this minimum number who would also be assigned other YMCA responsibilities. On the issue of student control, the state secretaries were unequivocal. While they allowed that students had the prerogative to form their own state councils linked to larger state committees, they also noted that it was "highly important that the State Student Council recognize that the State organization is the constitutionally prescribed agency of supervision for all Association work in its field."[45]

Students were incensed at both the tone and the content of the state executives' proposal, and they refused to accept the recommendations of the majority position as well. Porter, writing to Mott in the midst of these deliberations, noted that the increased power given to state and regional secretaries had become a significant threat to the democratic freedom of the student movement, "squeezing the life" out of the chapters. Porter suggested that the new system was creating a context within which positions could be given to men of "less spiritual leadership but more happiness in fitting into machinery of organization." While he acknowledged that the student proposal would sound radical to some, he saw it as the only hope for the continued vitality of the movement. "We are caught between two currents," he concluded. "It is literally true that if we please the students we are in danger of displeasing the secretaries. If we please the state and regional men we run the risk of losing out with the leaders of the student Associations and Councils." He urged Mott to intervene so as to preserve a student movement "with as much freedom as you have always insisted it must have."[46]

After multiple student proposals were rejected, the situation grew critical. In the Middle Atlantic region, a field council set up its own independent structure with paid secretaries, a self-contained budget, and policy-determining authority. At the same time, a number of associations in the southern colleges were threatening to form their own federation detached from the national and state hierarchies. Soon after, Porter and his entire national staff resigned, demanding that sole control be given to a National Council of Student Associations for personnel, program, and budget. Recognizing the gravity of the situation and the potential for the student organization to move completely outside of YMCA auspices, a "Committee of Eleven" was formed to present a new set of recommendations to the National Council.[47]

This committee recommended that the council establish a new Student Division, to be placed on par with the home, personnel, and foreign divisions of the YMCA. The field councils were to be designated as Advisory Counseling Committees of this National Student Division, providing input from local associations expressed through their regional representatives. While allowing for cooperation with state agencies, the Committee of Eleven placed responsibility for program development, budget, and supervision on the newly created student division and its committees. In fact, they noted bluntly that "nothing in the process of clearance through Regional or State Executives should operate to prevent or to obstruct the service or approach of National Student Secretaries to Student Associations." Instead, out of courtesy, students groups would "advise" state leaders about their program decisions. With the action of this committee and at Mott's personal request, Porter and his colleagues withdrew their resignations.[48] Exactly fifty years after an intercollegiate movement was officially endorsed, this movement gained the status of an independent YMCA division. It was now, as one student leader contended, a legitimate national student movement highlighting student initiative as a fundamental organizational principle. "I wonder if there is any place in American society," he concluded, "where the emphasis upon democracy . . . took firmer rootage than in the student movement."[49]

Educating for the Kingdom

While a democratic organizational structure promised to enhance the sense of a legitimate student movement, YMCA leaders also recognized that a new programmatic method would be required to fully liberate creative students to direct the organizational agenda. When Paul Super wrote his 1929 *Formative Ideas in the YMCA*, he noted that changing views of educational method were among the most important revisions of character-building processes in the 1920s.[50] Specifically, Super pointed to the importance of a group of liberal progressive educators who had assumed inordinate influence in the organization as experts in educational technique. William H. Kilpatrick's project method, the democratic educational theories of John Dewey and George Albert Coe, and the group discussion methodology popularized by Harrison Elliott were cited as new program staples, all having achieved centrality since the war. Although he acknowledged that the critical examination of old methods was "difficult and painful" for older leaders, Super suggested that the weight of opinion was rapidly shifting in favor of progressive educational techniques. "In fact," he contended, "there is perhaps no field of Association activity which is not being influenced by the findings of modern educational science, and it is also true that no field of study will yield Association leaders larger returns in usable and useful knowledge, skill and habits, than the study of the theory and methods of education."[51]

Super's evaluation was quite astute. In the 1920s, the YMCA as a whole became enamored with educational theory and methodology, seeking to utilize the ideas and practices recommended by progressive educators in order to foster democratic student initiative in creating a better world. That both religious and secular progressive educators should have exerted such a powerful influence on the "Y" is in many ways not surprising. Education and Religious Education professors from Teacher's College, Columbia University and Union Theological Seminary invested heavily in the YMCA during this era, and two held paid positions in the organization. Union professor Harrison Elliott, dubbed "Sunny" by YMCA workers, had been Bible study coordinator for the college YMCA movement and also served as a secretary in religious work, editor for Association Press, chairman of the Committee on Work with Boys of the National Council of the YMCA, and chairman of the Committee on Counsel and Guidance of the YMCA's Christian Citizenship program. Goodwin Watson, research director for the Religious Education Association and professor both at Union and Teachers College, had grown up within the YMCA system. As a former member of a Hi-Y club and a long-time participant in the boys' conferences and camps, Watson enthusiastically assumed the role of research director for the National Council of the YMCA. Coe and Kilpatrick remained intellectual "fathers" of the college YMCA throughout the period, speaking at conferences and writing copiously for association publications. The close ties between progressive educators and the YMCA ensured that their educational vision would gain a serious hearing.[52]

Progressive educators pressed secretaries to renounce the "propagandizing" methods that attended the transmission of Christian dogma in favor of an experimental religious attitude marked by democratic deliberation. The ideal, they noted, was an educational program characterized by scientific processes of problem solving and by democratic student collaboration in forging creative solutions to personal and global problems. With regard to methodological matters, it was John Dewey's seminal 1910 work, *How We Think*, that had the greatest influence on association leaders. In a comprehensive method he entitled "reflective thought," Dewey advocated a five-step procedure that included the "feeling" of a problem, the location and definition of that problem, the isolation of the problem into component parts, the suggestion of a hypothesis for solving the problem, deliberation with regard to probable outcomes, and, finally, the selection and testing of the hypothesis in the real world.[53] For Elliott, this scientific paradigm, exercised through democratic collaborative discussion groups, provided the path to creative social thinking in the college realm. He championed a method in which collegians would gather to consider personal and social problems, analyze data from multiple sources, formulate joint hypotheses, and test ideals through real-world action. Through a continual modification of ideas and the incessant testing of concepts in practice, the group would eventually decide on a solution that was greater (producing better consequences) than any individual response. Rather than applying adult authoritative principles, the group was attempting to develop adequate solutions to contemporary problems in light of the creative and democratic wisdom of youth.[54] It was through this model of "religious education" that the YMCA would seek to empower the voices of radical youth and operationalize its new liberal vision.

The focus on such methods revealed something very critical about the changing college YMCA, recognizable in the oft-repeated refrain that the organization was seeking to promote "creative" rather than "transmissive" education. Previously, the association had assumed a given religious truth and utilized education as a means of generating student willpower to act upon that truth. Now, however, the "right" and the "good" was less certain and education became more of a means of continual collaborative discernment and adaptation. Education was thus no longer a locus for the communication of fixed principles or the channel for religious and moral inspiration but was now rather a setting for creative inquiry, discussion, and the formation of new ideals.[55] The search for universally valid religious content, it seemed, was gradually giving way to a search for stable methods that might guide individuals in creative experimentation in an ever-changing world.[56] Many argued that this kind of religious methodology was precisely what the postwar generation of students desired. Resistant to authoritative pronouncements and suspicious of hypocritical proclamations from the so-called experts, 1920s youth were apt to be wary of the direct transmission of moral and religious truths. Students did not want to be told what to believe, YMCA leaders noted, but they wanted those with experience to provide facts and guide processes in which they could determine new solutions to global and personal problems.[57]

This progressive educational model merged perfectly with the newly embraced tenets of liberal Protestantism. Both sought to deemphasize the transmission of authoritative information and to emphasize instead the functional value of content derived from democratic inquiry. Both were indebted to evolutionary theory and to the scientific and democratic mindset it fostered. Both were optimistic about the role of education in shaping a world that was responsive to human intervention. Both also drew from a more optimistic theological anthropology. Waning beliefs in innate human sinfulness, combined with a growing appeal to human malleability, highlighted education over regeneration and creative inquiry over obedience. An unblinking trust in youth and human nature, in fact, was an absolute prerequisite of a program of collective group thinking. In order to place full confidence in discussion-based problem solving, leaders had to believe that students, when involved in such procedures, would really desire to move society in directions that would enhance democracy rather than special interest. This is in part why beliefs about human nature were so critical to the movement, and why many of the theories then gaining popularity in American culture were so dangerous to the basic beliefs of its leaders. If individuals were portrayed as inherently brutal (from World War I), irrational (from psychoanalysis), or sinful (from conservative religion), participatory group thinking would lose its power as a guiding force of renewal. Cooperative self-direction was the goal of collaboration, but if people were viewed pessimistically, self-direction would turn quickly to self-destruction.[58]

The gradual adoption of progressive educational theory as a methodological staple of the college movement revealed that the democratic purpose of a youth-initiated advance toward the Kingdom of God consisted of two mutually reinforcing elements. As described in the previous chapter, the liberal religious vision prompted a commitment to developing students who would work to secure justice for all people, regardless of race, class, nationality, or creed. A second component of the democratic moral vision, inextricably intertwined with this search for an equitable democratic society, was a push for developing and institutionalizing democratic procedures of inquiry and problem solving. YMCA leaders were convinced that a Christ-inspired social order would depend upon democratic deliberation rather than creeds and codes handed down from past and present authorities. Such a procedure would free youth from the tyranny of authoritative teachers and/or content and provide for a means of developing values appropriate and relative to concrete situations. Both descriptive (Christ-inspired social justice) and procedural (group thinking) forms of democracy were essential for the Kingdom of God.

Such a transformation was perhaps manifested most clearly in the gradual replacement of traditional Bible study with a variety of social problem "discussion groups" based on the new progressive educational models.[59] In Elliott's *How Jesus Met Life Questions*, published in 1920, there were a number of "life questions" to be considered, related to everyday conduct ("How should a Christian get even?" "What shall we do on Sunday?"), lifework ("What makes a person a success?" "What is our responsibility to the world?"), and social, economic, and world questions. Within each sphere, students were to isolate the problem, examine Jesus' attitudes toward the problem, consult other helpful resources, and develop hypotheses and action plans on the basis of their findings. "Earnest attempt has been made not to bias in any way the thinking and conclusions of those who use the book," Elliott suggested in his introduction. "There is no particular set of ideas regarding the life and teachings of Jesus which the author is attempting to set forth. The purpose is to get others to form ideas of their own rather than to adopt those which may appeal to him."[60] Similarly, in his *Building a New World*, liberal social gospel ideals and progressive educational methodology were blended seamlessly for collegians. This curriculum for college social problem discussion groups consisted of a series of questions related to America's place in World War I, foundations for developing a new democratic world, and the role of Jesus' thinking in these processes. Such questions as "What is the Real Issue of the War?" "What are the Chief Hindrances to Internationalism?" and "What is the Goal of Democracy?" were accompanied by "reference quotations" from presidential addresses, scholarly articles, books, and occasional biblical passages. Such quotations represented disparate opinions and perspectives on these issues, and students were to utilize these segments as resource materials within their open-ended conversations on these issues.[61] With these conflicting opinions before them, students were to forge their own creative solutions to personal and social problems.

Also helpful along these lines was the work of Dr. A. Bruce Curry, a professor of English Bible at Biblical Seminary in New York who had worked with Elliott to develop a series of Bible study materials for soldiers in World War I. Following this work, Curry was relieved of his teaching responsibilities between 1923 and 1925 in order to train leaders and assist student YMCA and YWCA groups attempting to apply the teachings of Jesus to personal and social problems. In his most popular study series, *Facing Student Problems*, Curry addressed campus problems such as academic honesty, male/female relationships, fraternities, and campus traditions, larger social issues related to war, patriotism, racial justice, and industrialsm, and more specifically religious questions such as prayer and the church. In every component, however, these studies reflected the aims and purposes of social and educational democracy. The section on fraternities, provocatively titled, "Shall Fraternities Be Abolished?" set students to thinking about fraternity life on campus in comparison to Jesus' own "fraternal" group of disciples. Looking at how Jesus would not allow such a group to erect barriers of exclusivity, students were asked to consider whether Christian students should boycott fraternities and fight for their elimination on the college campus.[62] Broader social issues were handled in similar ways. In the study entitled "How Radical Shall We Be?" students attempted to discern whether social action was best taken through radical or gradual means, looking at how Jesus lived as a radical in his day. At the end of the study, students were told to "Let your group plan some actual experiment in the direction of radical Christian action, and see it through together."[63] "If the group does not exist for the very purpose of committing itself to testing out new and old ideas, attitudes, and modes of conduct in the laboratory of daily life," Curry mused, "one wonders what is its justification for existence." "Some of us," he noted, "are ready to change the name of these groups from 'discussion groups' to 'life-experiment groups.' "[64]

Even topics such as sexuality were engaged utilizing the new approach. Gone were the "private talks" for men and other schemes designed to foster conviction of sin with regard to personal immorality. In their place were lectures and small groups presenting detailed factual information for group discussion. Select experts were invited to local campuses to present scientific talks on sexuality and sexually-transmitted diseases. Unlike earlier presentations in the White Cross Army and the evangelistic crusades of the 1890s and early 1900s, these discussions were couched in the language of social science and open for group deliberation. Highlighted by authors such as Frank N. Seerley and Winfred Scott Hall, this work was largely coordinated and popularized by Max Exner, who conducted similar work for the International YMCA at home and in the military. Armed with data and medical research, his approach was to promote objective and "intelligent discussion" of sexual issues, providing scientific facts that could secure informed personal and corporate decisions. Whatever the context, the goal was less an appeal to self-control than an educational focus on what Exner called "the hygiene of the imagination." Instead of seeking to press students to obey established Christian teaching in this area, the groups were to collectively determine proper boundaries between men and women. It was education and deliberation, not willpower, which would solve the collegiate "sex problem."[65]

It was clear that these groups provided contexts for students to think hard about personal and social questions, "turning their minds inside out" positing and implementing solutions emerging from their joint discussions.[66] In fact, the general secretary at Iowa State noted that "People who think that college students think of nothing deeper than the next good show or who to date for the next prom should be in some of these groups."[67] Yet in many cases, such groups were quite far removed from the Bible studies of an earlier era. While earlier voluntary studies within the movement had certainly allowed for the presence of extrabiblical materials to solve social problems, many of these new groups looked purely to the opinions of collected members or to external scholarly and popular resources. At Stanford, YMCA "bull sessions" of the 1920s provided for a "warm argument by a small group on a subject of common interest" to "stimulate independent creative thought" but there was no necessary religious content included in the discussions.[68] By 1928, the University of Illinois had dropped the name "Bible discussions" to designate such activities, choosing instead to call them "fireside forums." The change in name was reflective of current practice, since even before this time one leader suggested that "They call these Bible discussions, but nobody gives out a text, and there may be no assigned subject. There's no preaching." Similarly, at Washington State University, W.D. Weatherford noted that the new discussion forums were lively and well attended but lacking in explicit biblical content. "The constant tendency," he noted, "was to leave out of it all that was religious." The shift from Bible study to bull session, it seems, did indeed reflect a shift in the locus of authority.[69]

A New Niche

Bolstered by their new progressive educational consciousness, YMCA leaders regularly contrasted this new educational philosophy with that characterizing the collegiate classroom. What YMCA leaders perceived in so much of higher education was a type of indoctrination that left students unable to think independently about contemporary issues. On the one hand, students might simply accept authoritative teachings, failing to wrestle with alternative options. On the other hand, a more subtle emotional manipulation might also take place, generating passionate loyalty to principles in the absence of critical thinking. In either case, they argued vehemently that professors were not allowing students to generate moral ideals themselves, thus depriving them of personal ownership of these principles and removing the potent emotional concomitants developed through the value formation process. "Modern college work is largely a pouring in process," Elliott noted. "In the class room, through the lecture method, the student is told what is true and what to believe, and his examination tests him on how well he can repeat what he has been told. . . . He is robbed too largely of the joy of finding some things out for himself, of daring to have ideas of his own."[70] Bereft of independent thinking, students were taught to conform to outside expectations and to remain intellectually docile in their academic lives.[71]

In reality, it was precisely this sense of institutional educational failure that provided the key rationale for the YMCA's continued existence within the sphere of higher education. As a progressive educational movement, the YMCA, unlike the college classroom, could give students access to self-initiated, creative, and democratic discussions on critical personal, campus, and social topics.[72] In that sense, the association resembled a laboratory within which students would integrate their academic work with personal and global issues, developing creative and often radical solutions to problems and testing them in practice free from authoritative compulsion.

Porter reasoned, therefore, that the chapters were just as essential to the religious life of the universities as physics laboratories were to the scientific departments. Since educators revealed that learning best occurred in settings where students were interested and able to link data with experience, Porter optimistically pronounced that "If the Student Associations were disbanded in every college today, the inevitable law of student life and thought in every vital, spiritual college would in a few months recreate another student organization which, whatever name might be given to it, would have essentially the same guiding principles that this Student Association movement now has." Anyone who disagreed, he concluded, was "out of touch with contemporaneous educational thought."[73]

Such a reality, of course, gave the association a place even as the administration began assuming responsibility for many of its service and religious functions. Such intervention, Porter argued, did not "lessen one whit the need for a voluntary religious movement organized and led by the students. You may lead students to chapel, but you cannot make them religious."[74] In fact, he noted, the extension of administrative reach into the extracurricular lives of students actually made the YMCA more necessary than ever. If control were wrested completely away from students through overweening administrative intervention, students would lose their ability to gain valuable student-initiated experience in experimentation and practice in real-world settings. The YMCA, he stated, was necessary to "save the educational idea or purpose of the institution even against the methods or policies of the administration itself."[75] Porter's suggestions pointed to the new way of conceptualizing the role of the YMCA in this radically new climate. In the end, rather than defining the movement in terms of activities left uncovered by administrative or church groups, the most common means of defending the importance of the college association was through an appeal to student voluntarism and experiential education, the heartbeat of progressivism.[76] "The more radical and inclusive the plan of the administration itself," he suggested, "the more necessary is proving to be the need of a voluntary, student-controlled Movement."[77]

The character of that student movement was worked out within the programs and discussion groups of the postwar college YMCA. Throughout the 1920s, the organization gradually embraced a curriculum that encouraged student members to scientifically and democratically engage vexing social issues of war and internationalism, economic justice, and racial reconciliation. Increasingly, the ideal was set forth as the development of a society ruled by the social principles of Jesus and the Kingdom of God.[78] Sparking a "New Reformation" within Christendom and utilizing college students as the prophets of the future, the new educational aim was to produce "radicals according to Jesus!"[79]

Chapter Ten
The Curriculum of the Kingdom

By the 1920s, organizational leaders had determined that the traditional program of the prewar movement, while successful within its own context, was hopelessly inadequate for developing the social prophets of the next generation. If the new movement was to embrace its calling as a collegiate vehicle of the social gospel, many felt that a new curriculum would be required. First broached at the 1922 International YMCA Convention in Atlantic City, a proposal along these lines was enacted with the creation of an Advanced Program Commission (APC). While the exact purpose was rather vague, there was a direct call to reinterpret the Christian gospel "in its bearings on the individual and his relationships to social, industrial, racial, national, and international issues."[1] Members of the commission recognized that the placement of religion in the university had changed dramatically since the era before the war:

> The program of a decade ago represented a simple gospel concerned with the individual religion and more immediate personal habits and relationships of the students. It asked students to "get right with God thru personal religious experience" and then to "observe the Morning Watch," "attend Association meetings," and in similar ways develop their spiritual life and to express their religion in the simpler forms of social service. Then every person agreed as to what was involved in becoming and being a Christian. Now religion has come to be concerned with the wider social conditions on the campus, in racial and industrial affairs, and in international relations.[2]

While not neglecting "the personal readjustments of students" to the college environment, the movement, the report suggested, should utilize the bulk of its limited resources for "Christian social engineering," the development of revolutionary Christians who would offer a prophetic message of international, economic, and racial reconciliation for the Kingdom of God.[3]

The most immediate ramification of this new vision was the rapid decline of the Mott-era program. The core foundational emphases of Bible study, evangelism, prayer, and missionary service, initiated from the outset of the movement, were increasingly deemphasized throughout the 1920s. With regard to Bible study, while approximately 38,000 students were enrolled in classes in 1915, by 1930 this number had dropped to just over 4,000. While it is more difficult to ascertain the amount of evangelistic work being done nationwide, statistical accounts of conversions (later called "decisions for the Christian life") were indeed indicative of broader trends. While approximately 5,000 decisions were recorded in 1917, less than 1,000 were recorded just 10 years later. In terms of the emphasis on prayer and personal Bible study, observance of the so-called morning watch declined precipitously. While about 11,000 were recorded as taking part in the morning watch daily in 1917, this number was less than 4,000 by 1927. Similar trends were evident with regard to missionary work. While approximately 2,700 students volunteered for foreign missionary service in 1920 in the wake of the war, by 1928 this number had fallen sharply to a meager 252. By all statistical measures, these movement standbys were dramatically curtailed.[4]

Replacing these traditional staples were a host of new approaches that placed a primary emphasis on enhancing student social intelligence related to global and national social issues. In light of the Advanced Program Commission's recommendations, the themes receiving attention were those related to internationalism, economic justice, and racial harmony. Such topics were considered absolutely critical for a proper approach to democratic harmony and the Kingdom of God. Since these were areas in which antidemocratic tendencies were at work in generating separation and injustice, they served as helpful entry points for students ready to exert their revolutionary influence on the campus, in the nation, and across the globe.

"In Christ There Is no East or West"

Despite Mott's own international proclivities and the missionary work of the SVM, the general movement in the early years of the twentieth century was directed toward service to students and members of the local community. Porter, just a few months into his tenure, noted that, while there might have been a time when such local loyalties were enough to satisfy the "narrow interests" of the movement, the war had "evidently stabbed us wide-awake to an interest in international affairs."[5] While the early twentieth century campus had prompted a loyalty to American values and the self-contained collegiate experience, the growing YMCA conviction was that such provincialism was atavistic in an interdependent global society. In addition, betraying the changing religious orientation of the movement, many now viewed localism as a sign of collective sin. For J. Lovell Murray, education director of the SVM, the failure to cultivate international sensibilities reflected not only a pervasive ignorance of global trends but also a sinful parochialism that was peculiarly American in character, the quality of a nation that was brimming with self-importance. "The man is a moral ostrich," he concluded, "who buries his head in the sand on the cheap assumption that there is any particular Divine concern for his own particular nation."[6] Because the Kingdom of God had "no national boundaries," isolationism was portrayed as large-scale selfishness, cloaked under the guise of nationalist sentiment. While love of country was proper, Murray contended, nationalism was only Christian if subsumed within a broader allegiance to human brotherhood.

One of the factors most responsible for facilitating the shift away from parochialism was the campaign to raise monetary funds for various postwar relief efforts. Contributing $1.5 million in 1917 and a remarkable $3 million to the 1918 United War Fund, individual YMCA campus campaigns raised large sums for these causes. In 1918 alone, Princeton gave over $53,000 and Yale close to $73,000 for the YMCA United War Work Fund. Penn State men filled an ice house to earn money, and at Washington and Lee, money earmarked for a campus dance was directed to war relief efforts. Mott called this fundraising effort a "moral equivalent of war" for American students threatened by selfishness, luxury, and indifference to global human suffering.[7] Even more remarkable, and potentially controversial, was the joint YMCA/YWCA Student Friendship Fund, raised in the immediate aftermath of the war for students and professors in war-torn countries and prison camps. At a time when currency depreciation was a major problem in many European nations, the $1,013,124 raised was a major resource assisting reconstructive efforts in this region. Particularly in this era of "100 percent Americanism," some political leaders and college presidents were hesitant to support such direct assistance to recent belligerents. However, for YMCA leaders this effort was viewed as a tool to enhance student global awareness and international goodwill. By "sharing in the sufferings" of the students of central Europe, American collegians could demonstrate, at least on a student level, that political differences were not a barrier to the expressive love of the Kingdom of God.[8]

On the campuses, the effort to link Christian thought with international education was expansive. Stemming from a 1918 national effort to highlight "Christian World Democracy" as a curricular theme for the movement, the college YMCA also adopted an ongoing program of Christian World Education (CWE) to train students in international issues. Combining representatives of the YMCA and YWCA, the CWE committee wrote books, provided discussion outlines, planned conferences, generated bibliographies to stimulate student research, and even sponsored internationalist worship services that included such helpful hymns as "In Christ

There Is No East or West."[9] On over a hundred campuses throughout the 1920s, CWE institutes were held in which select speakers were invited to campus for several days in order to speak in classes, deliver lectures, and conduct informal discussion groups on global issues. Disappointed by Versailles and by the subsequent isolationist American foreign policy, college YMCA chapters across the country also worked hard in the 1920s to promote American involvement in both the League of Nations and the World Court, sponsoring mock multinational debates regarding participation. Such efforts did seem to heighten awareness and activism. The faculty at St. Bonaventure noted that student political interest had "come to life" as a result of the YMCA World Court Conference, while another school reported that the YMCA propaganda on these issues had forced them to reserve an entire section of the college library for international study and research.[10] Many, to be sure, must have taken comfort in the words of the Indiana University faculty member who acknowledged, "Most students and some faculty members have just discovered Europe. Go on informing them about it."[11]

In the midst of all of these issues, it was the movement against international warfare that generated the most interest. Despite their direct involvement with the war overseas and with the SATC on local campuses during World War I, the call for brotherhood and the Kingdom of God led to an increasing appeal to end all warfare. Porter himself was a pacifist and his convictions along these lines spread quickly to other leaders and students. Rooted both in Jesus' "reverence for human life and personality" and in the sense that democratization mandated dialogical (rather than military) procedures of international problem solving, the collegiate YMCA sought to expunge the escalating spirit of militarism within the national consciousness. At the 1922 meeting of the WSCF in Peking, China, delegates from all of the nations passed the "Peking Resolution," stating that the various movements of the federation should "face fearlessly and frankly, in the light of Jesus' teachings, the whole question of war and of those social and economic forces which tend to issue in war."[12]

While the attention to this issue may not have fulfilled the mandate to "focus their whole program on the study of and the working out of the implications of the Peking Resolution," students took part in a number of efforts along these lines.[13] Peace clubs were common offshoots of association chapters. At the University of Nebraska, the YMCA and YWCA together sponsored a "Protest Party" on the same night as the "Military Ball" in order to deflect attention from that event. In discussion groups, students studied the causes of war, analyzing its political and economic sources within the nationalist, imperialistic, and militarist dimensions of American culture. Others emphasized seeds of war on the campus itself, noting that the "spirit of selfishness, greed, suspicion, and hate" was fostered by fraternity exclusivity, athletic competition, and the glorification of American military heroes in history classes.[14] "The moment we stop to think of the spirit of war in our own lives," Porter warned students, "we realize that our hands are not entirely clean."[15]

These issues were also engaged "close to home" through YMCA discussions about the presence of ROTC units on the campuses. Tucker Smith, a general secretary of the New York University YMCA and later director of the Committee on Militarism in Education, noted that while 96 institutions provided courses in military training in 1912, 225 did so by 1927 at a cost of over 10 million dollars. The number of students enrolled in such courses increased 400 percent across this same period. While some opposed ROTC because of its compulsory nature, YMCA advocates challenged the very principle of such training, convinced that these courses fanned the flames of nationalism and encouraged jingoistic tendencies among impressionable youth. They noted that such courses never dealt with the causes of war, neglected themes related to alternative methods of ending warfare, exaggerated American virtues, and gave inaccurate, deprecatory pictures of other countries so as to stir up feelings of war hatred within students.[16] By 1926, the National Council of the Intercollegiate YMCA passed a resolution stating that, since ROTC units "retard the progress toward international good-will, and infringe upon freedom of speech and conscience," the YMCA registered formal protest against their presence on campus. For leaders, the choice was clear: "You may spend your lives as cannon fodder to support an imperialistic program in the world, or you may spend them behind the Cross of Jesus fighting imperialism on the home ground in behalf of the brotherhood of man."[17]

As time went on, particularly by the late 1920s and early 1930s, the YMCA placed much of its hope for creating a warless world on attempts to secure disarmament in the global community. In order to bolster the "fight for peace," YMCA leaders instituted both a campaign for international disarmament and an educational program designed to inform students about key international economic and political issues impinging upon nations' abilities to move in this direction.[18] Perhaps the most enduring aspect of this work came with the formation of the Intercollegiate Disarmament Council (IDC) at the 1931 conference of the WSCF. The council, formed in league with George Albert Coe's Committee on Militarism in Education, was designed to raise student awareness of disarmament issues in preparation for the 1932 World Disarmament Conference in Geneva, Switzerland. Much like the previous League of Nations and World Court programs, the disarmament campaign included educational aspects marked by lectures and discussions and a model disarmament conference where campuses assumed mock roles as various nations and sought to work with other groups to develop a disarmament treaty. They also sent Yale senior James Green to the World Disarmament Conference to represent the Christian student perspective, recommending the abolition of weapons, restrictions on preparation, budget limitations for arms, the elimination of compulsory ROTC, and "moral disarmament," an end to imperialistic ambitions. Green eloquently defended the college YMCA position:

> After contemplating the events preceding the catastrophe of 1914, we remain unconvinced of the wisdom of our predecessors. Fourteen years after the Armistice the glamour and heroism of that period fail to impress us, even when inscribed in gilt on stone memorials. The swords have lost their brilliance; the helmets and shiny buttons are tarnished. In fact the whole glorious temple of Mars has crumbled into ashes. We respect the noble war dead; but we question the judgment of those responsible for their death. . . . It is my generation which will be called upon to surrender all we consider worth while in life, in order to become targets for machine-gun bullets and victims for the latest poisonous gas. It is the men and women of my age who will be commanded to commit suicide. It is my generation which will be requested to destroy the best of human culture, perhaps civilization itself for causes which future historians will discover to be erroneous, if not utterly stupid. . . . We have thus lost interest in being prepared for cannon fodder.[19]

Working with the Committee on Militarism in Education, the YMCA earned the commendation of chairman Smith, who noted that the national peace movement depended heavily upon college secretaries and "those consecrated students who felt a greater loyalty to Jesus than to Mars."[20]

While much of the focus on internationalism was consumed with efforts to promote global peace, this impulse also prompted domestic efforts to expand YMCA ministry to international students on American campuses. An explosion in international student enrollment in American higher education, a growing YMCA international consciousness, and the diminishing of cross-cultural fears all combined to heighten interest in this sector of the campus community.[21] Increasingly, college YMCA leaders saw international student work less as a missionary or character-building force than as a strategic means of enhancing international harmony. YMCA leaders hoped that ministry with internationals would eliminate unhealthy stereotypes (Henry Wilson spoke of those who associated "banditry and bullfighting" with Mexican students or laundry businesses with the Chinese) and lay the foundation for expanding sympathy and mutuality.[22]

Those within the YMCA recognized that earlier directed attempts at evangelism and service had alienated many foreign students, especially those who resisted American attempts to impose Western religious and cultural values. For many, the new ideal was to encourage a universalized faith in human brotherhood expressed through the religious systems of each representative nation. At one YMCA-sponsored event, attendees proclaimed corporately the contributions of all the world religions, including Islam's "reverence in worship," Buddhism's "deep sympathy for the world's sorrow and unselfish search for the way of escape," Hinduism's "desire for contact with Ultimate Reality," and Confucianism's "belief in a moral order of the universe."[23] One Russian student noted that since the faith of the movement had merged so significantly with social justice concerns, foreign students were provided with a faith perspective that was essentially material rather than metaphysical in nature. "And though we do not say anything about Jesus," he commented, "we are working for the same things He worked for. He tried to give the poor a chance." Work with international students in many ways demonstrated the most concrete ramifications of religious pluralism in the movement.[24] Since the principles of Christian brotherhood were more important than formal Christianity, the Kingdom of God seemed to demand a more leveling approach.[25]

By the 1920s, the organization's new social gospel paradigm mandated a degree of reciprocity in such relations. While previous missionary labors had assumed an educational purpose of teaching other nations about the gospel, the postwar movement among international students more often emphasized the ways in which such students could teach Americans. Foreign students, one leader suggested, would "internationalize the minds of his hosts," opening student eyes to the cultural riches of other societies and blunting tendencies toward American exceptionalism. In order to accomplish this educational task, in fact, foreign students were enlisted to assist the associations in preaching the message of internationalism. Many groups gradually shifted from serving foreign students to promoting international forums where they could engage in dialogue about disparate cultures. Many chapters sent these students out on deputations to nearby colleges, schools, and churches to detail the beauty of their cultures and the desire for mutual understanding. In addition, such groups were able to "shock" American complacency by citing instances of discrimination on American soil. Rather than viewing these students as evangelistic projects for zealous proselytizers, it was clear that the new liberal mindset of the YMCA highlighted such individuals as important agents for concretizing the sometimes abstract internationalist message of the movement. These were the new "teachers."[26]

The Economy of the Kingdom

In addition to international issues, economic forces also drew the attention of the student YMCA in the 1920s and early 1930s. In the years before

World War I, concern for the poor had been an organizational staple, witnessed most clearly in the significant social service work conducted among American and foreign workingmen and in the general charity work done close to local campuses. However, in the years following the war, student concern for economic issues took on a more academic and policy-oriented cast. Not content any longer with service projects directed at those who were struggling financially, the new liberal social gospel perspective on such issues mandated a concern for the structural forces generating poverty, for the larger global economic realities generating corporate greed and inequity, and for the campus issues that laid a groundwork for such social ills. While the Depression obviously heightened such interest and lent a degree of urgency to social engineering along these lines, demonstrated concern for economic justice was already movement policy by the 1920s.

Approaches to these issues were manifold and diverse. Regularly, campus YMCA leaders invited speakers to come and address topics related to the battle between capital and labor. At the 1929 summer conference at Blue Ridge, North Carolina, for example, speakers representing both labor and owners presented two perspectives on the conflict in the textile industry.[27] Students at several universities joined forces with the League for Industrial Democracy and Socialist Club chapters in order to hold talks on economic issues. In addition, YMCA students, in cooperation with other radical campus organizations, held World Economics Conferences to deal with the global economy in preparation for the International Economic Conference in London in 1933. While the YMCA as a whole never officially adopted socialist policies or the political candidates espousing such positions, such economic ideals were strongly represented in these events.[28]

One of the most direct examples of the fight for economic justice came through a discussion series inaugurated in 1931 entitled *Toward a New Economic Society*. The product of a newly formed Economics Commission in the movement consisting of professors and representatives of the YMCA, YWCA, and the League for Industrial Democracy, this study was meant to provide students with information and questions related to economic justice on the college campus, in the nation, and in the larger world.[29] The curriculum stated unabashedly from the outset that the present economic system did not secure the greatest good for all people due to its tendency to promote "enervating wealth for a few and poverty and unemployment for many." The profit motive, the authors suggested, worked directly against the Christian value of love, leading "too often to the willingness to exploit labor with terrible human consequences, to fool the consumer politely and respectably, to extort concessions and privileges from the state by legal means." Flooded with statistics regarding the economic disparities between men and women, workers and employers, and those of different racial and ethnic groups, students were challenged to consider and rectify the economic sins of their world, fighting to replace the profit motive with a motive of democratic justice.[30]

Students were reminded, in very personal terms, that their position as students had significant economic implications. The authors demonstrated that students were often privileged to prolong their dependence because of the financial assistance of taxpayers (in state institutions) or endowment funds invested in farm mortgages and industry. They recommended that students attempt to recognize the "toiling hosts" whose labor made it possible for them to study, committing to use that education to benefit the community. Issues related to collegiate snobbery, fraternity luxury, money-driven athletics, and staff and faculty remuneration policies were also addressed, and students were encouraged to take an active role in facilitating justice through articles, petitioning on behalf of collegiate staff, and forming cooperatives.[31] Through all of this, the underlying message was that students must move beyond "incidental charity" to give careful consideration to the causes of poverty, economic injustice, and the "widest and most indirect consequences of our conduct." "If we make our money in an industry in which labor is underpaid, overworked, forbidden to organize, or left in a condition of perpetual insecurity," the curriculum suggested, "we are profiting by the exploitation of other persons." "We may never see those persons," the text allowed. "They may live in Gastonia or Danville or in the coal fields of Pennsylvania or they may be natives in Africa, but the geographical distance between us—whether we live in California or in New York—and our indirect relations with them constitute no moral reason for indifference on our part."[32]

In order to expand student exposure to such issues, practical means were also employed. While service projects with workingmen had been in place since 1910, more indicative of the new social gospel aim was the Summer Industrial Group (SIG) program. Initiated in Denver in 1920 under the leadership of Rocky Mountain college YMCA Director Ben Cherrington, the SIG program employed college men in industrial positions for the summer, directly introducing them to the vocational world of unskilled labor. The twenty-one men in this pilot program—many sociology, business, and engineering majors—secured jobs as drivers of milk wagons, machinists' helpers, carpenters' apprentices, street car conductors, and other manual labor positions in lumber yards, iron works, packing companies, and railroads. During their time of employment, these men received factory tours and attended meetings of the Central Trades and Labor Assembly, many joining the representative unions of their trades. On Tuesday and Thursday nights, the students gathered for seminars, led by insiders representing both capital and labor. On Sunday mornings, they discussed the relationship between religion and economic concerns. By 1921, SIGs were located in Portland, Omaha, Indianapolis, Minneapolis, and New York. In 1923, programs were started in Rochester, Chicago, Denver, and Atlanta, and by 1929–1930, Houston, Detroit, San Francisco, Kansas City, and Seattle had been added to the growing list of cities hosting this innovative student work program.[33]

While the students were never in a position to step completely into the experiences of common laborers—the short-term nature of the assignment almost guaranteed this—they were provided with a legitimate taste of the workingman's plight. They were able to enter into the violent Denver Tramway strike. In some cases, they lost their jobs due to cutbacks and had to face the reality of unemployment. They were treated as common laborers and thus given a sense of the hopelessness of individuals "chained to the wheels of industry." One noted that "much is to be gained in understanding by the student who has felt in his body the effect of high speed mechanical production, and who has tried to stretch a meager wage over a modest array of student wants." In many cases, students remarked at how difficult it was to experience the "severity and monotony" of industrial labor and to be labeled on the basis of job position rather than innate intelligence and character. As one observer suggested, "It has proved rather a startling disillusionment to some college men to discover how easily the intellectual can be disguised by a pair of overalls." Leaders of the program hoped that, while the students would likely one day serve as managers and owners rather than unskilled laborers, they would at least be given a perspective that would render them far more sensitive to the needs of the working class. Charles Elwood, professor of sociology at the University of Missouri, said it was one of the most hopeful experiments he had seen for eliminating labor unrest nationwide.[34]

Separate and Unequal? Racial Justice in the College YMCA

The final component of the new college YMCA "advanced program" was a concentrated focus upon racial issues. In contrast to the white student movement, African American college YMCA chapters continued to expand throughout the postwar period. After William A. Hunton's death in 1916, Jesse Moorland assumed national leadership as senior secretary of the YMCA's Colored Work department, leaving relative newcomers Channing Tobias and Max Yergan in charge of the student movement. By 1920, when the organization began including such chapters in national membership reports, YMCA chapters were present in 113 schools, enrolling 6,643 of the total 12,705 male students in these institutions (52 percent).[35] By 1924, the year Tobias replaced Moorland as Colored Work Department senior secretary, association chapters could be found in 123 out of the 200 black collegiate institutions, enrolling 8,000 students and making it the largest intercollegiate organization among African American collegians. After Tobias was replaced by William Craver in 1923, the movement's growth only escalated. In 1927, there were associations in 150 black colleges across the United States with 15,000 members and 4 separate summer conferences. Remarkably, this growth took place with minimal leadership. Despite continual pleas for extra assistance, the black student work was under the leadership of two national secretaries throughout the period. In addition, only three campuses—Hampton, Tuskegee, and Howard—possessed full-time campus secretaries.[36]

Even before the war, of course, the organization's social service emphasis had spawned studies of North American social problems, including race. After the war, however, the focus on racial issues both increased in intensity and changed in basic content and method. The war raised expectations for many African Americans returning from the front, both because of African American contributions to the war effort and because many hoped that the ideals of democracy defended on the battlefield would now be extended to the homeland.

In the immediate aftermath of the war, white student secretaries in the South spoke to black and white schools, churches, and organizations on racial problems and cooperated with the Southern Sociological Congress to give reports on issues such as "Child Welfare" and "Lynching." They also sponsored a variety of interracial forums in which representatives of black and white colleges met to discuss strategies for improved race relations.[37] The turning point for the student movement, however, came through the Indianapolis Convention in 1924. By the end of the conference, the assembled gathering set forth a series of resolutions committing the college chapters to interracial prayer and study groups, to encouraging the presence of African American speakers in YMCA conferences, to opposing organizations proclaiming white superiority, and to breaking down race divisions in fraternities, athletics, churches, and dorms. Students at the conference went on record as proclaiming that "racial discrimination and prejudice were unchristian" and that "the Negro should have equal opportunity for education, life and justice."[38]

Following the convention, the study of interracial issues became quite popular within the national movement and among local chapters. Movement leaders wrote a number of books and discussion syllabi on the issue, including *The Trend in the Races* by George Haynes in 1923 and *Race Relations and the Christian Ideal*, a piece compiled for the expressed purpose of helping students deal with the issues raised at Indianapolis. Discussion groups studying these works and J.H. Oldham's *Christianity and the Race Problem* attempted to delineate the ways in which the teachings of Jesus related to racial brotherhood, confirming the general sense that this was not just a social issue but also a matter of structural sin and disobedience to the larger claims of the Kingdom.[39] To actuate the spiritual forces to combat institutional and cultural racism, however, leaders were unequivocal in stating that traditional religion was inadequate. Oldham stated bluntly that "None of the social philosophies of the past, neither the conceptions of the mediaeval Church nor the ideas thus far evolved by modern Protestantism, are in the least adequate to deal with the new conditions." "If the Christian spirit is to exert a controlling influence on these modern developments," Oldham challenged, "it can only be by means of . . . fresh insights and conceptions that are still waiting to be born."[40]

The way in which Christianity could assist in this endeavor, many contended, was to elevate the conception of the Kingdom of God as an interracial brotherhood with no barriers. Because of the multiracial definition of the Kingdom, African Americans were to be viewed as fellow citizens fighting for common values. Oldham recommended that white students view their black colleagues "as the soldier views the comrade who serves the same flag with himself, and who dies for the same cause. In the Kingdom you, and your enemy, and yonder stranger, are one."[41] Pointing to the example of Jesus, many highlighted his equitable treatment of members of all races despite national Jewish prejudices that tended to alienate foreigners. In addition to his obvious friendliness to those of "alien races," however, Jesus also demonstrated in his very Jewish identity the ideal for race relations. As one noted, "When God found human affairs coming to such a state of narrowness and selfishness that only his coming in person could shock us into thinking, he came as a member of a despised race. In the Roman Empire, he came as a Jew. If he came to America, might he be born in a Negro cabin?"[42] With Jesus himself portrayed as an oppressed minority, the college YMCA now philosophically objected to the "separate but equal" doctrine and began battling for the brotherhood of all men under the fatherhood of God.

However, despite their social gospel inspired radicalism on racial issues, there were limits to this progressive work. First, in northern states, vigorous interest in racial issues proved to be the exception rather than the rule. While the "great migration" of blacks to northern cities had increased the number of African American students attending white colleges after World War I, students in these institutions demonstrated far less interest in these issues than their southern peers.[43] In addition, while discussions, forums, and stated resolutions on racial justice were quite generalized in the student movement, it quickly became clear that interracial practice was more difficult to achieve. The southern student conferences at Blue Ridge quickly became test cases for the practicality of theoretical commitments to racial integration. Blue Ridge had, from its origins, maintained a policy of racial exclusivity, forcing black students to initiate separate conferences at nearby King's Mountain. In the 1920s, an appeal was made to allow African Americans to attend Blue Ridge, provided that they agree to eat, sleep, and attend class in segregated areas of the grounds. Black students and their leaders were not impressed by these token gestures toward integration, suggesting that if white Christians found it necessary to build up barriers of separation while training for the task of reconstruction, they preferred not to get the training. It is not difficult to see how visible and forced segregation in this case was far more demeaning to black students than the voluntary separation at King's Mountain. For many blacks, racial independence seemed a better path than to suffer the humiliation of segregation in purportedly interracial settings.[44]

Such regional issues were compounded at the national level by tensions related to the organizational structure of African American student work. When the white student work was given the status of a division in 1927, black student chapters were not included in this new structure. Technically, because the African American chapters were located within the Colored Work Department of the Home Division, they were separated administratively from the white student movement. Leaders of the African American student movement began arguing vehemently that colored student work should be integrated with the larger Student Division. In part, they recognized that a move toward integration would allow financially strapped black associations access to the significant budgetary resources of the larger student division. In addition, leaders of black student work contended that African American students had more in common with white students than with other black YMCA departments. Most importantly, however, officers contended that there was a need to take action that would organizationally confirm the gospel of racial solidarity that was preached throughout the student division. The present setup, they argued, was a structural negation of the racial wisdom developed in the discussion groups of the larger movement, an outgrown shell militating against cherished ideals. Both to maintain integrity and as a "prophetic anticipation of the day when racial lines of all forms of Christian work will disappear," a unified merger seemed like the best option.[45]

Much to students' surprise, leaders in both the new student division and in the Colored Work department did not seem especially eager to integrate. When the white student movement was engaged in its contest to achieve divisional status, Porter asked Tobias to defer discussions of potential integration until after such status was secured because of potential resistance from leaders within the National Council. Many black leaders criticized this capitulation, the dean of men at Howard University stating that "if the Christianity of the YMCA cannot stand the test of discussing the colored students along with the white students, then the sooner that weakness in the organization is exposed, the better."[46] Other white YMCA leaders, especially those working in the South, were fearful that impetuous student radicalism on issues of racial integration would foment racial antagonism and eliminate the gains that had been secured by more gradualist means. R.H. King, associate executive secretary of the new YMCA Commission on Interracial Cooperation, suggested that such work had to proceed slowly: "There is fear that enthusiastic white and colored students who do not have to live with those problems day in and day out and whose relationship thereto is short-lived and largely academic and sometimes based largely on theory or sentiment may wreck a good and growing colored student and colored general work and may play havoc with the steady progress which has undoubtedly been made along interracial lines."[47] As long as white students were holding interracial forums, working to enhance black opportunities, and laboring to end abuses such as lynching, liberal YMCA leaders evinced a general support for their activities. When they attempted to transcend segregationist policies and Jim Crow practices, however, students often bumped up against the limits of liberal YMCA racial commitments.

At the same time that black students and their leaders were receiving this rather ambiguous response from national white YMCA leadership, they were also feeling pressure from black YMCA leaders to remain within the Colored Work department. In part, hesitation along these lines stemmed from legitimate fears that integration would result in white domination of black YMCA activities and the loss of significant leadership positions for African Americans. Deprived of their own structural haven, in other words, black students would be at the mercy of their oppressors.[48] Yet in addition to these concerns, legitimate in light of recent history, black leaders were also reticent to liberate black students because of issues of organizational survival. Since 105 of the 158 black associations in the nation were student chapters and since three of the eight Colored Work secretaries were engaged in student work, many feared the long-term consequences of such a move, especially since budget cuts were already threatening to decimate the administrative infrastructure of black YMCA work.[49] Philosophically committed to integration, they were also realistic about the need to maintain and bolster racially segregated programs to facilitate development in their own communities. Here was an issue where black students and their leaders came into conflict. Younger members bristled at gradualist notions, viewing them as a sign of tepid conviction. For older leaders, however, a chastened realism anchored in life experience at times fostered a rejection of radical integrationist approaches.[50]

In light of these events, many black students developed a cynicism toward YMCA "social Christianity." Advanced positions on racial issues were regularly articulated in conferences and written resolutions. In fact, student Colored Work Secretary John Dillingham noted in 1927 that "So frequently is race being discussed at National Conferences and Conventions that no one gets excited now and turns in the 'riot alarm' when the question is raised either on the platform or on the floor."[51] However, such radical pronouncements had not been manifested in equally radical procedures, and the disjunction between theory and practice was beginning to wear thin with black students. These matters came to a head at the National Faculty-Student Conference held in Detroit in 1930. While the conference had been set up as an interracial gathering, the Book-Cadillac hotel refused to allow black delegates to enter the coffee shop and restaurant and also segregated these students on a separate floor. When a similar event had taken place at the 1922 International YMCA Convention in Atlantic City, a student delegation had actually left the conference hotel in order to find new accommodations with their black colleagues. After meeting together to determine a response to this new inequity, however, the majority of students voted to remain in the hotel, utilizing only those facilities that were "equally available" to all delegates. In addition, they agreed to write a letter of protest to hotel administrators and to inform higher education institutions of the facts of the situation. A vocal minority of seventy-two students, both black and white, left the hotel and came back only for the general sessions, noting in a written statement that it was a "grave mistake" to register minimal protest against inhumane treatment. "Our failure in the past," the students noted, "has been that of committing ourselves to general principles while at the same time compromising in specific situations."[52]

For many black students and leaders, the action taken was simply a confirmation that the college YMCA was "all talk." In a follow-up report on the conference published by the *Christian Century*, a few African American critics noted that the Detroit experience might have long-term implications for the relationship between religion and race in the colleges. Because of Detroit, the periodical reported, influential black intellectuals chastised black students who thought religion could be of any assistance in solving the race problem in America. This conference, they noted, revealed that religion was nothing more than a "delusive reliance" for black Americans in that it "lavished sentimental phrases about brotherhood upon him while it has acquiesced in the prevailing attitude which deprives him of justice." The conference, the report concluded, "will be seized upon by certain Negro intellectuals as further evidence of the impotence of religion in the face of their race's great problem."[53]

It was not until 1933, in the midst of financial pressures spurred by the Depression, that the Colored Work department finally granted the students' request and permitted the integration of black students into the larger student work.[54] Progress in interracial policymaking notwithstanding, leaders could not obscure the fact that such gains were slow and in many cases short-lived. Active radicalism in racial issues proved to be more challenging than in areas of internationalism and economics. Racial barriers were a bit "closer to home" than these other issues, and students often faced powerful social forces of heritage and tradition. The stands taken by many in forging interracial dialogue, particularly in the South, were quite bold in their context. In addition, students were far more radical in this work than other departments of the YMCA. However, issues of race continued to serve as areas where official pronouncements and concrete practices were often misaligned.[55]

The Consequences of Radicalism

As historian Paula Fass has suggested, while moral rebellion was quite universal in the colleges and universities of the 1920s, social and political radicalism was far less prevalent. When it did exist, she notes, such radicalism was often a "Christian phenomena," bathed in a version of Christian sociology that highlighted both national reform and internationalism. Students influenced by these trends, she suggested, "often turned to social reform and politics just as their less spiritual brothers and sisters turned to football and jazz."[56] Contemporary observers identified this orientation among postwar YMCA students. Morton King, professor at Vanderbilt University, recalled that "in my experience the campus Y was the most 'liberal' organization, indeed influence, at Vanderbilt and probably on many campuses at that time. Its posture was that of liberal Protestantism. Heavy emphasis was on the 'social gospel,' addressing the issues of war-peace, racial justice, poverty/economic justice." Harry Ward noted that the student Christian movement had "noticeably increased the tendencies in the younger generation toward a prophetic ethical religion," while Norman Thomas, socialist candidate for president, suggested that "the student Christian association in a great many of our colleges and universities is a genuine center of intellectual curiosity. . . . More than once I have been impressed with the bravery with which secretaries of the student YMCA and the student officers of it have stood out for their convictions even at cost to themselves."[57] Not all campus chapters, of course, followed national prescriptions and embraced radical pedagogy. Yet despite the fact that some chapters were "quite unaware of contemporary issues," the "corporate impression" of the movement was certainly one of Kingdom-inspired social radicalism.[58]

Yet despite the fact that such a shift had been undertaken for the purpose of enhancing relevance to the campus, the radicalization of the college

YMCA had distinct ramifications. In an era defined by "100 percent American" and anti-Bolshevist sentiment, many university leaders across the country began to denounce the social reform biases of the movement, claiming that local chapters had traded useful service for "troublemaking." Some campus leaders accused YMCA leaders of socialist tendencies, a charge which was probably true of a vocal minority of movement leaders but unrepresentative of the movement as a whole. One chapter leader noted that the YMCA men leading the fight against ROTC were labeled as "reds," "bolshevists," and "un-American." The army and navy club at the University of Michigan spoke of YMCA antiwar advocate Sherwood Eddy as a "preacher of sedition."[59] At this same university, conservative faculty members ridiculed the organization for its radicalism, suggesting that this posture amounted to a denial of purpose. "Cut out the politics, war and peace, economic planning, and sociology," commented Assistant Professor of Mathematics Norman Anning, "and concentrate on just two things: education because all beginners need help concerning how to study . . . religion because religion is written into your name and charter and no other campus group has a better right to talk about it."[60]

The YMCA was also criticized for its new alliances in the college setting. Early in the 1920s, the national college movement urged cooperation with groups such as the Young Democracy, a student movement that stood for "thoroughgoing democracy in government, in education, in industry, and in international and industrial relations and for searching inquiry into the basic causes of war for the purpose of their eradication." The connections to the YMCA student movement were strong, and chapters collaborated with the organization on special projects, particularly related to antiwar protest and the World Court. In addition, while YMCA national and campus leaders spoke appreciatively of the work of the National Student Federation, they cooperated far more actively with the League for Industrial Democracy and various socialist clubs, chiefly on economic issues. League leaders were invited to address students, and college YMCA groups worked together with this organization in a joint Economics Commission in the early 1930s.[61]

These cooperative relationships were always guarded. The Advanced Program Commission, recognizing the helpful work already undertaken by joint ventures, warned that the organization must not lose or dilute its distinctively Christian message. Kenneth Scott Latourette at Yale acknowledged that joint efforts were "immensely hopeful," but he also stated that "unless they are Christian they will breed selfishness, cynicism, class warfare, and a type of radicalism which takes a scornful attitude toward existing institutions." Our "Youth Movement," he suggested, "must be actuated by the spirit of Christ."[62] Such a statement reveals what was in actuality a policy of partial cooperation with such groups, joining in matters of mutual interest but resisting radical overlap so as to maintain a distinct mission.

Yet even if such a policy existed, there were cases of "guilt by association." At the University of California at Berkeley, for example, the organization was criticized by campus leaders for allowing the radical National Student League to use the YMCA's Stiles Hall for its meetings. In reality, leaders of the YMCA at Berkeley had officially declared that they would not cooperate on peace issues with explicitly communist groups since "the objectives and methods of such an organization as the National Student League are so fundamentally different from ours." At the same time, committed to issues of free speech and tolerance for minority viewpoints, the association did allow such groups to use their facilities as long as meetings were not "inciting violence" or planning "overt acts" (YMCA Secretary Harry Kingman noted that they desired to protect "real freedom of speech" but not to protect "bad manners"). The YMCA came under attack from the Joint Americanism Committee and other conservative groups for this policy, but they remained steadfast that one of the reasons for the YMCA's existence was to provide opportunities for the discussion of religious, political, and economic questions "which cannot be permitted on the campus of a state university."[63]

In many cases, this growing tension was also fueled by the changing relationship between YMCA secretaries and their local institutions. Secretaries in the prewar era were important campus spiritual leaders, assuming a shepherding role in these institutions and providing an important link between students and faculty/administrators. The altered emphases of the intercollegiate movement, however, served to complicate this symbiotic relationship. The shift from activist service to intellectual engagement initiated a commensurate change in expectations for general secretary qualifications, replacing character, personal magnetism, and spiritual warmth with a desire for academic expertise. The YMCA Commission on the Training of Student Secretaries mandated in 1924 that student secretaries possess a bachelor's degree plus additional training in at least a representative sampling of subjects germane to their position such as philosophy of religion, religious education, history of Christianity, sociology, anthropology, psychology, economics, political science, and social hygiene. As of 1925, about 35 percent of all local secretaries possessed graduate degrees, an increase of 100 percent over the totals in 1915. By the time of the next large-scale study in 1933, a full 52 percent had earned a graduate degree.[64]

Whether by training or disposition, secretaries were apt to be champions of liberal social causes. Despite the fact that a majority of secretaries in the 1920s hailed from the more conservative southern and central regions, surveys in the early 1930s revealed that more read magazines dealing with social and political problems (141) than religious (89), and financial (4) titles.[65] With regard to perspectives on social issues, many declared themselves "absolute" pacifists, and 70 out of 106 polled in 1933 registered a fundamental opposition to war. At the same time, in answer to the question, "Do you make it a rule to speak to others about becoming a Christian?" twenty-nine said they did, thirty-six that they did not, and ten that they did so occasionally. In many cases choosing such positions over church vocations because of the liberal academic environment, secretarial interests certainly paralleled those of the larger movement at this time.[66]

Because of these views, secretaries were content to exist outside of the authoritative purview of university leaders, free from the structural and intellectual constraints imposed upon those in more guarded positions. Over 90 percent of the secretaries polled in 1928 argued that placing or keeping secretaries on the collegiate payroll would limit their "progressive" stances and inhibit the freedom of inquiry characteristic of the new student movement.[67] YMCA leaders at the University of Minnesota and University of Denver noted that the experimentalism characterizing the movement would be jeopardized were they somehow linked officially to the campus structure. The likelihood of a ban on radical speakers was enough to discourage this proposed plan for the Minnesota leader.[68] At Denver, the secretary suggested that because of the conservative nature of university presidents, the result of affiliation would be straightforward: "Would not be free to experiment in social theory. Could not be radical."[69] Thus, YMCA secretaries, formerly the most explicit points of connection between the college chapters and campus leaders, now registered a somewhat oppositional and counterinstitutional stance vis-à-vis these administrative officers. Coming to their positions older and with higher educational achievements, newer secretaries were less willing to submit to college leaders' policies, preferring an independent position true to the "prophetic minority" ideal.

This oppositional stance was increasingly problematic. In 1922, for example, the administrators of Northwestern University and Purdue University both issued statements denouncing campus chapters for their "subversive" activities. The Northwestern administrative statement was typical of the growing disenchantment between local clubs and their sponsoring institutions: "We are definitely dissatisfied with the Association because it seems not to be performing its appropriate functions as a religious organization, but is absorbed with social and economic issues in which it takes a rather extreme leftist position. Many business leaders have argued against them on this front."[70] As a result of their distaste for the new functions of the association, the Northwestern administration withdrew the considerable subsidy that had been given to the campus organization for secretarial salaries. When the secretary was released because of insufficient funding, the position was filled by a university-sanctioned chaplain. Similarly, at Purdue in 1927, campus leaders simply disbanded the association both because of its stand against the campus ROTC and its tendency to "produce in students a disdain for those in authority." A campus chaplain and dean of students were hired to cover the work of this organization.[71] As one historian put it, the "new convictions gained by students on controversial issues outran the development at the same time of skill in making their convictions effective without endangering the harmony and unity of campus life."[72] For many administrators, the risks were simply too substantial.

These shifts also alienated many of the conservative Christian students on the campuses. Many understood that the new social gospel approach would segregate the movement from the more popular domains of campus life, particularly fraternity members and athletes. However, it was also true that, in their enthusiasm for expanding religion to incorporate social topics, many had disparaged traditional offerings and therefore threatened a conservative religious constituency. As Gary Scott Smith has contended, as long as social service was viewed as a way of applying the biblical command to love one's neighbor, conservatives and liberals in the Progressive era could work harmoniously together for social causes and the amelioration of social ills. However, once liberals began developing a conscious theology of social activism that both diminished the importance of traditional evangelical themes and redefined conservative doctrinal tenets, the platform for common work was markedly reduced. The transformation of views on sin, salvation, and the Bible crossed lines that conservatives were unwilling to sanction.[73] When YMCA groups spoke of Jesus as a social reconstructionist rather than a personal Savior and glossed over significant moral issues, they were perceived as exerting a corrosive influence. When they began admitting members of all faiths and those devoid of explicit faith, conservatives found little common ground to stand upon.

The loss of Bible study, prayer, and missionary themes highlighted denominational groups as more appropriate settings for some conservative Christian students. Baptists, for example, were far more likely to affiliate with the new Baptist Student Union because of the reaction against the liberal YMCA. One historian of the Southern Baptist Convention, in fact, asserted that the chief reason for the urgent formation of this organization was the "conservative student reaction swinging away from interdenominational radicalism."[74] More importantly, the increasingly liberal stance of the college YMCA served as the impetus for the formation of new, conservative interdenominational student movements. In the mid-1920s, Princeton conservative J. Gresham Machen led students at his institution to establish a League of Evangelical Students (LES) in order to "rise up a student protest against modern unbelief and to take a stand for the defense and propagation of the gospel of everlasting salvation through the sacrificial death of God's only begotten Son." Linked to Westminster Theological Seminary, this organization formed about sixty chapters around the country and initiated a new student religious magazine, *The Evangelical Student.* However, it was InterVarsity Christian Fellowship that emerged as the most popular conservative Christian response to the liberalizing YMCA. Initiated as a reaction against the liberal college YMCA in Great Britain, InterVarsity chapters were birthed in the United States in the 1930s for similar reasons. Two Wheaton College (IL) graduates, Stacey Woods and Charles Troutman, were commissioned to travel among colleges and universities in order to organize the new movement. In 1938, exactly eighty years after the college YMCA was birthed at the University of Michigan, conservative Christians, many of whom were former YMCA members, developed this new movement to channel religious enthusiasm. By 1946, InterVarsity was operating on 277 campuses worldwide, boasting a student membership of 10,000.[75]

In addition, the rise of Protestant neo-orthodoxy in the mid-1930s proved to be a challenging context for the liberal religious and educational vision of the college YMCA. Neo-orthodoxy was, more than anything, a reaction to the excesses of liberal theology in light of the tragic reality of world events. The Depression squelched much of the optimism characteristic of the progressive faith in science and democracy. Belief in evolutionary progress and the inevitable growth of the Kingdom of God seemed both quaint and blind in a world giving birth to fascism and Nazi ethnic cleansing. Likewise, the confidence in pure scientific method and democratic deliberation seemed increasingly illusory in a society that revealed base motives in economic dealings and the spread of Freudian conceptions of irrational and unconscious impulses. In short, neo-orthodox theorists accused liberals of painting a distorted picture of reality, substituting glib hopes of global democratization and generalized altruism for the tragic reality of human sin and suffering. These theorists contended for a return to a belief in a transcendent God, a serious consideration of the reality of sin, and a more fulsome confidence in the authority of Scripture.[76]

Although some liberal leaders attempted to dismiss this reaction as a "recrudescence of fundamentalism," they quickly ascertained that neo-orthodoxy was a more formidable opponent. By the mid-1930s, in fact, many collegians were attracted to this posture. In a world threatened by war and economic collapse, many argued that students were again desirous of an authoritative religion that could provide a "dependable standard of values" and "some sustaining world-view."[77] In a 1935 conference on "what kind of religion we believe in," the National YMCA Student Council stated that the Christian faith espoused by the association was identified "too closely with the surrounding culture." Attempting to "prove itself compatible with modern life and science," it had "made science the real revealer of truth" and compromised the importance of "theology and tradition." Speaking highly of the neo-orthodox message, the council noted that the YMCA possessed "too optimistic a view of man," suggesting that the liberal mantra of "respect for personality" had blunted "critical judgment." Criticizing the identification of the Kingdom with "certain social, economic, and political philosophies," one student suggested that "Christianity is not only a spirit and an ethic, it is a stupendous personal religion. . . . When Christianity is seen not merely as man's wistful attempt to march Godward, but as God's supreme and victorious action in marching manward, a new light has begun to shine."[78]

Many older leaders and theorists within the college YMCA, including Harrison Elliott and George Albert Coe, reacted strongly against these themes, disparaging neo-orthodox critics as "other-worldly theorists" who were concerned more for an individual's mystical relationship with God than with the real and pressing struggle for the Kingdom of God on earth. Judging this movement as a knee-jerk reaction to global crises, Elliott accused these theorists of seeking false security in divine authority, substituting religious tyranny and a theologian-inspired demagoguery for the more difficult work of democratic deliberation.[79] What is perhaps ironic about this stance was the degree to which it appeared to reverse leaders' faith in youth. While liberals in the 1920s held up youth as the hope of the world, they now dismissed the antiliberal proclivities of the rising generation as whims of immaturity. Instead of portraying the next generation as fresh and inventive, liberals now described them as misguided souls, clinging on to security rather than embracing the freedom of scientific methodology.

The glaring inconsistency of a progressive movement resistant to change and youthful contributions seems less difficult to comprehend when it is acknowledged as a philosophical dilemma within liberal progressivism itself. This generational conflict demonstrated the difficulty in sustaining both the descriptive and procedural dimensions of the democratic platform. Leaders desired to provide students with methods that would be experimental, allowing them to make contributions on the basis of democratic inquiry. At the same time, they desired to inculcate what they perceived to be the liberal democratic values of the Kingdom of God. This conflict between trust in the youth-led inquiry process and a desire to inculcate "advanced social positions" was a deep and intrinsic fissure within the very spirit of liberal progressivism. In reality, liberals' commitment to democratic scientific methodology, rooted in student initiative and social intelligence, meant that decisions might be made contrary to the most "radical" social solution. The fact that leaders were not always willing to receive the democratically determined conclusions of youth revealed that there were limits to the freedom of science.[80]

The decline of the YMCA chapters on university campuses was thus in many ways a product of its success. The program to develop an organization radically appropriate to the changing religious and university world made perfect sense. It was, in essence, an attempt to remain relevant to its clientele and to its context, both of which were in a state of flux after World War I. But this revision of the association "mission" to appeal to its apparently shifting "market" was more complex than it first appeared. For one thing, the market, in its broadest sense, was never monolithic. The student YMCA had to appeal to students, but its very livelihood depended on relationships with university administrators and church leaders as well. Even the student culture was not monolithic and indeed revealed that conservatism could be found not only in "popular" students but also in students who were actively involved in religious communities. For an "in-between" institution that had previously held a steady but tenuous position at the intersection of these constituencies, the decision to embark on a program reorientation resulted in enhanced opposition to program themes from administrators, church leaders, and both popular and conservative religious students.

This is not to say that the YMCA plan to restructure itself somehow "backfired." Many leaders were not convinced that growth was their highest calling. One leader in the late 1920s suggested that "Whether in any given situation the alternative to this is an Association that is larger or smaller numerically than formerly is not important so long as the Association becomes increasingly more catholic, gathering into itself the broad reaches of Christian insight and experience, regardless of church affiliation."[81] In this sense, the transformation, what one called the creation of a "prophetic minority," was certainly an internal "success." At the same time, such tensions ensured that the movement would indeed be characterized by a minority consciousness on the campus.

Conclusion

The intercollegiate YMCA was in many ways a transitional institution between the evangelical piety of the 1870s and the growing secularism of the early twentieth century. Birthed in the revivalist fervor of the 1850s and channeling this energy to the colleges, the religious, social, and institutional changes of the next decades pushed the associations to embrace new ideals and new practices that coordinated more effectively with campus life. The transformation of the college YMCA across its three core time periods can be viewed from a number of unique vantage points. From a theological angle of vision, the shifting philosophy of the movement can be characterized as a transition from conservative evangelicalism (1858–1888) to liberal evangelicalism (1888–1917) to a pluralistic liberal Protestantism (1917–1934). The individualistic gospel, concerned primarily for the salvation of individual souls, marked Wishard's era and gave the YMCA a central place within the larger tenor of Protestant revivalism. All "social work" was taken up with gospel proclamation, bathed in the sentiment that mass personal conversions would foster a larger social renewal. The conservative social gospel of the Mott era, alternatively, highlighted social study and service in addition to personal work. A gospel for social as well as personal regeneration, this emphasis was nonetheless directed at character reformation in the community. Service, in other words, was directed not at large-scale social change but rather at the moral improvement of individuals and groups in the fourfold life. Finally, the liberal social gospel of the post–World War I era pushed into new areas of emphasis. More concerned with social than individual sin, leaders adopted a framework in which students would focus upon joint intellectual inquiry for the purpose of stimulating democratic social justice within the broader culture. This shift from an individualistic gospel to a conservative social gospel to a liberal social gospel was a clear source of the larger transformations of this student movement.

These religious changes generated corresponding alterations in many of the Christian themes and practices that characterized this movement. Jesus, for example, was defined in Wishard's era as a personal Savior, the source of divine regeneration through his atoning and sacrificial death. In the years leading up to World War I, however, he was described more often as an exemplar of fourfold character and manly service, the ultimate representation of a muscular faith. In the 1920s, this same Jesus was defined as a social radical, a brilliant thinker whose principles carried the force of a social revolution. Thus, through dramatic changes in philosophy and policy, the college YMCA could maintain an unbroken connection to "Jesus' way of life" even as the meaning and purpose of his ministry was completely redefined. The Bible was transformed from a sacred text containing apologetic arguments and evangelistic directives into one that contained pragmatic assistance for character and service during the Mott era. Then, moving into the postwar period, the Bible was viewed more self-consciously as a source of principles for radical social reconstruction. During both of these latter two eras, the Bible was increasingly described as one helpful resource among many rather than as a text that possessed unique revealed truths. In addition, while the early movement embraced and communicated the entire text to students, the later years saw a more exclusive concentration on the Gospels and the life of Jesus, largely ignoring the Old Testament.

These redefinitions altered the ways in which movement leaders conceptualized the task of developing "Christian" students. In many ways, the changing religious orientation of the college YMCA fostered a parallel change in the domain of student life serving as the focal point of programmatic efforts. The early movement directed its program squarely at the human soul, seeking the divine work of the Holy Spirit to bring an immediate and radical conversion. The muscular Christianity of the Mott era, on the other hand, focused upon student behavior, actions within the fourfold life, and in social service venues. The postwar era then altered this emphasis, giving primary consideration to the student mind. Intellectual labor through the scientific method, in this sense, was the central locus of Christian responsibility in the attempt to bring about the Kingdom of God on earth. Even the desired career trajectories for students reflected this transformation. While leaders in the Wishard and early Mott eras wanted students to pursue explicitly ministerial and soul-saving careers in pastoral, missionary, and YMCA occupations, before World War I other vocations, including social work and engineering, appeared equally "Christian" in emphasis. By the postwar era, few spoke of full-time ministry professions. Instead, they focused upon jobs in which intellectual labor and policymaking were central. This significant evolution—essentially a shift from heart to hands to mind—was a critical means of demonstrating the practical effects of theological changes.

In addition to these theological changes, this era in YMCA history can also be described in terms of organizational changes. To use Paul Super's descriptive terms, the YMCA was transformed from a movement to an institution between 1888 and 1917, only to return to a movement consciousness after World War I. The early movement had a definite vision of spiritual purpose in the attempt to evangelize collegians, the surrounding communities, and the world, all the time possessing little institutional machinery. During the Mott era, however, the chapters became institutions on the campuses, largely through the construction of full-service buildings, the hiring of paid staff, and the assumption of student service responsibilities. Once these responsibilities were removed, the college YMCA actively sought to regain a movement consciousness by empowering student initiative, democratic practices, and a radical purpose in establishing the Kingdom of God on earth. In some ways, this larger shift was related to the status of the chapters on the campuses. A minority status, the reality of both the early and the later periods, coordinated well with a movement sensibility, challenging students to pursue a righteous cause that was oppositional to the campus mainstream. Between 1888 and 1917, however, the YMCA was a functional participant in the campus mainstream, enjoying the prestige of centrality within both student culture and the larger administrative functions of the schools. Such a position was clearly an impetus to a growing sense of institutional presence on the campuses of this era. Thus, despite major differences in theology and aim, the Wishard and Porter eras might have possessed a similar organizational self-understanding, especially when contrasted with the unique success of the two decades preceding World War I.

In the end, both the movement and institutional dimensions of the college YMCA in these years played a role in fostering the secularization of higher education. From the methodological angle, the YMCA promoted the ability of institutions of higher learning to place the responsibility for student religious care outside of required courses and activities. For faculty, the presence of such an organization meant that they could pursue their specialized fields in the spirit of scientific free inquiry without reference to creedal religious concerns. It also meant that they could serve as personal ministers and mentors to students in the extracurricular YMCA. What this did was to encourage a variety of dualisms between facts and values, public and private truths, and sacred and secular spaces. Essentially, the academic curriculum became the context for factual inquiry, a secular sphere shorn of the subjective religious values that might inhibit scientific objectivity. The extracurriculum, on the other hand, was viewed as a fitting place for the values of one's private faith to be cherished, nurtured, and expressed. Here was the sacred space where pious faculty and students together could deepen their commitments to Christian faith. Administrators could survey their campuses and herald the fact that these institutions were brilliantly succeeding on both fronts. Classwork was radically scientific and designed to promote knowledge production free from the baggage of religious presuppositions. At the same time, students were regular in church attendance, flocking to YMCA evangelistic meetings, and forging the greatest Bible study movement ever witnessed on the campuses. This division of labor seemed to be bearing tremendous fruit for both science and religion.

Such a smooth transfer of functions, however, might have had paradoxical results. By adopting the role of moral and religious leadership in the colleges and universities, a role that institutions of higher education were then gradually abandoning, the YMCA may actually have helped to marginalize the previous consensus that these institutions ought to provide moral training. In fact, the success of the YMCA may have assisted in the removal of religion from the university while its decline in the 1920s might have prompted a renewal of this traditional goal (i.e., in the form of university chaplains and religious directors). Ironically, the YMCA, through its "success" on college campuses and its adoption of the character-building functions of the traditional liberal arts college, helped to ease the transition to modern secular education.

For the students who took part in the organization, the reality of ideological secularization was perhaps even more substantive. Liberal Protestant innovations were clearly directed to the intellectual credibility and cultural relevance of the Christian faith, especially critical for many within cutting-edge university contexts. At the same time, the provisions of this evolving faith stance did promote secularization in some fairly direct ways. The first might be characterized as a broad shift within the organization to a consequentialist ethic that was often described as a move from "belief" to "life." During the Mott era, the emphasis upon character and service (as contrasted with evangelism and apologetic certainty) clearly moved in this direction. As long as students were growing in the various aspects of the fourfold life and taking part in humanitarian service to others, the source of such results was of little concern to movement and chapter leaders. In other words, promoted by the pragmatist sensibilities of this period, while the ends of Christian growth were clearly stated, the means of achieving these ends mattered little. Such a reality was evidenced in the alteration in membership standards within the movement, many chapters denouncing evangelical church membership criteria anchored to expressed and specific creedal foundations. The fact that the Bible was now viewed as only one helpful resource among many also demonstrated this reality. The words of Scripture were now judged by the same evaluative criteria of any other published work, rooted not in a revelatory source but rather in its capacity to produce lives of character and service in the world. If belief served life, then it was indeed heralded for its value. If life could be secured without belief, however, there was little reason to label this an inferior outcome.

Of course, such an ethic was a clear impetus for the growing tolerance for religious pluralism in the YMCA. As consequences were highlighted over the belief-oriented sources of action, the door was opened to members of other faiths, regardless of their adherence to evangelical or even Protestant creeds. Nonevangelical Protestants, Catholics, Jews, and even those without faith could thus all proceed similarly toward ideals of character and service. In addition, both missionary and service work soon mandated a sensitivity to the cultural religious perspectives of ministry recipients. Leaders gave clear directives in service work to avoid directive proselytizing and to allow industrial laborers and others to aspire to character within the context of their own faiths. While missionary work always maintained more consciousness of Christian influence, even this became more a matter of sparking character development in the fourfold life rather than seeking direct conversion to the Christian faith. These changes were always made with an eye to expanding the reach and influence of the Christian message to larger numbers of students and more expansive dimensions of human life. Such growing inclusivity, however, made it increasingly difficult to generate boundaries that took seriously the reality and particularity of religious beliefs. Of course, religious pluralism (the removal of Protestant hegemony) cannot be equated with secularization. But to the degree that such pluralism was an indication that religious perspectives were no longer important within the movement, they depict a distinct secularizing trend.

Ideological secularization was also facilitated by the organization's changing educational vision. The method of Deweyan procedural democracy, defined as it was in the context of collaborative group decision making, did theoretically allow for a variety of perspectives to be included "around the table." At the same time, it became increasingly clear that conservative religious students were not equal players. The strong faith in scientific method espoused within the postwar movement did not leave much of a place for those who promulgated revelatory ways of knowing. In other words, students could join in the process of collaborative decision making, but only if they were willing to adopt the scientific epistemological ground rules embraced by liberal Protestants. By eliminating those who were not scientifically "enlightened" from the participatory process, these liberals enacted a kind of epistemological hegemony that, while certainly viewed as a liberation from a priori argumentation, also practically excluded certain perspectives from gaining a legitimate hearing. Thus, while the prewar college YMCA facilitated the methodological secularization of the university, the postwar YMCA chapters hastened an ideological secularization that divorced religious expression from its doctrinal roots even as it denigrated traditional religiosity because of its antidemocratic sentiments. Designed to combat secularization, the YMCA became one of the tools for its realization on American college and university campuses in the early twentieth century.

Notes

Introduction

1. Woodrow Wilson, "The Significance of the Student Movement to the Nation," in *APJRM*, 168.
2. Francis Patton, "The Significance of the Student Movement to the Church," *Int* 25, no. 4 (January 1903): 81.
3. For example, see Lawrence R. Veysey, *The Emergence of the American University* (Chicago: University of Chicago Press, 1965); Frederick Rudolph, *The American College and University* (Athens, GA: The University of Georgia Press, 1962); John Brubacher and Willis Rudy, *Higher Education in Transition: A History of American Colleges and Universities*, 4th ed. (New York: Harper & Row, 1958); Helen Horowitz, *Campus Life: Undergraduate Cultures from the End of the Eighteenth Century* (Chicago: University of Chicago Press, 1987); and Harry E. Smith, *Secularization and the University* (Richmond: John Knox Press, 1968).
4. See, for example, Veysey, *The Emergence of the American University*; Rudolph, *The American College and University*.
5. See George M. Marsden, *The Soul of the American University* (New York: Oxford University Press, 1994); Julie Reuben, *The Making of the Modern University: Intellectual Transformation and the Marginalization of Morality* (Chicago: University of Chicago Press, 1996); Jon H. Roberts and James Turner, *The Sacred and the Secular University* (Princeton: Princeton University Press, 2000).
6. From Marsden's perspective, liberal Protestantism eased the transition from the seemingly innocent methodological perspective to the more insidious ideological variety that prevented religious themes from entering the university "marketplace of ideas." Julie Reuben, while not discounting Marsden's argument, does complicate his perspective by looking more specifically at how religion was conceived in relation to the search for truth at major universities in this era. Roberts and James Turner in *The Sacred and the Secular University* focus more specifically on university faculty in this transition, looking at the sciences and the humanities.
7. George M. Marsden and Bradley J. Longfield, eds., *The Secularization of the Academy* (New York: Oxford University Press, 1992), 29.
8. John R. Mott, *The Students of North America United* (New York: The International Committee of Young Men's Christian Associations, 1903), 15.
9. In works on the history of higher education, the YMCA is addressed only briefly if at all. Despite his rather engaging analysis of the "mind of the undergraduate," historian Lawrence Veysey in *The Emergence* neglects the student YMCA as an important component of student life. Frederick Rudolph's *The American College*

and University helpfully places the YMCA within the context of Progressivism but offers only a single paragraph on the movement's contributions (362). Even Helen Horowitz, whose *Campus Life* is devoted to undergraduates, ignores this organization. As historian P.C. Kemeny summarized, "Standard treatments . . . have neglected the critical role that the YMCA had in the daily lives of students and the importance that this organization played in many modern colleges' and universities' efforts to perpetuate their historic Protestant mission." See P.C. Kemeny, *Princeton in the Nation's Service: Religious Ideals and Educational Practice, 1868–1928* (New York: Oxford University Press, 1998), 163–164. By far the most helpful treatment in this genre is James Burtchaell's *The Dying of the Light: The Disengagement of Colleges and Universities from their Christian Churches* (Grand Rapids: Eerdmans, 1998). While Burtchaell's aim is to describe the waning relationship between institutions and their founding churches, he mentions the YMCA frequently as a platform for student religious expression.

10. Clarence P. Shedd, *Two Centuries of Student Christian Movements* (New York: Association Press, 1934); William H. Morgan, *Student Religion during Fifty Years: Programs and Policies of the Intercollegiate YMCA* (New York: Association Press, 1935). In his larger history of the YMCA, C. Howard Hopkins devoted two chapters to the student movement and its relationship to the parent organization, deriving much of his information from a 1947 Yale dissertation by Harry M. Philpott. See C. Howard Hopkins, *A History of the YMCA in North America* (New York: Association Press, 1951); H.M. Philpott, "A History of the Student YMCA, 1900–1941" (Ph.D. diss., Yale University, 1947).
11. Marsden and Longfield, *The Secularization of the Academy*, 16–17.

Chapter One Organizing a College Revival

1. Rudolph, *The American College and University*, 88–101.
2. Ibid., 86–109; Roger Geiger and Julie Ann Bobulz, "College as It Was in the Mid-Nineteenth Century," in Geiger, ed. *The American College in the Nineteenth Century* (Nashville: Vanderbilt University Press, 2000), 80–90.
3. See especially David Allmendinger, *Paupers and Scholars: The Transformation of Student Life in Nineteenth Century New England* (New York: St. Mortin's Press, 1975), 8–19; 91–94; Allmendinger, "The Dangers of Ante-Bellum Student Life," *Journal of Social History* 7 (Fall 1973): 75–83.
4. Horowitz, *Campus Life*, 27. See also Phillip Greven, *The Protestant Temperament: Patterns of Childrearing, Religious Experience, and the Self in Early America* (New York: Knopf, 1977).
5. Donald G. Tewksbury, *The Founding of American Colleges and Universities Before the Civil War: With Particular Reference to the Religious Influences Bearing upon the College Movement* (New York: Arno Press, 1932), 16–28; Natalie A. Naylor, "The Ante-Bellum College Movement: A Reappraisal of Tewksbury's Founding of American Colleges and Universities," *History of Education Quarterly* 13 (Fall 1973): 261–274.
6. As historian David Potts has documented, denominationalism might have actually grown later in the century when improved transportation allowed for a more religiously homogenous clientele to seek out affiliated institutions "across the miles."

David Potts, "American Colleges in the Nineteenth Century: From Localism to Denominationalism," *History of Education Quarterly* 11 (1971): 363–380.

7. Walter P. Rogers, *Andrew D. White and the Modern University* (Ithaca: Cornell University Press, 1942), 82. On this theme of a broadly Protestant constituency, see W. Bruce Leslie, *Gentlemen and Scholars: College and Community in the Age of the University, 1865–1917* (University Park: Pennsylvania State University Press, 1992), 96.
8. Marsden, *The Soul of the American University*, 84–93.
9. On natural theology, see Reuben, *The Making of the Modern University*, 30–35, 50–53; Roberts and Turner, *The Sacred and the Secular University*, 22–24, 32–33.
10. Reuben, *The Making of the Modern University*, 19–21, 88–90; Roberts and Turner, *The Sacred and the Secular University*, 20–21, 43–49, 107–122; Rudolph, *Curriculum: A History of the American Undergraduate Course of Study Since 1636* (San Francisco: Jossey-Bass Publishers, 1981), 39–42, 90–93.
11. Frederick Marsden, *The Soul of the American University*, 90–93, 103–106. See also Mark A. Noll, "The Revolution, The Enlightenment, and Christian Higher Education in the Early Republic," in Joel A. Carpenter and Kenneth W. Shipps, eds. *Making Higher Education Christian: The History and Mission of Evangelical Colleges in America* (Grad Rapids: Christian University Press, 1987), 56–76.
12. Bradley J. Longfield, "From Evangelicalism to Liberalism: Public Midwestern Universities in Nineteenth-Century America," in Marsden and Longfield, eds. *The Secularization of the Academy*, 47–48. See also Shedd, *Two Centuries of Student Christian Movements*, 80–84.
13. According to Frederick Rudolph, University of Illinois students lodged their protests through foot stamping, Bible theft, and the release of stink bombs. Rudolph, *The American College*, 79.
14. Shedd, *Two Centuries*, 18–25.
15. Ibid., 18–43; Cornelius H. Patton and Walter T. Field, *Eight O'Clock Chapel* (Boston: Houghton Mifflin Company, 1927), 209.
16. Shedd, *Two Centuries*, 55–58.
17. Ibid., 69.
18. Kathryn T. Long, *The Revival of 1857–1858: Interpreting an American Religious Awakening* (New York: Oxford University Press, 1998), 7. See also Timothy Smith, *Revivalism and Social Reform: American Protestantism on the Eve of the Civil War* (Gloucester, MA: P. Smith, 1957).
19. See Hopkins, *A History of the YMCA in North America*, 18.
20. Long, *The Revival of 1857–1858*, 61.
21. Hopkins, *History of the YMCA*, 25.
22. Ibid., 19, 36–38.
23. Orr, *Campus Aflame*, 67, 83. Orr cites major revivals in every part of the country, claiming that 5,000 came to faith in the Midwest alone between March and May of 1858.
24. Shedd, *Two Centuries*, 44–45.
25. Rudolph, *The American College*, 83.
26. Philip Alexander Bruce, *History of the University of Virginia, 1819–1919: The Lengthened Shadow of One Man*, vol. 3 (New York, Macmillan, 1922), 138.
27. Hugh McIlhany, "The Founding of the First Young Men's Christian Association Among Students," *Alumni Bulletin of the University of Virginia* 2,

no. 1 (January 1909): 50; Matthew Leggett, "The Evolution of the Young Men's Christian Association of the University of Virginia: 1858 to 1968," *Magazine of Albemarle County History* 58 (2000): 33.

28. Adam K. Spence, "A History of the Founding of the Young Men's Christian Association at the University of Michigan," 2 (A.K. Spence Papers, Bentley Historical Library, University of Michigan, 1870).
29. Shedd, *Two Centuries*, 95–98; Victor Wilbee, "The Religious Dimensions of Three Presidencies in a State University" (Ph.D. diss., University of Michigan, 1967), 114–119. Robert Weidensall is very clear that Michigan should be given credit as the first college YMCA chapter because of its stable evangelical basis. Clarence Shedd has posited that the University of Virginia should be given credit because of its quick allegiance to the movement and more developed program. See Weidensall, "The Early History of the College Young Men's Christian Association Work, and the Part the Virginia University Association Had in It," 7–8. Folder "Early History," Student Work, Kautz Archives; Shedd, *Two Centuries*, 94–102.
30. Shedd, *Two Centuries*, 104.
31. Ibid., 103–110. See also *Proceedings of the Thirteenth Annual Convention of the Young Men's Christian Associations of the United States and British Provinces* (New York: Executive Committee, 1868), 101. Robert E. Lee was later made an honorary member of the association. See Wishard, "Virginia College Conference," *CB* 8, no. 3 (December 1885): 1.
32. *Proceedings of the Thirteenth Annual Convention*, 88–89.
33. Ibid., 101–104; Hopkins, *History of the YMCA*, 271–274.
34. Weidensall, "The Early History," 13.
35. Ibid., 11–12.
36. Rudolph, *Curriculum*, 60.
37. Rudolph, *The American College*, 250–258. Horowitz, *Campus Life*, 69–70, 115.
38. Geiger, *The American College*, 127–152.
39. Cited in Geiger, *The American College*, 87.
40. See Leslie, *Gentlemen and Scholars*, 111–113; Allmendinger, *Paupers and Scholars*, 103, 113, 117.
41. Joseph R. DeMartini, "Student culture as a change Agent in American Higher Education: An Illustration from the Nineteenth Century," *Journal of Social History* 9 (June 1976): 526–541; Horowitz, *Campus Life*, 36–39.
42. *Proceedings of the Thirteenth Annual Convention*, 137.
43. Shedd, *Two Centuries*, 105–107.
44. Weidensall, "The Early History," 14; Hopkins, *History of the YMCA*, 273–274.
45. *Proceedings of the Fifteenth Annual Convention of the Young Men's Christian Associations of the United States and British Provinces* (New York: Executive Committee, 1870), 64.
46. Weidensall, "The Early History," 62, 66.
47. Ibid., 30, 40–49, 94; C.K. Ober, *Exploring a Continent* (New York: Association Press, 1929), 113.
48. Weidensall, "The Early History," 28–185; Shedd, *Two Centuries*, 116–117.
49. Robert Weidensall, "College Work, 1877," 40. Robert Weidensall Papers, Box 1, Kautz Archives. By this time, chapters had been formed in 8 state universities, 16 colleges, 5 state normal schools, 5 denominational colleges, and 1 military institute.

50. John R. Mott, *The Intercollegiate Young Men's Christian Association Movement* (New York: YMCA, 1895), 2; Weidensall, "College Work, 1877," 38.
51. Luther Wishard, *The Beginning of the Student's Era in Christian History: A Reminiscence of a Life* (New York: Association Press, 1890), 11–12, 18–25; David B. Lowry, "Luther D. Wishard (1854–1925): Pioneer of the Student Christian Movement" (Department of History, Princeton University, 1951, photocopy), 5. Folder "Luther Wishard," Kautz Archives.
52. Marsden, *The Soul of the American University*, 197. For more on McCosh's ideals, see J. David Hoeveler, *James McCosh and the Scottish Intellectual Tradition* (Princeton: Princeton University Press, 1981).
53. Kemeny, *Princeton in the Nation's Service*, 57.
54. Wishard, "Day of Prayer for Colleges," *CB* 1, no. 3 (January 1879): 1. See also, "The Day of Prayer for Students," *Int* 10, no. 3 (January 1888): 21; Lowry, "Luther Wishard," 2–3; Kemeny, *Princeton in the Nation's Service*, 108.
55. Kemeny, *Princeton in the Nation's Service*, 107.
56. Wishard developed a life-long friendship with Woodrow Wilson at this time. See C.K. Ober, *Luther Wishard: Projector of World Movements* (New York: Association Press, 1927), 25. While affiliation of the Princeton chapter took place at this time, full membership was delayed until 1883 because of these issues.
57. Wishard, *The Beginning of the Student's Era*, 74–75.
58. "Minutes of the Philadelphian Society, 1875–1880 (May 19, 1877)." SCARP, Box 2, Folder 7; "Minutes of the Philadelphian Society, 1875–1880 (March 27, 1877)," SCARP, Box 2, Folder 7; Ober, *Luther Wishard*, 32–35.
59. From 1877 to 1887, this journal was called the *College Bulletin*. In 1887, it was enlarged and continued for four years under the name of *The Intercollegian*. The next two years it was edited by the International Committee as the "Intercollegian Department" of the *Young Men's Era*. Then the *Young Men's Era* became *Men* for five years and maintained a "College Department." In 1898, *The Intercollegian* was reinstated as the official organ.
60. Wishard, *The Beginning of the Students' Era*, 80–81; "The First Year of the College Work," *CB* 1, no. 3 (January 1879): 1–2. See also "Review of Last Year's Work," *CB* 3, no. 1 (October 1880): 2.
61. James B. Reynolds, Samuel Fisher, and Henry B. Wright, *Two Centuries of Christian Activity at Yale* (New York: G.P. Putnam's Sons, 1901), 106–107.
62. Earl H. Brill, "Religion and the Rise of the University: A Study of the Secularization of American Higher Education, 1870–1910" (Ph.D. diss., The American University, 1969), 557; Wishard, "College Visitation," *CB* 2, no. 5 (January 1880): 2.
63. Ober, *Luther Wishard*, 52; Wishard, "The Association in Harvard," *Int* 9, no. 1 (January 1887): 2; Reynolds, Fisher, and Wright, *Two Centuries of Christian Activity at Yale*, 107–108, 212–213, 222–223.
64. David B. Potts, *Wesleyan University: Collegiate Enterprise in New England* (New Haven: Yale University Press, 1992), 108.
65. Wishard, *The Beginning of the Student's Era*, 90.
66. Ibid., 19; "Membership of the College Association," *CB* 8, no. 5 (February 1886): 18–19. See also Wirt Wiley, *History of Y.M.C.A.—Church Relations in the United States* (New York: Association Press, 1944), 17–18; Richard C. Morse,

Relation to the Churches of the North American Young Men's Christian Associations (New York: The International Committee of Young Men's Christian Associations, 1906).

67. "Membership of the College Association," 19. See Wiley, *History of Y.M.C.A.—Church Relations*, 17–18. In 1869, 60% of chapters already possessed such a test, while 30% had a "good character" test. See Hopkins, *A History of the YMCA*, 363–364.
68. Wishard, "An Outline of the Work of the College Young Men's Christian Association (1885)," 10. SCARP, Box 12, Folder 5; "Membership of the College Association," 19.
69. James Thomas Honnold, "The History of the Y.M.C.A. of the University of Wisconsin, 1870–1924" (master's thesis, University of Wisconsin, 1954), 35.
70. Ibid., 38–44. On Bascom, see also J. David Hoeveler, "The University and the Social Gospel: The Intellectual Origins of the 'Wisconsin Idea,' " in Lester F. Goodchild and Harold S. Wechsler, eds. *The History of Higher Education*, 2nd ed. (Boston: Simon and Schuster, 1997), 234–246.
71. Honnold, "The History of the Y.M.C.A. of the University of Wisconsin," 46–47.
72. "Correspondence, March 24, 1888," *The Philadelphian* vol. 2 (May 1888): 454. SCARP, Box 13, Folder 4.
73. Weidensall, "College Work, 1877," 5. Robert Weidensall Papers, Box 1, Kautz Archives. Weidensall noted that he tried "several times" to place the University of Minnesota chapter on an evangelical basis but failed each time.
74. On the reasons for this growth, see Lynn D. Gordon, *Gender and Higher Education in the Progressive Era* (New Haven: Yale University Press, 1990), 16–26. See also Barbara Miller Solomon, *In the Company of Educated Women* (New Haven: Yale University Press, 1985), 64.
75. Shedd, *Two Centuries*, 190–191. Weidensall actually noted that he had tried to form a Young Women's Christian Association at the University of Wisconsin in 1871. The low number of women at the University made this an impossibility. Weidensall, "College Work, 1877," 4.
76. Wishard, *The Beginning of the Students' Era*, 159.
77. Cited in Harold W. Hannah, *One Hundred Years of Action: The University of Illinois YMCA, 1873–1973* (n.p., 1973), 18.
78. Wishard, *The Beginning of the Students' Era*, 159–160.
79. Hopkins, *History of the YMCA*, 293; Wishard, *The Beginning of the Students' Era*, 160–161.
80. Shedd, *Two Centuries*, 202.
81. Ibid., 209–210.
82. Floyd Loveland, "Secretary's Report," *The Bulletin of the Cornell University Christian Association* (October 1895): 33–34. AYSD, Box 53, Folder 766; Clarence P. Shedd, *The Church Follows Its Students* (New Haven: Yale University Press, 1938), 183.
83. C. Grey Austin, *A Century of Religion at the University of Michigan* (Ann Arbor, MI: University of Michigan, 1957), 4–5; William Walker, "The Students' Christian Association from 1883 To 1889," 3. Office of Ethics and Religion Records, 1860–1991, University of Michigan, Box 1, Folder "Historical Materials"; "Shall We Separate?" *The Monthly Bulletin of the University of Michigan* 7, no. 17 (April 1896): 108.

84. President Bascom lent a theological rationale, noting that "In Christ there is neither male nor female and to restore the division is to expel the master." Honnold, "The History of the Y.M.C.A. of the University of Wisconsin," 50.
85. Ober, *Luther Wishard*, 71.
86. Wishard, "The Intercollegiate Work," *CB* 2, no. 2 (October 1879): 2.
87. See Wishard, "First Five Years' Work of the College Secretary," *CB* 5, no. 1 (October 1882): 4; "First Decade of the Intercollegiate Work," *Int* 10, no. 1 (September 1887): 5.
88. Having been a student member of the Williams YMCA chapter, Ober went on to serve as the general YMCA secretary for the state of Massachusetts before joining Wishard in the national college department. Ober, *Luther Wishard*, 55; Wishard, "First Decade," 5.
89. Wishard, *The Beginning of the Students' Era*, 90.
90. C.M. Davies "California," *CB* 1, no. 5 (March 1879): 4; Wishard, "Pacific Coast Tour," *Int* 9, no. 3 (May 1887): 18.
91. Wishard, "The Outlook," *CB* 7, no. 4 (February 1885): 1.
92. Shedd, *Two Centuries*, 109.
93. YMCA *Yearbook*, 1882–1883, 32. See also Nina Mjagkij, *Light in the Darkness: African-Americans and the YMCA, 1852–1946* (Lexington: The University Press of Kentucky, 1994), 33–34.
94. Hopkins, *History of the YMCA*, 216.
95. Weidensall, "The Early History," 78–79. See also Weidensall, "The College Work, 1877," 17.
96. The white Student Department was located administratively in the Home Division of the American YMCA.
97. Cited in Mott, "The College Man's Religion at the Beginning of the Twentieth Century," *The Sunday School Times* 43, no. 3 (January 19, 1901): 1–2. Located in JRMP, Box 137, Folder 2213.

Chapter Two Saving Collegians to Save the World

1. Wishard, "College YMCA Constitution," *CB* 3, no. 6 (March 1881): 2. See also "The Organization of the College YMCA," *CB* 2, no. 5 (January 1880): 1–2.
2. Wishard, "Continual Service," *CB* 4, no. 4 (February 1882): 1.
3. Hopkins, *History of the YMCA*, 283.
4. Wishard, "The College Problem," *CB* 1, no. 6 (April 1879): 2. See also "The Great Fact in the Religious Life of Our Colleges Today," 2. JRMP, Box 143, Folder 2365.
5. Wishard, "Day of Prayer for Students," *CB* 3, no. 4 (January 1881): 1.
6. Wishard, "The Work of the Past Year," *CB* 4, no. 6 (April 1882): 1.
7. College Session of the International Convention of Young Men's Christian Associations, "College Department of the Young Men's Christian Associations: The Opportunities in College Life For Making Religious Impressions upon Young Men" 1. AWSCF, Box 77, Folder 624.
8. Wishard, "Now or Never," *CB* 1, no. 6 (April 1879): 4. See also "Day of Prayer for Colleges," *CB* 1, no. 3 (January 1879): 1; "College Revivals," *CB* 5, no. 2 (November 1882): 3.

9. College Session of the International Convention of Young Men's Christian Associations, "College Department," 4. See also Wishard, "Day of Prayer for Students," 1.
10. Wishard, "The College Problem," 2. Among Christian students, Wishard suggested, less than one-third had identified with any Christian organization.
11. Wishard, "New Converts," *CB* 6, no. 5 (February 1884): 1.
12. C.K. Ober, "Advantages of Union with the International Organization," *CB* 4, no. 3 (January 1882): 2.
13. Ibid.
14. Wishard, "Why You Should Have a College Association," *CB* 2, no. 5 (November 1880): 3. See also Wishard, "An Outline of the Work of the College Young Men's Christian Association," 5–6.
15. Wishard, "The Intercollegiate Young Men's Christian Association Movement," *CB* 1, no. 1 (November 1878): 2. See also Ober, *Exploring a Continent*, 70.
16. Morgan, *Student Religion during Fifty Years*, 35–36. See also "Correspondence," *The Philadelphian* (December 1887), 158–160. SCARP, Box 13, Folder 4; *Proceedings of the Twenty-Fourth Annual Convention of the Young Men's Christian Association* (New York: Executive Committee, 1879), 44; Wishard, "What Then?" *CB* 1, no. 6 (April 1879): 3.
17. "Correspondence, September 29, 1887," *The Philadelphian* (November 1887): 78–79. SCARP, Box 13, Folder 4. See also "An Historical Sketch of the Rutgers College Y.M.C.A," 7. Records of the Young Men's Christian Association of Rutgers College, Record Group 48/14/01, Special Collections and University Archives, Rutgers College.
18. Wishard, "The Outlook," *CB* 7, no. 4 (February 1885): 1; Ober, *Luther Wishard*, 43; Wishard, "Much Land Yet to be Possessed," *Int* 10, no. 2 (November 1887): 13.
19. Wishard, *CB* 6, no. 1 (October 1883): 2.
20. Wishard, *CB* 7, no. 1 (November 1884): 2; Wishard, "First Decade of the Intercollegiate Work," *Int* 10, no. 1 (September 1887): 5; "Much Land Yet to be Possessed," 13.
21. *The Dartmouth* 14 (February 10, 1893): 139; *Bowdoin Orient* 28 (February 8, 1899): 193. Cited in Brill, "Religion and the Rise of the University, 555–556.
22. Wishard, "Day of Prayer for Students," *CB* 3, no. 4 (January 1881): 1. See also Shedd, *Two Centuries*, 165; Wishard, "Day of Prayer for Colleges," *CB* 2, no. 5 (January 1880): 1.
23. Wishard, "Day of Prayer for Colleges," *CB* 2, no. 5 (January 1880): 1; "Rhode Island," *CB* 2, no. 5 (January 1879): 2; "Virginia," *CB* 2, no. 5 (February 1879): 4.
24. Wishard, "Methods of Bible Study," *CB* 3, no. 3 (December 1880): 2; "Bible Study," *CB* 7, no. 4 (February 1885): 2; "The College Association Bible Class," *CB* 2, no. 4 (December 1879): 1.
25. Wishard, "Bible Study at Amherst," *CB* 5, no. 3 (December 1882): 3.
26. Wishard, "Bible Study," 3; "The College Association Bible Class," 2; "Report of the Fifth Annual Conference of the Young Men's Christian Associations of New England Colleges at Yale University, February 18–20, 1887," 20–26. Student Work, Kautz Archives; Morgan, *Student Religion during Fifty Years*, 85.
27. Wishard, "The Bible Training Class," *CB* 8, no. 1 (October 1885): 2.

28. Wishard, "Bible Study," 3; Morgan, *Student Religion during Fifty Years*, 14–16.
29. Wishard, "The Association Room," *CB* 3, no. 5 (February 1881): 3.
30. Minutes of Philadelphian Society Meeting, January 19, 1878, 141. SCARP, Box 2, Folder 7. See also *Int* 10, no. 1 (September 1887): 7.
31. Hannah, *One Hundred Years of Action*, 59–60.
32. *APJRM*, 18. See Wishard, "Day and Week of Prayer for Young Men," *CB* 6, no. 1 (October 1883): 3.
33. George Adam Smith, *The Life of Henry Drummond* (New York: Doubleday and McClure, 1898), 140–145, 152.
34. "Deputation Work at Yale," *Int* 10, no. 3 (January 1888): 19.
35. "Scotch University Men in the American Colleges," *Int* 10, no. 1 (September 1887): 6–7; George Taylor, "The Knox College Movement," *Int* 12, no. 6 (March 1890): 91. Amherst followed a similar pattern. "Correspondence, January 16, 1888," *The Philadelphian* (January 1888): 220. SCARP, Box 13, Folder 4.
36. "College Items," *CB* 5, no. 2 (November 1882): 4.
37. Minutes of Philadelphian Society Meeting, October 21, 1876, 109–111. SCARP, Box 2, Folder 7.
38. Wishard, *The Beginning of the Students' Era*, 110. See also Ober, *Luther Wishard*, 44.
39. T.J. Shanks, ed. *A College of Colleges* (Chicago: Fleming Revell, 1887), 12–14; Ober, *Luther Wishard*, 44.
40. Shanks, ed. *A College of Colleges*, 12–13; Wishard, "A Word Concerning Missions," *CB* 1, no. 5 (March 1879): 4.
41. Wishard, "The College Missionary Meeting," *CB* 3, no. 2 (November 1880): 2. See also "Missionary Revival," *CB* 3, no. 2 (November 1880): 1.
42. Wishard, "A Visitor from England," *CB* 8, no. 2 (November 1885): 3.
43. Wishard, "How to Promote Missionary Spirit in College," *CB* 1, no. 5 (March 1879): 2.
44. Wishard, "The College Missionary Meeting," 2; "Development of Missionary Interest," *CB* 6, no. 2 (November 1883): 1.
45. See Wishard, "Missionary Meeting Topics," *CB* 3, no. 5 (February 1881): 2; "The Monthly Missionary Meeting," *CB* 2, no. 6 (February 1880): 2.
46. Wishard, "Work among Chicago Medical Students," *CB* 3, no. 6 (March 1881): 3. See also "Medical Students' Missionary Conference," *CB* 5, no. 5 (February 1883): 1; "Medical Students' Reception in Chicago," *CB* 5, no. 3 (December 1882): 3.
47. Wishard, "Inter-Collegiate Correspondence," *CB* 2, no. 7 (March 1880): 2.
48. Ibid., 2; Wishard, "Inter-Collegiate Work," *CB* 6, no. 3 (December 1883): 2–3.
49. Wishard, "Inter-Collegiate Work," 3; "Amherst College," *CB* 4, no. 6 (April 1882): 4.
50. In the first year it was issued, nearly 2,000 students purchased tickets. Ober, "The College Vacation Ticket," *CB* 5, no. 3 (December 1882): 1; "The College Vacation Ticket," *CB* 6, no. 7 (April 1884): 4.
51. Wishard, "Third College Conference," *CB* 4, no. 1 (November 1881): 1; "Fourth International Conference of Students," *CB* 5, no. 7 (April 1883): 3.
52. "Report of the Fifth Annual Conference of the YMCAs of New England Colleges," 10.

53. Shedd, *Two Centuries*, 181–182.
54. Ober, *Luther Wishard*, 63.
55. Wishard, *The Beginning of the Students' Era*, 99.
56. Ibid., 103–104.
57. Mott, "The Influence of Dwight L. Moody on the Student Movement," *Int* 22, no. 4 (January 1900): 87–88; Wishard, "Work in Oxford University," *CB* 5, no. 4 (January 1883): 1–2; "Mr. Moody at Harvard," *Int* 9, no. 1 (January 1887): 5; John R. Mott, "The Greatness of Moody," *Association Men* 25, no. 5 (February 1900): 151.
58. Wishard, *The Beginning of the Student's Era*, 146. Moody had perhaps four years of formal schooling. See Lyle W. Dorsett, *A Passion for Souls: The Life of D.L. Moody* (Chicago: Moody Press, 1997), 35.
59. "College Students' Summer School for Bible Study," *CB* (April 1886): 1; Ober, *Luther Wishard*, 89.
60. "College Young Men's Christian Association Encampment," *Int* 9, no. 2 (March 1887): 11; "College Students' Summer School," *Int* 9, no. 1 (January 1887): 7; Shanks, *A College of Colleges*, 215–216.
61. "Delegations," *Philadelphian Bulletin* 1, no. 4 (February 1892): 1; "College Students' Encampment," *Int* 9, no. 3 (May 1887): 20–21; "Tents at Northfield," *Int* 10, no. 5 (May 1888): 34. See also James F. Findlay, *Dwight L. Moody: American Evangelist, 1837–1899* (Chicago: University of Chicago Press, 1969), 352.
62. Shanks, ed. *College Students at Northfield, or A College of Colleges, no. 2* (New York: Revell, 1888).
63. Shedd, *Two Centuries*, 250; Findlay, *Dwight L. Moody*, 351–352; Shanks, *College Students at Northfield*, 76–81. This fight to maintain orthodoxy was complicated by Moody's own eclectic theology and tendency to work together with ministers across a broad range of theological perspectives. Drummond was criticized, for example, for his views regarding the Bible and for his attempted reconciliation of biblical truth and evolutionary theory in *Natural Law in the Spiritual World* (1883). See Wishard, *The Beginning of the Student's Era*, 155.
64. T.J. Shanks, *College Students at Northfield*, 92–93.
65. *Addresses and Papers of John R. Mott*, 10.
66. Timothy C. Wallstrom, *The Creation of a Student Movement to Evangelize the World* (Pasadena: William Carey International University Press, 1980), 33–36; Robert P. Wilder, *The Great Commission: The Missionary Response of the Student Volunteer Movements in North American and Europe* (London: Oliphants, Ltd., n.d.), 13–18; Dwayne G. Ramsey, "College Evangelists and Foreign Missions: The Student Volunteer Movement, 1886–1920" (Ph.D. diss., University of California, 1988), 5–7.
67. This quote, often attributed to John R. Mott, was first used by Wilder and the Princeton Foreign Missionary Society. See Wallstrom, *The Creation of a Student Movement*, 35.
68. Wilder, *The Great Commission*, 18.
69. Ramsey, "College Evangelists and Foreign Missions," 12.
70. Wishard, *The Beginning of the Students' Era*, 123.

71. Shedd, *Two Centuries*, 269–272. See also Wallstrom, *The Creation of a Student Movement*, 50–51.
72. Hopkins, *History of the YMCA*, 304. See also Wishard, "Students in Conventions," *Int* 10, no. 4 (March 1888): 28.
73. C. Howard Hopkins, *John R. Mott, 1865–1955: A Biography* (Grand Rapids, MI: Eerdmans, 1979), 304.
74. The leaders included John R. Mott (YMCA), Nettie Dunn (YWCA), and Robert Wilder (ISMA).
75. Suzanne de Dietrich, *Fifty Years of History: The World Student Christian Federation* (Geneva: World's Student Christian Federation, 1995), 17.
76. Funded by philanthropists John Wanamaker and J.V. Farwell, Wishard spent time in England, France, Germany, and Japan promoting student work and recruiting foreign student delegates for Northfield. Wishard went on to serve in a number of missionary organizations until finally, in 1904, he turned his attention to business, serving as president of Wishard Securities Company. See "Luther Deloraine Wishard" in *Fifty Years of Princeton '77: A Fifty-Four Year Record of the Class of 1877 of Princeton College and University* (Princeton: Princeton University Press, 1927), 143–144. See also Hopkins, *History of the YMCA*, 328–330.
77. Weidensall, "The Early History of the College Young Men's Christian Association Work," 14–15.
78. Long, *The Revival of 1857–1858*, 124–125.
79. As J. David Hoeveler has maintained, the revivalist fervor in antebellum colleges was often accompanied by reformist zeal. Hoeveler, "The University and the Social Gospel: The Intellectual Origins of the 'Wisconsin Idea,' " in Goodchild and Wechsler, eds. *The History of Higher Education*, 2nd ed., 235; Mark A. Noll, *The Scandal of the Evangelical Mind* (Grand Rapids: Eerdmans, 1955), 64.
80. Honnold, "The History of the Y.M.C.A. of the University of Wisconsin, 10.
81. "Twenty-Fifth Anniversary of the Michigan University Association," *CB* 5, no. 5 (February 1883): 4.

Chapter Three Expanding the Boundaries

1. Hopkins, *John R. Mott, 1865–1955*, 10–12.
2. Ibid., 17.
3. Basil Mathews, *John R. Mott* (London: Student Christian Movement Press, 1934), 53.
4. Ibid., 48. See also Hopkins, *John R. Mott*, 23; de Dietrich, *Fifty Years of History*, 17; Wishard, *The Beginning of the Student's Era*, 103.
5. *APJRM*, 11.
6. *The Association Bulletin* 3, no. 3 (December 1887): 3. See also W.S. Sloan, *A White Cross Question* (New York: YMCA, 1885).
7. Hopkins, *John R. Mott*, 41. See also *APJRM*, 7; Ober, *Exploring a Continent*, 22–24; Morris Bishop, *A History of Cornell* (Ithaca, NY: Cornell University Press, 1962), 269; *APJRM*, 12–13.
8. Mathews, *John R. Mott*, 48; de Dietrich, *Fifty Years of History*, 17; Hopkins, *John R. Mott*, 45.

9. Ober, *Exploring a Continent*, 77.
10. Ibid., 80.
11. Fred L. Norton, ed. *A College of Colleges* (Chicago: Fleming Revell, 1889), 9.
12. *APJRM*, 115, 156. See also "An After-Dinner Speech of John R. Mott," 1892. JRMP, Box 143, Folder 2364.
13. *APJRM*, 77.
14. Veysey, *The Emergence of the American University*, 263–341. See also Geiger and Bobulz, "The Crisis of the Old Order," in Geiger, ed. *The American College in the Nineteenth Century*, 264–276.
15. Richard Hofstadter and C. DeWitt Hardy, *The Development and Scope of Higher Education in the United States* (New York: Columbia University Press, 1952), 31. Growth came from a variety of sources, including not only traditional undergraduates, but also graduate students, professional students, and those attending summer sessions. See Roger L. Geiger, *To Advance Knowledge: The Growth of American Research Universities, 1900–1940* (New York: Oxford University Press, 1986), 15.
16. Burton J. Bledstein, *The Culture of Professionalism: The Middle Class and the Development of Higher Education in America* (New York: W.W. Norton and Company, Inc., 1976), 90, 98. In the 1870s and 1880s, many students were coming from academies and collegiate preparatory departments, but by the 1890s the colleges were receiving increasing numbers from the public high schools. See Leslie, *Gentlemen and Scholars*, 111, 205; Geiger, *To Advance Knowledge*, 266–267.
17. Horowitz, *Campus Life*, 50.
18. See Robert H. Wiebe, *The Search for Order, 1877–1920* (New York: Hill and Wang, 1967), 112. While B.A. students expanded by 64% between 1890 and 1904, other undergraduate degrees reflecting more technical and practical pursuits rose by an estimated 304%. See Geiger, *To Advance Knowledge*, 13–14.
19. Herbert Lyman Clark, "The Trouble at College," *Int* 23, no. 8 (May 1901): 176–177.
20. Robert E. Speer to John R. Mott, April 24, 1902, 2. JRMP, Box 84, Folder 1509.
21. L.H. Miller, "The Relation of the Association to the Faculty," *Int* 28, no. 8 (May 1906): 182–183.
22. Hannah, *One Hundred Years of Action*, 14.
23. Geiger, *To Advance Knowledge*, 11.
24. J.D. Dadisman, "The Association as a Factor in the Solution of the Moral Problems of the Institution," in *College Leadership* (1909), 84–85. AYSD, Box 68, Folder 925.
25. Marsden, *The Soul of the American University*, 107. See also Mott, "It Is Time to Reap," *Int* 26, no. 4 (January 1904): 77.
26. Roberts and Turner, *The Sacred and the Secular University*, 88.
27. Francis Patton, "The Contribution of the Association to the Moral and Religious Life of Universities and Colleges," in *The Jubilee of Work for Young Men in North America* (New York: The International Committee of Young Men's Christian Associations, 1901), 127–128.
28. Robert E. Speer to John R. Mott, April 24, 1902, 2. JRMP, Box 84, Folder 1509.

29. Mott, *The Students of North America United,* 15.
30. At Princeton, the faculty membership fee was three dollars in 1898. See "Circular to Faculty on Faculty Membership" (1898), 52. SCARP, Box 3, Folder 6.
31. D.G. Hart, *The University Gets Religion* (Baltimore: The Johns Hopkins University Press, 1999), 21.
32. Reuben, *The Making of the Modern University,* 29–30, 57–60, 66, 72, 119–124. See also Marsden, *The Soul of the American University,* 113–121.
33. Marsden, *The Soul of the American University,* 156–159; Roberts and Turner, *The Sacred and the Secular University,* 28.
34. Required courses in natural theology were expunged from many college catalogs, and many of the new prestigious universities initiated in these years failed to offer any course with this title. Moral philosophy also passed away in this era, often replaced by specialized courses in the new social sciences. See Robert and Turner, *The Sacred and the Secular University,* 29–31, 107–122; David Hollinger, "Inquiry and Uplift: Late Nineteenth-Century American Academics and the Moral Efficacy of Scientific Practice," in Thomas L. Haskell, ed. *The Authority of Experts: Studies in History and Theory* (Bloomington: Indiana University Press, 1984), 147.
35. Marsden, *The Soul of the American University,* 152–153. See also Longfield, "From Evangelicalism to Liberalism," 49.
36. Kemeny, *Princeton in the Nation's Service,* 17–22; Albert Perry Brigham, *Present Status of the Elective System in American Colleges* (New York, n.p., 1897), 361.
37. "Chaplain's Report," 201; Presidents' Papers, 1889–1925, Box 71, Folder 3, University of Chicago Archives; Reuben, *The Making of the Modern University,* 119–124; Longfield, "From Evangelicalism to Liberalism," 49–53; Winton U. Solberg, "The Conflict between Religion and Secularism at the University of Illinois, 1867–1894," *American Quarterly* 18 (1966): 183–199; James B. Angell, "Religious Life in Our State Universities," *Andover Review* 13 (April 1890): 365–372. Oberlin was a typical illustration. In 1892, students were permitted to substitute attendance at YMCA meetings for Sunday evening services. Students were then relieved of requirements to attend a second Sunday service (1898), morning prayers (1901), and any church service (1906). John Barnard, *From Evangelicalism to Progressivism at Oberlin College, 1866–1917* (Columbus: Ohio State University Press, 1969), 103.
38. Reuben, *The Making of the Modern University,* 122.
39. David Starr Jordan, *The Call of the Twentieth Century: An Address to Young Men* (Boston: Beacon Press, 1903), 72–73.
40. Mott to A.A. Stagg, October 20, 1892, 2. Presidents' Papers, Box 71, Folder 3, University of Chicago Archives.
41. Ibid., 3–4.
42. "The Organization of Religious Work at the University in the Year of 1892–3," Presidents' Papers, 1889–1925, Box 55, Folder 2, University of Chicago Archives.
43. Mott to Harper, October 4, 1892, 2–3. Presidents' Papers, Box 71, Folder 4, University of Chicago Archives.
44. Mott to Stagg, October 20, 1892, 10. Presidents' Papers, Box 71, Folder 3, University of Chicago Archives.

45. Mott to Stagg, November 28, 1892, 2. Presidents' Papers, Box 71, Folder 3, University of Chicago Archives.
46. "The Organization of Religious Work at the University in the Year of 1892–3," 2. See also J. Lawrence Laughlin to Frederick T. Gates, December 18, 1892, 1. Presidents' Papers, Box 55, Folder 2, University of Chicago Archives.
47. See especially Charles Henderson, "Religious Work in the University," in *The President's Report*, 1892–1902, 371–386; 1898–1899, 143–147. University of Chicago Archives; John M. Coulter to H.P. Judson, February 1, 1912, 1–2. Presidents' Papers, 1889–1925, Box 71, Folder 4, University of Chicago Archives.
48. "Chapel Assembly," *The University of Chicago Weekly* (1901): 742. University Presidents' Papers, University of Chicago Archives.
49. Charles Henderson to William Rainey Harper, August 12, 1896, 1. Presidents' Papers, 1889–1925, Box 38, Folder 10, University of Chicago Archives.
50. Henderson to Harper, May 18, 1896, 1. Presidents' Papers, Box 38, Folder 10, University of Chicago Archives.
51. Ibid., 2.
52. Ibid.
53. Reuben, *The Making of the Modern University*, 127.
54. Ibid., 131.
55. Mott, *The Students of North America United*, 15. If only active members are considered, approximately 18% of male students in 1901 were members of the YMCA. See also Henry D. Sheldon, *Student Life and Customs* (New York: D. Appleton and Company, 1901), 274.
56. *APJRM*, 132–133.
57. Mott, "Religious Conditions of the Colleges of the Pacific Coast," in *APJRM*, 134. See also Mott, "Remarkable Conference in the Webfoot State," *Young Men's Era* (April 14, 1892): 464.
58. Ibid., 128.
59. "The Summer Conferences," *Int* 12, no. 4 (January 1890): 54–55; "Again 'tis Northfield," *Int* 11, no. 4 (March 1889): 5–6; *APJRM*, 53.
60. Shedd, *Two Centuries*, 324. See also Luther Wishard, "Foreign Missionary College Work," *CB* 6, no. 7 (April 1884): 1.
61. See Harrison Elliott, *Theological Students and the Student Movement* (New York: Association Press, 1913), 11–12.
62. *APJRM*, 149–153. See also Geiger, *To Advance Knowledge*, 13–15.
63. See *APJRM*, 147–154.
64. Ibid., 204–208.
65. Ibid. See also Elliott, *Theological Students and the Student Movement*, 3–9; William F. McDowell, *Dangers that Beset Theological Students* (New York: International Committee of Young Men's Christian Associations, 1903).
66. See Elliott, *Theological Students and the Student Movement*, 15.
67. Moorland graduated as valedictorian of Howard University, studying in the Theological Department while serving as secretary of the African American Association in Washington, D.C. Owen Pence, *The YMCA and Social Need: A Study of Institutional Adaptation* (New York: Association Press, 1939), 59.
68. Hunton to Mott, n.d. Kautz Archives. See also Pence, *The YMCA and Social Need*, 59.

69. W.A. Hunton, "The Association Movement among the Colored Colleges," *Int* 21, no. 5 (February 1899): 110–111.
70. In 1896, 41 of 60 black associations were found in colleges. See Hunton, "Association Work among Colored Young Men," *Int* 27, no. 2 (November 1904): 34–36.
71. *APJRM*, 114; Mathews, *John R. Mott*, 103; Pence, *The YMCA and Social Need*, 60.
72. *APJRM*, 35, 38, 52.
73. Brockman served as southern student secretary between 1892 and 1898 after graduating from Vanderbilt University. He assumed leadership of the SVM during Mott's extensive travels and later served as a missionary to China and general secretary of the National Committee of Chinese YMCAs from 1901 to 1915. Beaver, a graduate of Penn State in 1890 and son of the governor of Pennsylvania, became an assistant state college secretary in Pennsylvania before serving as a national secretary for medical, professional, and metropolitan institutions. Hopkins, *John R. Mott*, 50, 101, 528–529; H.M. Philpott, "A History of the Student Y.M.C.A." (Ph.D. diss., Yale University, 1947), 22; *APJRM*, 37.
74. Mott, "The American Inter-Collegiate Young Men's Christian Association." JRMP, Box 145, Folder 2407.
75. See, for example, Angell, "Religious Life in Our State Universities," 365–372; Glenn C. Altschuler, *Andrew D. White—Educator, Historian, Diplomat* (Ithaca, NY: Cornell University Press, 1962), especially 200–216; W.R. Harper, *The Trend in Higher Education* (Chicago: University of Chicago Press, 1905), 373.
76. Marsden, *The Soul of the American University*, 177. See also Patton and Field, *Eight O'Clock Chapel*, 222.
77. Mott, *The Students of North America United*, 19.
78. Mott, "The College Man's Religion at the Beginning of the Twentieth Century."
79. Ibid.
80. Robert E. Speer to Mott, April 24, 1902, 2. JRMP, Box 84, Folder 1509.

Chapter Four Building a Campus Presence

1. Henry S. Canby, *Alma Mater: The Gothic Age of the American College* (New York: Farrar & Rinehart, 1936), 24. See also Leslie, *Gentlemen and Scholars*, 111, 205.
2. Woodrow Wilson, "What Is College For?" in Arthur Link ed., *The Papers of Woodrow Wilson*, vol. 19 (Princeton: Princeton University Press, 1975), 344–346.
3. Ralph Henry Gabriel, *Religion and Learning at Yale* (New Haven: Yale University Press, 1958), 209.
4. Owen Johnson, *Stover at Yale* (New York: Collier Books, 1912), 148. See also Joseph Kett, *Rites of Passage: Adolescence in America, 1790 to the Present* (New York, Basic Books, 1977), 175–176.
5. At the University of Delaware, association leaders continually found themselves battling to convince students that the YMCA chapter was governed by Delaware men for the prosperity of the campus rather than the YMCA national headquarters. J.H. Mitchell, "The Beginning of the Y.M.C.A. at Delaware—Reminiscences," *The Alumni News* 2, no. 2 (April 1916): 7–9.

6. *The Bulletin of the Cornell University Christian Association* (October 1895): 31.
7. H.H. Horne, "The Place of the Student Young Men's Christian Association," *Int* 25, no. 9 (June 1903): 195. Some of those "representative students" criticized the YMCA for its pious posturing and exacting religious requirements. A student at Knox College spoke of YMCA members as a "pious little enclave" that marred Friday night entertainment with its "devotional sessions." See Hermann R. Muelder, *Missionaries and Muckrakers: The First Hundred Years of Knox College* (Urbana: University of Illinois Press, 1984), 227.
8. Ibid.
9. "Whitmore Lake Conference, 1913," 64. Office of Ethics and Religion SCA Records, Box 5, Folder "Cabinet Meetings 1911–1915," Bentley Historical Library, University of Michigan.
10. Honnold, "The History of the Y.M.C.A. of the University of Wisconsin, 126.
11. A. Bruce Minear, "Maintaining Bible Class Attendance," *Int* 27, no. 3 (December 1904): 62–63.
12. Bruce Barton, "Out of the 'Y' and in again," *The Outlook* 79 (1904): 257. See also Muelder, *Missionaries and Muckrackers*, 244.
13. Ibid. See also Thomas Le Duc, *Piety and Intellect at Amherst College, 1865–1912* (New York: Arno Press, 1969), 142.
14. Frank V. Slack, *The Interesting and Enlisting of Fraternity Men in the Work of the Student Young Men's Christian Association* (New York: Young Men's Christian Association Press, 1908), 5.
15. Ibid., 8–9; Conference of Secretaries of the Student Department of the International Committee, September 7–9, 1905, 3. AYSD, Box 41, Folder 603.
16. Ibid., 6–7, 10, 17. See also H.H. Horne, "The Place of the Student Young Men's Christian Association," 195.
17. Such a plan was undertaken in the late nineteenth century by the Massachusetts Institute of Technology, Northwestern University, Purdue University, University of Illinois, University of Wisconsin, University of Michigan, University of Montana, and University of Kansas.
18. John R. Mott, *College Association Buildings* (New York: College Series no. 302, 1891), 5. See also Neil McMillan, Jr., "America's Experience in Student Young Men's Christian Association Buildings," 6; Pence, *The YMCA and Social Need.*
19. Mott, *College Association Buildings*, 13. JRMP, Box 143, Folder 2362. See also Harry F. Comer, "The Equipment and Operation of a Student Young Men's Christian Association Building." Folder "Student Secretaries," Student Work, Kautz Archives.
20. Mott, *College Association Buildings*, 11, 13. See also W.W. Berry, "Religious Life in the University of Tennessee," *Int* 24, no. 9 (June 1902): 209.
21. Executive Committee of the University of Oregon, "Prospectus of the Proposed Building for the Young Men's and Young Women's Christian Associations of the University of Oregon." AWSCF, Box 76, Folder 620; *APJRM*, 109; Report of the President of the YMCA of Rutgers College, for year ending March 13, 1912, 10. Records of the Young Men's Christian Association of Rutgers College, Record Group 48/14/01, Special Collections and University Archives, Rutgers College. See also James Edward Gilson, "Changing Student Lifestyle at the University of Iowa, 1880–1900" (Ph.D. diss., University of Iowa, 1980), 87–88; Mott, *College Association Buildings*, 15.

22. Bates College, "Annual Report to the President 1912–1913," 89. Muskie Archives and Special Collections Library, Bates College; Robert C. McMath, Jr., *Engineering the New South: Georgia Tech, 1885–1985* (Athens, GA: The University of Georgia Press, 1985), 111; Mott, *College Association Buildings*, 15.
23. Gilson, "Changing Student Lifestyle at the University of Iowa," 87–88; Advisory Committee of the Young Men's Christian Association of Iowa State College, "Prospectus," 7.
24. Norton, *A College of Colleges*, 10.
25. Mott, *College Association Buildings*, 9. See also McMillan, "America's Experience," 5. University building expansion in these years was minimal. According to the U.S. Department of Education, the total value of campus buildings went from $297,153 in 1910 to $495,920 in 1920. By 1930, however, the value of such buildings stood at $1,490,014.
26. "Murray Hall, the Princeton College Association Building," *Int* 9, no. 2 (March 1887): 1.
27. Both Dodge Hall and Earl Hall were donated in memory of William Dodge's son, Earl, who died at the age of 25. "Dodge Hall at Princeton," *Int* 23, no. 3 (December 1900): 65; "Dedication of Earl Hall," *Int* 24, no. 7 (April 1902): 172–173; "Dedication of Madison Hall, University of Virginia," *Int* 28, no. 1 (October 1905): 73–74; "The Cornell Christian Association," *Cornell Alumni News* (January 24, 1900): 2. JRMP, Box 150, Folder 2499; "Union College Association Building," *Int* 24, no. 1 (October 1901): 15–16; F.B. Rankin, "The New Association Building at the University of North Carolina," *Int* 29, no. 8 (May 1907): 195; McMath, Jr., *Engineering the New South*, 110; Austin, A *Century of Religion at the University of Michigan*, 28–29; Graduate Committee of the Yale Young Men's Christian Association, *Dwight Hall, Yale University, Its Origin, Erection, and Dedication* (New Haven: Tuttle, Morehouse & Taylor, 1887), 7–21.
28. Eugene Levering to the Trustees of the Johns Hopkins University, May 1, 1889. AWSCF, Box 76, Folder 619.
29. Charles Brewster, *A History of the Dartmouth Christian Association* (Ann Arbor, MI: Edwards Brothers, 1927), 12–13; Gilson, "Changing Student Lifestyle at the University of Iowa," 86–88; Patricia Sleezer, "The Religious Struggle at the State University of Iowa." Yale Divinity School, 1942, seminar paper. University of Iowa Archives. See also Harold Reinhart, "The History and Activities of the Iowa State College Young Men's Christian Association, 1890–1947 (n.d.)," 1. Iowa State University Archives, YMCA Records, Record Group 22/10/00/03.
30. John R. Mott, *How to Secure a College Association Building* (New York: The International Committee of the Young Men's Christian Association, 1892).
31. "The Virginia Polytechnic Institute Building," *Int* 24, no. 4 (January 1902): 85–86; "A Students' Movement." AWSCF, Box 76, Folder 618; Peterson, *The New England College*, 85; "A Band of Men," *Int* 9, no. 3 (May 1887): 23.
32. Rankin, "The New Association Building at the University of North Carolina," 195; Allen, "A History of the Young Men's Christian Association at the University of Virginia," 214; Hugh McIlhany, "Dedication of Madison Hall, University of Virginia," *Int* 28, no. 1 (October 1905): 74. Luxurious by campus standards, Barnes Hall was valued at $55,000, Dwight Hall at $60,000,

Murray-Dodge Hall at $100,000, and Earl Hall at $175,000. Competition clearly reigned in this arena, and schools were often driven to exceed the elegance of buildings at peer institutions. For typical examples, see "Student Association Building Campaigns," *Int* 27, no. 3 (December 1904): 54–55; "Progress in Building Movements," *Int* 27, no. 2 (November 1904): 36–38.

33. Honnold, "The History of the Y.M.C.A. of the University of Wisconsin," 97–101; See also Mayer N. Zald, *Organizational Change: The Political Economy of the YMCA* (Chicago: University of Chicago Press, 1970), 32.
34. W.D. Weatherford, "Shall the College Associations Have Their Own Buildings in Our Colleges?" 53. Folder "Student Secretaries," Student Work, Kautz Archives.
35. Rankin, "The New Association Building at the University of North Carolina," 196.
36. Graduate Committee of the Yale Young Men's Christian Association, *Dwight Hall*, 13; "Report of C.H. Lee, President of the Cornell University Christian Association," *The Association Bulletin* 4, no. 7 (April 1889): 86–87.
37. Pence, *The YMCA and Social Need*, 130; Zald, *Organizational Change*, 34. According to Zald, the number of secretaries nationally escalated from 178 in 1880 to 5,076 in 1919.
38. *The YMCA Yearbook*, 1887, p. xli.
39. For information on other local advisory boards, see, for example, Bates College, "Annual Report to the President 1911–1912," 82. Muskie Archives and Special Collections Library, Bates College; John H. Safford, "The Advisory Committee of the College Association," *Int* 29, no. 9 (June 1907): 222–223. Many "advisory boards" became "boards of directors" when buildings were secured.
40. The University of Toronto also had an early secretary. See Shedd, *Two Centuries*, 168–169. Schools often began by hiring part-time staff, securing full-time positions only when the position's usefulness had been verified. Charles D. Hurrey, *The General Secretary of a Student Young Men's Christian Association* (New York: Young Men's Christian Association Press, 1908), 3.
41. Allen, "A History of the Young Men's Christian Association at the University of Virginia," 182.
42. Honnold, "The History of the Y.M.C.A. of the University of Wisconsin," 88.
43. Hurrey, *The General Secretary*, 4–7. See also A.S. Johnstone, "The Spiritual Life of the Secretary," in *College Problems: A Study of the Religious Activities of College Men* (Nashville: Publishing House of the Methodist Episcopal Church, South, 1907), 64.
44. H.L. Heinzman, "The General Secretary in Relation to Student Life and Affairs." AYSD, Box 68, Folder 925.
45. W.D. Weatherford, *Student Secretaries in Training* (New York: Young Men's Christian Association Press, 1910), 9.
46. George Stewart, *Life of Henry B. Wright* (New York: Association Press, 1925), 14–16.
47. Reynolds, Fisher, and Wright, *Two Centuries of Christian Activity at Yale*, 236.
48. Weatherford, *Student Secretaries in Training*, 10–11. See also Heinzman, "The General Secretary," 169–170.
49. As late as 1910, among eighty-three secretaries surveyed, only a "very few" had been in the work longer than three years and twenty-six noted that they

intended to remain in the post no more than two or three years. Hurrey, *The General Secretary*, 16.

50. Milton C. Towner, ed. *Religion in Higher Education* (Chicago: The University of Chicago Press, 1931), 299–301.
51. *Int* 12 no. 1 (October 1889): 2.
52. Arthur Reed Campbell, "Yale's Unadvertised Side," Private Book of the General Secretary, 1897–1900, 9. SCARP, Box 3, Folder 6; E.O. Jacobs, "Work for the Entering Class," *Int* 31, no. 9 (June 1909): 225–226; J.C. Prall, "Work for New Students the Opportunity of the Year," *Int* 21, no. 8 (May 1898): 179–180; "State University," *CB* 6, no. 7 (April 1884): 4; Allen, "A History of the Young Men's Christian Association at the University of Virginia," 152, 175.
53. A.F. Jackson, "Reaching New Students—Some Practical Methods," *Int* 25, no. 9 (June 1903): 207–208. See also "Work for the Entering Class at Yale," *Int* 21, no. 1 (October 1898): 18–19.
54. James L. McConaughy to various employers, October 16, 1912. Bowdoin College Archives; "Rutgers College YMCA Cabinet Meeting, Feb. 4, 1918," 2. Records of the Young Men's Christian Association of Rutgers College, Record Group 48/14/01, Special Collections and University Archives, Rutgers College; Annual Report, Young Men's Christian Association University of Southern California, 1908–1909, 2. University Archives, University of Southern California.
55. *Int* 11, no. 2 (November 1888): 1. See also "Northwestern University YMCA Annual Report—1898," 6. Northwestern University Archives; "Report of the Committee on Employment for Students, 1910–1911," Student Work, Kautz Archives. "Old Journalism Kennel Now Open," *The Montana Kaimin* 12, no. 11 (November 24, 1914): 2.
56. Hannah, *One Hundred Years of Action*, 26–27.
57. "The Burden for the Oriental Students in Our American Colleges," in "Notes of the Fall Conference of the Secretaries of the Student Department of the International Committee, 1909." AYSD, Box 41, Folder 605; Committee on Friendly Relations among Foreign Students, *The Unofficial Ambassadors*, 3; Program of International Friendship Committee, University of Illinois YMCA, 1930–1931. University of Illinois YMCA Archives, Record Series Number 41/69/322, Box 12; Committee on Friendly Relations among Foreign Students, "Friendly Relations among Future Leaders," 1920. CPSP, Box 40, Folder 393.
58. L.T. Savage, "Religious Life in a Typical State University of the West," *Int* 22, no. 5 (February 1900): 105.
59. Robert Eskline, "Changing Student Religion at the State University of Iowa: 1880–1920" (Ph.D. diss., University of Iowa, 1976), 187.
60. "Response to Survey on the Place of the Student YMCA in the University: Oklahoma and Colorado," Student Work, Kautz Archives.
61. "The Contribution of the Association to the Moral and Religious Life of the Colleges," 136. In *The Jubilee Convention* of *1927*, Student Work, Kautz Archives.
62. Kemeny, *Princeton in the Nation's Service*, 163; Gabriel, *Religion and Learning at Yale*, 209; Reynolds, Fisher, and Wright, *Two Centuries of Christian Activity at Yale*, 123.

63. Henry Seidel Canby, *College Sons and College Fathers* (New York: Harper & Brothers, 1915), 13. See also H.A. Hunt, "Y.M.C.A. Department," *William and Mary College Monthly* 9, no. 6 (April 1900): 271.
64. Lewis Sheldon Welch and Walter Camp, *Yale, Her Campus, Class-Rooms, and Athletics* (Boston: L.C. Page and Company, 1899), 55.
65. Ibid., 59–61. See also Sheldon, *Student Life and Customs*, 280.
66. Edwin Starbuck, "How Shall We Deepen the Spiritual Life of the College," *Religious Education* 4 (April 1909): 84–85.
67. W.W. Dillon, "A Spiritual Awakening in Every College," *Int* 25, no. 4 (January 1903): 85–86. See also Heinzman, "The General Secretary," 169–170; *APJRM*, 26.

Chapter Five A Muscular Faith

1. Sinclair Lewis, *Elmer Gantry* (New York: Hardcourt, Brace, and Company, 1927), 36–37.
2. Ibid.
3. Mark Schorer, *Sinclair Lewis: An American Life* (New York: McGraw Hill, 1961), 74.
4. Clifford Putney, *Muscular Christianity: Manhood and Sports in Protestant America, 1880–1920* (Cambridge: Harvard University Press, 2001).
5. Gail Bederman, " 'The Women Have Had Charge of the Church Work Long Enough': The Men and Religion Forward Movement of 1911–1912 and the Masculinization of Middle-Class Protestantism," *American Quarterly* 41, no. 3 (September 1989): 432.
6. Bird Baldwin, "The Boy of High School Age: The Moral and Religious Development of the Adolescent Boy," *Religious Education* 8 (1913): 23. See also David Macleod, *Building Character in the American Boy: The Boy Scouts, The YMCA, and Their Forerunners* (Madison: University of Wisconsin Press, 1981), 43.
7. Richard Hughes, "The Churches and the Religious Problem in State Universities," *Religious Education* 10, no. 2 (April 1915): 184. See also William Rainey Harper, *The Trend in Higher Education* (Chicago: University of Chicago Press, 1905), 197.
8. Josiah Strong, *The Times and Young Men* (New York: Baker & Taylor, 1901), 179–180.
9. Lawrence J. Dennis, *From Prayer to Pragmatism: A Biography of John L. Childs* (Carbondale, IL: Southern Illinois Press, 1992), 20.
10. John R. Mott, *The Future Leadership of the Church* (New York: Student Department Young Men's Christian Association, 1908), 85–86. See also T.W. Graham, "The Importance of Recruiting Strong Men for the Ministry," in *Report of the Lake Forest Summer School 1911* (New York: Association Press, 1911), 131–135. AYSD, Box 68, Folder 925.
11. The books in the series, all published in New York (1909) by the Student Young Men's Christian Association, included George Angier Gordon, *The Claims of the Ministry upon Strong Men*; Edward Increase Bosworth, *The Modern Interpretation of the Call to the Ministry*; William Fraser McDowell, *The Right*

Sort of Men for the Ministry; Phillips Brooks, *The Minister and His People*; Woodrow Wilson, *The Minister and His Community*; Arthur Steven Hoyt, *The Call of the Country Church*; Walter William Moore, *The Preparation of the Modern Minister*; Bosworth, *The Weak Church and the Strong Man*; Charles Edward Jefferson, *The Minister as Preacher.*
12. Gordon, *The Claims of the Ministry upon Strong Men*, 12–13.
13. Bosworth, *The Modern Interpretation of the Call to the Ministry*, 11. See also William F. McDowell, "The Claims of the Ministry upon College Students," *Int* 26, no. 8 (May 1904): 173.
14. T.J. Jackson Lears, *No Place of Grace: Antimodernism and the Transformation of American Culture, 1880–1920* (New York: Pantheon Books, 1981), 26–32. Female school teachers constituted 59% of all teachers in 1870, but that number had escalated to 86% by 1920. See G. Stanley Hall, "Feminization in School and Home," *World's Work* 16 (1908): 10238.
15. A number of authors have recently addressed the crisis of masculinity in terms of economic shifts. See Margaret L. Bendroth, *Fundamentalism and Gender, 1875 to the Present* (New Haven: Yale University Press, 1993), 17; Bederman, "The Women Have Had Charge of the Church Work Long Enough," 432–465; E. Anthony Rotundo, *American Manhood: Transformations in Masculinity from the Revolution to the Modern Era* (New York: BasicBooks, 1993), 248.
16. Bederman, "The Women Have Had Charge of the Church World Long Enough," 435–438.
17. For popular books, see, for example, Carl Case, *The Masculine in Religion* (Philadelphia: American Baptist Publication Society, 1906); Jason Pierce, *The Masculine Power of Christ* (Boston: Pilgrim Press, 1912); Jasper Massee, *Men and the Kingdom* (New York: Fleming Revell, 1912). The Men and Religion Forward Movement was also a direct result of this enthusiasm, promising "more men for religion and more religion for men." See Putney, *Muscular Christianity*, 99–126; L. Dean Allen, *Rise up, O Men of God: The Men and Religion Forward Movement and the Promise Keepers* (Atlanta: Mercer University Press, 2002).
18. Kett, *Rites of Passage*, 173. On Roosevelt, see John Higham, *Writing American History: Essays in Modern Scholarship* (Bloomington: Indiana University Press, 1970), 78–88.
19. Henry Churchill King, *How to Make a Rational Fight for Character* (New York: Young Men's Christian Association Press, 1911), 5–6.
20. Ibid., 14–15.
21. Ibid.
22. Ibid., 8–9, 16–17.
23. Ibid., 22, 27.
24. Young Men's Christian Association of Oberlin College, Annual Report 1908–1909, 5. Oberlin College Archives, Record Group 29, YMCA/YWCA, Folder 1902/03–1915/16.
25. Hurrey, *The General Secretary of a Student Young Men's Christian Association*, 9.
26. "Topics for Religious Meetings," *Int* 25, no. 6 (March 1903): 130–131.
27. Clay Shepard to Parents, November 25, 1900. Edgar Raymond Shepard Letters, 1900–1915 (SR 6/6/6/12), Oregon State University Archives.

28. Annual Report, Young Men's Christian Association University of Southern California, 1908–1909, 2–3. University Archives, University of Southern California.
29. Peter I. Berg, "A Mission on the Midway: Amos Alonzo Stagg and the Gospel of Football" (Ph.D. diss., Michigan State University, 1996), 24–79; Erin A. McCarthy, "Making Men: The Life and Career of Amos Alonzo Stagg, 1862–1933" (Ph.D. diss., Loyola University, 1994), 56–76, 86–96; Amos Alonzo Stagg, *Touchdown!* (New York: Longmans, Green and Co., 1927), 130–131.
30. See "Topics for Religious Meetings," *Int* 25, no. 6 (March 1903): 130–131; "Topics for Religious Meetings," *Int* 26, no. 6 (March 1904): 140–141; "2nd Ledger, May 6, 1876–May 5, 1926: Minutes," 457. Records of the Young Men's Christian Association of Rutgers College, Record Group 48/14/01, Special Collections and University Archives, Rutgers College. Young Men's Christian Association of Oberlin College, Annual Report, Oberlin College Archives, Record Group 29, YMCA/YWCA, Folder 1906/1907, 4–5.
31. It was often more difficult to secure regular attendance than to secure "enrollment." Bible study secretary Neil McMillan estimated that only about 30% of students continued more than one year, statistics revealing that larger schools had more difficulty achieving commitment. "Striking Achievements and Plans of Student Bible Study," *Int* 30, no. 9 (June 1908): 210.
32. Mott to the Presidents of the College Associations of North America, October 1, 1897. JRMP, Box 144, Folder 2369. See also Bible Study Department, "Some Facts Relating to Bible Study among Students," *Int* 26, no. 9 (June 1904): 210; Cooper, "Campaign for Fifty Thousand College Men in Bible Classes," *Int* 30, no. 1 (October 1907): 14–15.
33. L. Wilbur Messer, *Christ as a Personal Worker: Topics and Methods for Workers' Bible Training Classes* (New York: YMCA International Committee, 1891); H. Clay Trumbull, *Individual Work for Individuals: A Record of Personal Experiences and Convictions* (New York: YMCA International Committee, 1901); Mott, "Bible Study," 3; "The Bible Class," *Int* 11, no. 3 (January 1889): 8.
34. "Report of the Fifth Annual Conference," 30. Student Work, Kautz Archives.
35. Robert E. Speer, *Studies in the Gospel of Luke* (New York: International Committee of YMCA, 1892); *Studies in the Book of Acts* (New York: International Committee of YMCA, 1892); Wilbert Webster White, *Thirty Studies in Jeremiah* (New York: International Committee of YMCA, 1896); *Inductive Studies in the Twelve Minor Prophets* (Chicago: Fleming H. Revell, 1892); Thomas Wakefield Goodspeed, *William Rainey Harper* (Chicago: University of Chicago Press, 1928), 75; "Inductive Bible Studies on the Books of Samuel," *Int* 11, no. 5 (May 1889): 11–12; Morgan, *Student Religion during Fifty Years* , 44, 53.
36. John R. Mott, *The Morning Watch* (New York: International Committee of Young Men's Christian Association, 1898), 5. JRMP, Box 137, Folder 2208. See also Mott, *The Bible Study Department*, 8.
37. Mott, *The Bible Study Department*, 16–23.
38. Harper, "What Can Universities and Colleges Do for the Religious Life of Their Students?" in *The Aim of Religious Education: The Proceedings of the Third Annual Convention of the Religious Education Association* (Chicago, 1905), 37–38; "President Harper on Bible Study," *Int* 27, no. 8 (May 1905): 182–183.
39. Ibid. See also Harper, *The Trend in Higher Education*, 373.
40. Mott, *The Bible Study Department*, 17–18.

41. Ibid., 10; Ernest Burton, "Betterments in Bible Study in the Colleges," *The Association Monthly* (October 1907): 9–11, 16.
42. O.E. Brown, "Relation of Association Bible Study to Curriculum Work," *Int* 32, no. 3 (December 1909): 59–63.
43. Mott, "Bible Study," 129; *APJRM*, 95; Reuben, *The Making of the Modern University*, 109–112; Marsden, *The Soul of the American University*, 205–215.
44. Herbert W. Gates, "What We Have a Right to Expect from Bible Study and How the Objective May Be Attained," in *College Leadership* (1909), 57–61. AYSD, Box 68, Folder 925.
45. Austin Evans to Neil McMillan, February 8, 1911. AYSD, Box 56, Folder 806. See also Clayton S. Cooper, *Things Which Make a Bible Class Effective* (New York: Young Men's Christian Association Press, 1908), 1.
46. Clayton S. Cooper, "The Bible Awakening among North American College Men," 9. AWSCF, Box 65, Folder 530.
47. Clayton S. Cooper, *College Men and the Bible* (New York: Association Press, 1911), 130–131; *The Bible and Modern Life* (New York: Funk and Wagnalls Co., 1911), 139–143. See also "The International Student Bible Conference, Columbus, Ohio, October 22–25, 1908," 1–2. AWSCF, Box 65, Folder 530.
48. Based upon Stevens and Burton's *A Harmony of the Gospels*, these materials, published in 1896 as *Studies in the Life of Christ*, sold 8,500 copies in their first year. n.a., *This One Thing: A Tribute to Henry Burton Sharman* (Toronto: Thistle Printing, 1959), 27–36. Other popular texts at this time included: Henry Burton Sharman, William Arnold Stevens, and Ernest De Witt Burton, *Studies in the Life of Christ* (1896); Edward Increase Bosworth and Ernest De Witt Burton, *Studies in the Acts and Epistles* (1898); Wilbert Webster White, *Studies in Old Testament Characters* (1900); Wilbert Webster White, *Old Testament Records, Poems, and Addresses* (1900); Edward Bosworth, *Studies in the Teachings of Jesus Christ and His Apostles* (1903); Edward Bosworth, *Studies in the Life of Jesus Christ* (1904), Robert A. Falconer, *The Truth of the Apostolic Gospel* (1904); William H. Sallmon, *Studies in the Life of Paul for Bible Classes and Private Use* (1896); *Studies in the Life of Jesus for Bible Classes and Private Use* (1897); *Studies in the Parables of Jesus Recorded by Matthew* (1906); *Studies in the Miracles of Jesus, Recorded by Matthew* (1903).
49. In 1902 alone, the association sold 14,302 volumes of the new Bible study materials. *APJRM*, 55, 104. H.B. Sharman, "A Four Years' Cycle of Bible Study: Its Plan and Advantages," *Int* 21, no. 1 (October 1898): 10–11.
50. Reynolds, Fisher, and Wright, *Two Centuries of Christian Activity at Yale* , 124; Gabriel, *Religion and Learning at Yale*, 197; William H. Sallmon, *How We Built up the Bible Study Department at Yale* (New York: The International Committee of Young Men's Christian Associations, 1893), 13. AWSCF, Box 66, Folder 532.
51. "The Man I Have Never Seen." AYSD, Box 69, Folder 933.
52. Cited in Susan Curtis, "The Son of Man and God the Father: The Social Gospel and Victorian Masculinity," in Mark Carnes and Clyde Griffen, eds. *Meanings for Manhood* (Chicago: University of Chicago Press, 1990), 72–75.
53. Harry Emerson Fosdick, *The Manhood of the Master* (New York: Association Press, 1913), 9, 13, 89, 96, 161.
54. Ibid., 173.

55. On the committee itself, see "The Denominations' Part in the College Voluntary Study Courses," 3. AYSD, Box 58, Folder 821. See also "College Voluntary Study Series: A Series of Graded Studies for Use in Non-Curriculum Classes in Sunday Schools and Student Christian Associations." AYSD, Box 58, Folder 821; Harrison Elliott, "Voluntary Bible Study: Its Place in the Religious Education of Students," *Religious Education* 7 (February 1913): 715–716.
56. Elliott, "Voluntary Bible Study," 713–714; Tentative Report of Commission to Make Suggestions on Texts Best Suited for Voluntary Bible Study in Educational Institutions (1912), 14–15. AYSD, Box 57, Folder 807.
57. "College Voluntary Study Series," 10; "A Standard Program of Voluntary Study," 3.
58. Gates, "What We Have a Right to Expect from Bible Study," 59.
59. "A Suggested Curriculum for Voluntary Study Groups in Colleges and Universities: Report of a Committee of the Council of North American Student Movements, 1912." AYSD, Box 58, Folder 821.
60. Minutes of Meeting of Committee to Outline Work to Be Covered in Voluntary Bible Study in Educational Institutions, November 30, 1912, AYSD, Box 57, Folder 807. See also the Council of North American Student Movements, Bible Study Prospectus: Bible Study Texts Recommended for 1913–1914 for Voluntary Bible Study in Colleges and Universities. AYSD, Box 58, Folder 821.
61. Harrison Elliott and Ethel Cutler, *Student Standards of Action* (New York: Association Press, 1914); "Student Standards of Action," *NAS* 3, no. 1 (October 1914): 34.
62. J. Lovell Murray and Frederick M. Harris, *Christian Standards of Life* (New York: National Board of the Young Women's Christian Associations, 1915).
63. Cooper, "The Bible Awakening among North American College Men," 4.
64. W.D. Weatherford, "Report to the Student Committee of the International Committee of the Young Men's Christian Associations for the Year Ending August 31, 1909," 1. AYSD, Box 41, Folder 601.
65. "Commission on Bible Study and Mission Study," 54. Report of the Lake Forest Summer School, 1910. AYSD, Box 68, Folder 925; Sallmon, *How We Built up the Bible Study Department at Yale*, 5. Increasingly, faculty and general secretaries were called upon to head up weekly "normal classes" designed for the training of student Bible study leaders and periodic "Bible study institutes" for more significant training. By 1908, there were about 140 regular normal classes operating and approximately 250 institutes held around the country. See "The Faculty Man and Student Bible Study," 192; *Int* 31, no. 2 (November 1908): 32–34; Clayton S. Cooper, "The Bible Institute," *Int* 27, no. 2 (November 1904): 38–39; Clayton S. Cooper, *The Training of Bible Teachers* (New York, Young Men's Christian Association Press, 1910), 5.
66. Such a move did not go unchallenged. The North American Council of the YMCA accused Elliott of minimizing the Bible's role by making it an "indirect" rather than a "direct" source of authority. See E.A. Corbett, "Letters to the Editor," *NAS* 3, no. 5 (February 1915): 224. For Elliott's response, see Elliott, "Action of North American Council—April 16 and April 21, 1914." AYSD, Box 57, Folder 808.
67. Frederick M. Harris, "Letters to the Editor," *NAS* 3, no. 6 (March 1915): 314.

68. See "The Washington Convention," *Int* 30, no. 3 (December 1907): 50–52. See also "Summary Discussion on the Basis, Northfield," September 8, 1912, 1. AYSD, Box 53, Folder 761; The Resolutions Adopted at the Washington Convention of 1907, 3. AYSD, Box 53, Folder 763.
69. "The Resolutions Adopted at the Washington Convention," 1, 2, 6; Pence, *The YMCA and Social Need*, 122.
70. "The Washington Convention," 51–52; Andersen, "The Practical Effects of the Washington Resolution," 3. AYSD, Box 53, Folder 763.
71. W. Richard Ohler to Mr. Moody, December 4, 1909, 1–2. AYSD, Box 53, Folder 763. For a similar case, see also "Shall We Separate?" 110.
72. Ibid. See also Gerald Ray Krick, "Harvard Volunteers: A History of Undergraduate Volunteer Social Service Work at Harvard" (Ph.D. diss., Boston University, 1970), 54–55; Harvard Christian Association, 1902–1903 Annual Report, 1. Harvard University Archives.
73. "Report of the Subcommittee Appointed by the Committee of Thirty-Three to Find Facts Concerning the Basis of Association Active Membership and Control," 9. AYSD, Box 53, Folder 762.
74. "Brown University," 2. AYSD, Box 53, Folder 765; Brill, "Religion and the Rise of the University," 558–561.
75. Brill, "Religion and the Rise of the University," 559–560. See also I.E. Brown, "The Evangelical Basis: Its Interpretation and Administration," in *College Leadership*, 17–21.
76. "Summary Discussion on the Basis, Northfield," 1.
77. Hugh A. Moran, "The Rhodes Scholarship and the Student Christian Movement," *Int* 30, no. 9 (June 1908): 201.
78. French also completed statues of Joseph Henry and Benjamin Franklin at Palmer Hall, John Harvard in Cambridge, Alma Mater at Columbia University, and Abraham Lincoln at the Lincoln Memorial in Washington, D.C. See Alexander Leitch, *A Princeton Companion* (Princeton University Press, 1978).
79. Rudolph, *The American College*, 372.
80. Putney, *Muscular Christianity*, 23; David O. Levine, *The American College and the Culture of Aspiration, 1915–1940* (Ithaca, NY: Cornell University Press, 1986), 76, 123–126; Burton J. Bledstein, *The Culture of Professionalism: The Middle Class and the Development of Higher Education in America* (New York: W.W. Norton and Company, Inc., 1976), 135, 254–255.
81. Hannah, *One Hundred Years of Action*, 23.

Chapter Six A Conservative Social Gospel

1. "Commission on Evangelism," 61–62. AYSD, Box 68, Folder 925.
2. E.T. Colton, "The Cardinal Student Sin," *Int* 24, no. 8 (May 1902): 184–185.
3. A commissioned army officer in the Civil War, Sayford took part in gospel temperance work before becoming involved in the YMCA. See S.M. Sayford, *Origin and Progress of Mr. Sayford's Work among Students* (Boston: Press of T.O. Metcalf & Co., 1891). AWSCF, Box 64, Folder 521.

4. S.M. Sayford, "In Training," *Int* 12, no. 4 (January 1890): 57.
5. Sayford, *Origin and Progress*, 36, 38–39; Sayford, "Overcoming Temptation," *Int* 23, no. 3 (December 1900): 54.
6. Hannah, *One Hundred Years of Action*, 13; "Y.M.C.A. Department," *William and Mary College Monthly* 8, no. 2 (December 1898): 72–73.
7. See Sayford, *Origin and Progress*, 15. See also *APJRM*, 73, 103; "Mr. S.M. Sayford," *Burlington Free Press and Times* (December 10, 1890). AWSCF, Box 64, Folder 521; "Mr. Sayford's Fall Campaign among the Colleges, 1897." AWSCF, Box 64, Folder 521.
8. Mott, *The Students of North America United*, 21.
9. "Special Address to Men Only," *Madison Hall Notes* 3, no. 7 (October 19, 1907): 1; "Christ on Sin—Which?" *Madison Hall Notes* 3, no. 8 (October 26, 1907): 2.
10. David Stass Jordan, "A Christian Man,"*Young Men's Era* 18, no. 26 (July 30, 1892): 819.
11. Mott, "Will-Power and Religion," 40, 44. JRMP, Box 138, Folder 2230.
12. H.B. Wright, "The Spiritual Awakening at Yale University," *Int* 22, no. 7 (April 1900): 155–156; "Revivals—What Hinders? What Helps?" *Int* 23, no. 4 (January 1901): 75–76.
13. Bederman, " 'The Women Have Had Charge of the Church Work Long Enough', 456.
14. Ohio State University *Makio*, 1898, 239. See also E.T. Colton, "Some Recent Spiritual Awakenings: Leland Stanford Junior University," *Int* 23, no. 6 (March 1901): 134–135; H.B. Wright, "Essentials in the Preparation for a Spiritual Awakening," *Int* 23, no. 4 (January 1901): 79.
15. Ferenc Szasz, "The Stress on Character and Service in Progressive America," *Mid-America: An Historical Review* 63, no. 3 (1981): 145–156.
16. Neil V. Salzman, *Reform and Revolution: The Life and Times of Raymond Robins* (Kent, Ohio: The Kent State University Press, 1991), 109. For other helpful works on Robins, see William Hard, *Raymond Robins' Own Story* (New York: Harper and Brothers, 1920); James A. Martin, "Raymond Robins and the Progressive Movement: The Study of a Progressive Reformer, 1900–1917" (Ph.D. diss., Tulane University, 1975).
17. In January and February of 1916, for example, Robins and Childs visited twelve colleges with a total attendance of 46,645. For more details on crusades, see Galen M. Fisher, *Public Affairs and the YMCA: 1844–1944* (New York: Association Press, 1948), 82–83; Progress in Southern Student Christian Life, 1915–1916, 1. AYSD, Box 64, Folder 896. Childs later served as a Midwest fieldworker for the Student YMCA.
18. On this theme, see Charles Sheldon, *In His Steps* (Philadelphia: H. Altemus, 1899).
19. Shailer Mathews, *The Social Teachings of Jesus* (New York: The Macmillan Company, 1917), 209, 224–225; Francis G. Peabody, *Jesus Christ and the Social Question* (New York: Grosset and Dunlap Publishers, 1900), 117–118; Peabody, *Jesus Christ and the Christian Character* (New York: The Macmillan Company, 1906), 197.
20. By 1912, 36,580 students, only half of whom were missionary candidates, were enrolled in YMCA mission study classes. See, for example, "Mission Study for

1903–1904," *Int* 26, no. 1 (October 1903): 17; J. Lovell Murray, "Mission Study Plans for 1907–1908," *Int* 30, no. 1 (October 1907): 15.

21. Harlan Beach, *Dawn on the Hills of T'ang*, rev. ed (New York: Young People's Missionary Movement, 1907), 141.
22. Robert E. Speer, *South American Problems* (New York: Student Volunteer Movement for Foreign Missions, 1912), especially 73–81. See also James S. Dennis, *Social Evils of the Non-Christian World* (New York: Student Volunteer Movement for Foreign Missions, 1897), especially 112–132.
23. Le Duc, *Piety and Intellect at Amherst College*, 143.
24. Mott, *The Future Leadership of the Church*, 43, 45.
25. Jeremiah W. Jenks, *Social Significance of the Teachings of Jesus* (New York: Association Press, 1911), 100–112.
26. Report of Immigration Conference of Student Christian Associations, Massachusetts and Rhode Island, Boston, October 20, 1911, 3, 5. AYSD, Box 62, Folder 862.
27. The copies sold were 30,000, and an estimated 50,000 students joined together in study groups to analyze the book prior to 1920. Jesse Moorland was a bit reticent to fully endorse the book since Weatherford did not consult the African American community in his work. See Mjagkij, *Light in the Darkness*, 103.
28. W.D. Weatherford, *Negro Life in the South* (New York: Association Press, 1911), 12.
29. Ibid., 15–18.
30. Ibid., 8–10.
31. Ibid., 6.
32. Ibid., 13–14.
33. Ibid., 4, 15.
34. Hopkins, *John R. Mott*, 419–420.
35. See also Booker T. Washington, "What the White Race Can Do for the Black," *NAS* 3, no. 8 (May 1915): 350–352.
36. W.A. Hunton, "The Association Movement among the Colored Colleges," *Int* 21, no. 5 (February 1899): 110–111.
37. Hunton, "Association Work among Colored Young Men," *Int* 27, no. 2 (November 1904): 34–36.
38. Ibid.
39. Martin Marty, *Righteous Empire: The Protestant Experience in America* (New York: The Dial Press, 1970).
40. Peabody, *Jesus Christ and the Social Question*, 117.
41. In some cases, historians have argued that service and urban reform were simply convenient tools to attract men back to the church. In fact, Bederman has suggested that muscular Christians "were inspired less by an awakened social conscience than by a search for manly church work to counter the churches' feminization." While she is correct in stating that muscular Christians saw service as a form of Christianity more conducive to manly ardor, service was inextricably intertwined with the larger YMCA goals of character building and moral reform. To argue that social service was simply a tool to masculinize Christianity is to place too much credence in the singular potency of gender

anxieties. See Bederman, "The Women Have Had Charge of the Church Work Long Enough," 455–456; Janet Forsythe Fishburn, *The Fatherhood of God and the Victorian Family: The Social Gospel in America* (Philadelphia: Fortress Press, 1981), 32; Curtis, "The Son of Man and God the Father," in Carnes and Griffen, eds. *Meanings for Manhood*, 67, 77; Putney, *Muscular Christianity*, 39–44, 73–98.

42. See Brubacher and Rudy, *Higher Education in Transition*, 166; Clayton Cooper, *Why Go to College* (New York: The Century Co., 1912), 58.
43. See Rudolph, *The American College and University*, 366–368.
44. Ibid., 356.
45. Ibid., 362.
46. These new impulses were recognizable in many domains of American Protestantism. The Federal Council of Churches, formed in 1908, assumed a strong bent in the direction of social work, creating a Commission on the Church and Social Service even before adopting a Commission on Evangelism. In addition, the Men and Religion Forward Movement (MRFM) of 1911–1912, led by YMCA evangelist Fred B. Smith, gave close attention to social service in domains of education, temperance, prison reform, sanitary inspections, playground construction, and race harmony. See Marty, *Righteous Empire*, 182; Gary Scott Smith, "The Men and Religion Forward Movement, 1911–1912," in Martin Marty, ed. *Protestantism and Social Christianity* (New York: K.G. Saur, 1992), 166–193.
47. See Porter in R.H. Edwards and A.M. Trawick, eds. *Clubs and Other Work with Boys* (New York: Association Press, 1914). AYSD, Box 68, Folder 928.
48. Campbell, "Yale's Unadvertised Side," 9.
49. "Deputation Committee," *The Association Record* 5 (July 1907): 25. See also Philip Alexander Bruce, *History of the University of Virginia, 1819–1919*, vol. 5 (New York: Macmillan, 1922), 248, 255.
50. "Student Evangelistic Deputation Work," c. 1912. Dartmouth Christian Association Records, Special Collections, Dartmouth College. Box 2.
51. "General Secretary's Book, 1897–1898," 9–22. SCARP, Box 12, Folder 10. See also Richard H. Edwards, "The Securing and Training of Social Workers." AYSD, Box 41, Folder 59; "City Mission Work at Yale," *Int* 24, no. 8 (May 1902): 193–194.
52. Porter, *Clubs and Other Work with Boys*, 17.
53. "City Mission Work at Yale," *Int* 24, no. 8 (May 1902): 193–194; Thomas Perry, "Philanthropic Work at Harvard," *Int* 24, no. 8 (May 1902): 192; W.W. Berry, "Religious Life in the University of Tennessee," *Int* 24, no. 9 (June 1902): 209–210. See also Burtchaell, *The Dying of the Light*, 42; David B. Potts, *Wesleyan University: Collegiate Enterprise in New England* (New Haven: Yale University Press, 1992), 203.
54. Fred H. Rindge, *Educational Classes and other Service with Workingmen* (New York: Association Press, 1915), 4.
55. Ibid. Many used Peter Roberts's 1909 text *English for Coming Americans* (New York: Association Press, 1912).

56. Ibid., 23; "Commission on Service," Report of the Lake Forest Summer School, 34. AYSD, Box 68, Folder 925.
57. Rindge, "Interesting Students in Industrial Problems and Practical Service." AWSCF, Box 74, Folder 600.
58. Rindge, "The Industrial Service Movement," *Int* 38, no. 9 (June 1919): 5–6.
59. Oliver F. Cutts, "The Alumni Work," *Report of the Lake Forest Summer School 1911* (New York: Association Press, 1911), 136–146. AYSD, Box 68, Folder 925. See also letter from Oliver Cutts to Secretary, Stanford University, February 4, 1914. Department of Special Collections and University Archives, Stanford University.
60. Rindge, *Educational Classes*; Rindge, "The Industrial Service Movement," 5–6; "The College Man's Opportunity," 4.
61. J. Bruce Byall, "The Christian Settlement and Summer Camps of the University of Pennsylvania Association," *Int* 26, no. 9 (June 1904): 205–207; "The University of Pennsylvania Christian Settlement and Summer Camps." AWSCF, Box 77, Folder 629.
62. William K. Selden, *The Princeton Summer Camp, 1908–1975* (Princeton: Princeton Education Center, 1988), 24.
63. Ibid., 22.
64. Clayton Moore, "Religious and Philanthropic Work of the Associations," *Int* 13 (1892): 42–46. See also Perry, "Philanthropic Work at Harvard," 192.
65. Richard H. Edwards, "The Securing and Training of Social Workers," 1. AYSD, Box 41, Folder 59.
66. Ibid., 6. The YMCA also desired to see alumni taking part in social service and organized committees to facilitate this through the Russell Sage Foundation. See Department of Special Collections and University Archives, Stanford University. "Progress of Religious Work among the Students of the University of Michigan." Office of Ethics and Religion SCA Records, Box 6, Folder "SCA Miscellaneous (3)," Bentley Historical Library, University of Michigan.
67. Francis Peabody, "The Religion of a College Student," *The Forum* 31 (1901): 451. On this theme, see also Shailer Mathews, "The National Significance of the Religious Life of State Universities," in Charles J. Galpin and Richard H. Edwards, eds. *Church Work in State Universities, 1909–1910, Report of the Third Annual Conference of Church Workers in State Universities* (Madison, WI: n.p., May 1910), 15.
68. Cope, "Newer Ideals of Religious Education in the Universities," 18, 21.
69. Charles F. Thwing, "Some Changes in the Emphases in College Life," *Int* 21, no. 9 (June 1899): 194.
70. Charles W. Gilkey, "The Devotional Meeting—What Shall We Do with It?" *Int* 27, no. 6 (March 1905): 137. See also J.H. Kirkland, "Requisites for Success in Christian Work in College," *Int* 21, no. 8 (May 1899): 175.
71. Reuben, *The Making of the Modern University*, 57–60; Marsden, *The Soul of the American University*, 119.
72. Burtchaell, *The Dying of the Light*, 824; Marsden, *The Soul of the American University*, 117.
73. Cooper, *Why Go to College*, 130.

74. SCA, *Religious Thought at the University of Michigan* (Ann Arbor, MI: The Inland Press, 1893), 3–14, 63, 65, 68.
75. Ibid., 3.
76. The YMCA president at Harvard in 1903 noted that, in its philanthropic work, the organization could "bring into cooperation" members of the Catholic Club, the St. Paul's Society, and even "that broader class of men belonging to no organization, who yet wish to join in work that furthers social brotherhood." See Phillips Brooks House 1902–1903 Report, 2. Harvard University Archives.
77. "Report From the Department of Philosophy," 6. Christian Association Records, University of Pennsylvania Archives, Box 1, Folder 9; William O. Milton, "Settlement Work at the University of Pennsylvania," *Int* 24, no. 8 (May 1902): 192–193.
78. Rindge, *Educational Classes*, 55.
79. Ibid., 36.
80. Ibid., 69.
81. Shailer Mathews, "Social Service Not a Substitute for Religion," *NAS* 3, no. 5 (February 1915): 204–208.
82. William Smith Pettit, "The Function of a College Association," *Int* 28, no. 6 (March 1906): 148.

Chapter Seven A Shrinking Sphere

1. Frederick P. Keppel, *The Undergraduate and His College* (New York: Houghton Mifflin Company, 1917), 191.
2. George E. Peterson, *The New England College in the Age of the University* (Amherst: Amherst College Press, 1964), 85.
3. See "The Man Who Surprised Harvard and Captured English University Men" and "Porter, David R.," Kautz Archives.
4. Levine, *The American College and the Culture of Aspiration*, 27.
5. Honnold, "The History of the Y.M.C.A. of the University of Wisconsin, 168.
6. Levine, *The American College*, 26–32.
7. Ibid., 29. See also Charles F. Thwing, *The American Colleges and Universities in the Great War, 1914–1919* (New York: The Macmillan Company, 1920), 55–84. Complying with federal requirements, colleges and universities received approximately nine hundred dollars to cover the cost of each individual enrolled.
8. Charles Brewster, *A History of the Dartmouth Christian Association* (Ann Arbor, MI: Edwards Brothers, 1927), 17; William Howard Taft, ed. *Service with Fighting Men: An Account of the Work of the American Young Men's Christian Associations in the World War*, vol. 1 (New York: Association Press, 1922), 403–405; A.J. Elliott, "The Association's Opportunity in the S.A.T.C.," *Int* 36, no. 3 (December 1918): 5–6. On the nature of service, see "The Service of the 'Y' with the Student Army." University of Illinois Archives, Record Series Number 41/69/322, Box 11; "What the S.A.T.C. is Like," *Int* 36, no. 3 (December 1918): 7–8.

9. "The Place and Function of Christian Associations in the Present University Situation," 1929, 7. Student Work, Kautz Archives.
10. See Paul Super, *What Is the YMCA?* (New York: Association Press, 1922), 24; YMCA *Yearbook* (New York: Association Press, 1934), 13. See also York Lucci, Report of a Study of the Student Work of the YMCA, 13. Columbia University Bureau of Applied Social Research, 1950. AWSCF, Box 78, Folder 634.
11. After slight gains between 1915 and 1920, the losses multiplied for the next twenty years. Sixty-nine college YMCAs were actually founded and terminated between 1920 and 1940. Among states, only Ohio had losses less than 25% in these two decades. Most regions lost between 25 and 45% of their chapters. See "What of the Future of Student YMCA's?" Report of the Commission on Student Work to the National Council of the Young Men's Christian Associations, October 1941, 15–17. AYSD, Box 43, Folder 631.
12. Paula Fass, *The Damned and the Beautiful: American Youth in the 1920s* (Oxford: Oxford University Press, 1977), 58–61, 81–85.
13. Keppel, *The Undergraduate and His College*, 68–69; Veysey, *The Emergence of the American University*, 266; Levine, *The American College*, 49–51.
14. Levine, *The American College*, 24.
15. Ibid., 39. The totals in 1930 do not include the 277 junior colleges that were now recorded in national tallies. On growing religious diversity, see Marcia Graham Synnott, *The Half-Opened Door: Discrimination and Admissions at Harvard, Yale, and Princeton, 1900–1970* (Westport, CO: Greenwood Press, 1979), 130–133; Edward Boyer, "Religious Education in Colleges, Universities, and Schools of Religion," *Christian Education* 11 (October 1927): 27; Dan A. Oren, *Joining the Club: A History of Jews at Yale* (New Haven: Yale University Press, 1985).
16. McMillan, Jr., "America's Experience," 8.
17. "Report of Sub-Committee #4 to the Meeting of the Advance Program Commission, September 11, 1924," 1. AYSD, Box 40, Folder 577.
18. "The Place and Function of Christian Associations in the Present University Situation," 7.
19. On this theme, see also Marsden, *The Soul of the American University*, 340.
20. Brubacher and Rudy, *Higher Education in Transition*, 330.
21. Ibid., 332.
22. Reuben, *The Making of the Modern University*, 255; William DeVane, *Higher Education in Twentieth Century America* (Cambridge: Harvard University Press, 1965), 58–60.
23. Scott J. Peters, *The Promise of Association: A History of the Mission and Work of the YMCA of the University of Illinois, 1873–1997* (Champaign, IL: University YMCA, 1997), 34.
24. "Abuse of Madison Hall," *Madison Hall Notes* 3, no. 12 (November 23, 1907): 2; Ira Baker, "Local History of the Y.M.C.A." CPSP, Box 77, Folder 715.
25. McMillan, Jr., "America's Experience," 10.
26. Scott J. Peters, *The Promise of Association*, 33–34.
27. Local commercial amusement enterprises—often constructed in more elaborate fashion than the association buildings—also diminished the need for such facilities on the campus.

28. Honnold, "The History of the Y.M.C.A. of the University of Wisconsin," 105, 123, 184. See also Robert Cooley Angell, *The Campus: A Study of Contemporary Undergraduate Life in the American University* (New York: D. Appleton and Company, 1928), 163–164; Prof. Ira I. Baker, "Prof. Baker Gives History of Y in 52 Years' Growth," *The Y's Indian* (February 20, 1925): 3–4.
29. See, for example, "Union Denies Activity in YMCA Row," *Michigan Daily* 22, no. 56 (December 7, 1911): 1. See also "YMCA Must Go or Give up Present Field," *Michigan Daily* 22, no. 57 (December 8, 1911): 1, 3; "Think Y.M.C.A. Needs Radical Overhauling," *Michigan Daily* 22, no. 54 (December 5, 1911): 1, 4.
30. McMillan, Jr., "America's Experience," 10; "University of Washington's New Building," *Int* 40, no. 9 (June 1923): 25; "General Secretary's Report to the Chairman and Members of the Board of Directors, Young Men's Christian Association at the University of Illinois" (March 14, 1928), 14. University of Illinois Archives, Record Series Number 41/69/322, Box 5.
31. Brubacher and Rudy, *Higher Education in Transition*, 331–336.
32. Statistics in the postwar era revealed that from one-half to two-thirds of college men and from one-fourth to one-third of college women were working. While the best estimation is that only 15% of all students were financially self-supporting during the 1920s, the number of students seeking odd jobs continued to climb. Fass, *The Damned and the Beautiful*, 134–135; *Report of the Faculty-Student Committee on the Distribution of Students' Time, January 1925* (Chicago: University of Chicago Press, 1925), 21.
33. See Kenneth S. Latourette, "Preparing Him for His College," *Int* 38, no. 1 (October 1920): 14; Elmore McKee, "Putting Him to Work," *Int* 38, no. 1 (October 1920): 14–15; "Why a Leakage?" *Int* 42, no. 3 (December 1924): 24.
34. Latourette, "Preparing Him for His College," 14; Young Men's Christian Associations, *American Students and Christian Living: A Year in the Life of a Movement, 1929–1930* (New York: The Student Division, 1931), 23.
35. *American Students and Christian Living*, 20–23; Latourette, "Preparing Him for His College," 14.
36. "The University of Michigan." CPSP, Box 77, Folder 720, 38.
37. Thwing, "The Effect of the War on Religion in College," *Religious Education* 13, no. 4 (August 1918): 271.
38. "Report of the Provisional Student Division Committee, June 24, 1928," 5. Student Work, Kautz Archives.
39. The Council of Christian Associations, *Education Adequate for Modern Times* (New York: Association Press, 1931), 195.
40. Porter, *The Church in the Universities*, 53.
41. Only an estimated one in twenty-five students at public universities enrolled in such courses. See Marsden, *The Soul of the American University*, 335–337. See also Hart, *The University Gets Religion*.
42. The Student Christian Association at the University of Michigan, "Religious Forces at Michigan," 2. Office of Ethics and Religion SCA Records, Box 6, Folder "SCA Miscellaneous (3)," Bentley Historical Library; Memorandum to President Burton on Social and Religious Welfare of the Student Body, February 18, 1924, 4–5. University President's Papers, 1889–1925, Box 46,

Folder 28, University of Chicago Archives. On this movement, see also Hart, *The University Gets Religion*, 67–90.

43. Council of Christian Associations, "The Place and Function of Christian Associations in the Present University Situation," 7.
44. "Report of the Provisional Student Division Committee," 5.
45. Committee of Student Secretaries, Central Region, *A Study of the Present Position of the Student Young Men's Christian Association in Relation to Higher Education* (New York: Association Press, 1925), 13.
46. Austin, *A Century of Religion at the University of Michigan*, 36–39, 46.
47. Report to President Hibben of Special Committee Appointed to Study Activities and Scope of the Philadelphian Society, December 31, 1926, 5, 11–12. SCARP, Box 14, Folder 6; Kemeny, *Princeton in the Nation's Service*, 214–219.
48. *Christian Work in State Universities* (n.p., 1918), 45; David R. Porter, ed. *The Church in the Universities* (New York: Association Press, 1925), 23.
49. "Report of the Committee of Six," *Religious Education* 1 (February 1907): 212. While not as much a source of direct competition, the Jewish Hillel Foundation also initiated chapters in the 1920s.
50. Joseph W. Cochran, "Conditions and Plans for the Religious Welfare of Students in Universities," *Religious Education* 5, no. 2 (June 1910): 116–117.
51. Richard C. Hughes, "The Church and the College Student," *Religious Education* 7, no. 4 (October 1912): 395.
52. Joseph W. Cochran, "Preparation for Leadership," *Religious Education* 5 (1911): 123.
53. Wallace Stearns, "Moral and Religious Training in the Universities and Colleges in the United States," *Religious Education* 2 (1908): 207–208.
54. "Student Associations and the Church," *Int* 41, no. 3 (December 1923): 28; Summary of the Cleveland Conference, Cleveland, Ohio, March 19, 1915, 10. CPSP, Box 58, Folder 524; F.K. Sanders, "The Most Serious Criticisms Regarding the Relation of the Association to the Church—Founded or Unfounded," September 1913. AYSD, Box 41, Folder 598.
55. Cochran, "Preparation for Leadership," 123–124.
56. Galpin and Edwards, eds., *Church Work in State Universities*, 67.
57. See Wiley, *History of Y.M.C.A.—Church Relations in the United States*, 50–51. See also Elmore McKee, "The Gap Between the Student Associations and the Church," *Int* 40, no. 5 (February 1923): 4–5.
58. "The Student Christian Movement in a New Christian World," 4. Student Work, Kautz Archives. See also Rev. Tissington Tatlow, *The Relation of the Student Christian Movement to the Church* (London: Students Christian Movement, 1913), 7–8.
59. "Report of the National Assembly of Student Secretaries, 1930," 25–26. Student Work, Kautz Archives.
60. Towner, *Religion in Higher Education*, 211. See also "Advance Program Commission, Report of Sub-Committee II," 21. AYSD, Box 39, Folder 572.
61. Shedd, *The Church Follows Its Students*, 76.
62. Porter, *The Church in the Universities*, 22–23, 27; Shedd, *The Church Follows Its Students*, 169–173.

63. Porter, *The Church in the Universities*, 42 ; William G. McDowell, "Relating Students to the Church," *Int* 38, no. 6 (March 1921): 5.
64. Shedd, *The Church Follows Its Students*, 66; Hugo Thompson, "The Student Pastorate and the Christian Associations," February 1930. AYSD, Box 41, Folder 602.
65. Porter, *The Church in the Universities*, 52–66. Administrators typically allowed such courses to count for credit (up to ten hours), providing that the instructor hold a Ph.D. from a respected university and that the courses be of sufficient rigor to conform to college standards.
66. John R. Mott, "A Policy of Co-operation in Meeting the Religious Needs of State Universities," in Galpin and Edwards, eds. *Church Work in State Universities*, 58–68.
67. Summary of the Cleveland Conference, 11–13; Summary of the Second Cleveland Conference, Hotel Hollenden, Cleveland, Ohio, November 23, 1916. AYSD, Box 41, Folder 602.
68. Ibid., 5; Summary of the Second Cleveland Conference, 5.
69. Ibid., 6–7.
70. David R. Porter, *Student Associations and the Church* (New York: International Committee of Young Men's Christian Associations, 1915); Mott, "Report of Progress in Co-operation between the Representatives of the Churches and the Christian Associations Since the Cleveland Conference," 1–6. AYSD, Box 41, Folder 602.
71. Gerald C. Smith, "High Lights in a Year of Work," *Int* 39, no. 7 (April 1922): 11; Fred M. Hansen, "Relating Students to the Church," *Int* 38, no. 6 (March 1921): 5; David R. Porter, "An Adventure in Church Unity," *Int* 36, no. 6 (March 1919): 10.
72. Thomas Evans, *The Church at Work in the Universities* (New York: Association Press, 1915), 7, 26–27, 32–33.
73. Richard Henry Edwards, *Cooperative Religion at Cornell University: The Story of United Religious Work at Cornell University, 1919–1939* (Ithaca, NY: Cornell Cooperative Society, 1939), 13.
74. Ibid.
75. "A Discussion in Regard to the Relations of the Student YMCA to the Various Branches of the Church, arranged for by Dr. Mott at the University Club, Chicago, February 28, 1925," 6–7. AYSD, Box 45, Folder 641.
76. Edwards, *Cooperative Religion at Cornell University*, 14–15.
77. Porter, *The Church in the Universities*, 12–13, 36–38, 40–44.
78. By the mid-1930s, such groups were formed at Ohio State, Duke University, the University of Miami, Purdue University, the University of Nebraska, the University of Washington, Louisiana State University, the University of Tennessee, Vanderbilt University, the University of South Carolina, and the University of Oregon. "The Development of an Inclusive Student Movement." AYSD, Box 60, Folder 842.
79. George W. David, "Significant Recent Changes in the Student Field," *Christian Education* 21 (December 1937): 110. By 1938, all summer conferences sponsored by the Christian associations were inclusive of both men and women, and a number also included representatives of the local denominational student groups.

80. Thomas Wesley Graham, "The Present Situation and Outlook among Students," 9–10. Kautz Archives.
81. On this theme, see Honnold, "The History of the Y.M.C.A. of the University of Wisconsin," 124.
82. The Council of Christian Associations, *Education Adequate for Modern Times*, 195.

Chapter Eight In Search of a Youthful Religion

1. "Report of the Provisional Student Division Committee to the General Board at Meridale Farms, June 24, 1928," 6. Student Work, Kautz Archives.
2. Fass, *The Damned and the Beautiful*, 137.
3. "Report of the General Secretary to the Board of Directors of the Philadelphian Society, April 8, 1924," 2–3. SCARP, Box 4, Folder "Scrapbook."
4. Ibid., 3.
5. Thomas St. Clair Evans, "The Religious Crisis at Princeton University," *Presbyterian* (July 25, 1918): 6–7.
6. L.C. Douglas, "The Campaign at Pennsylvania State College," *NAS* 2, no. 7 (April 1914): 350.
7. Shoemaker befriended Buchman while both were in China as missionaries. See Kemeny, *Princeton in the Nation's Service*, 202–203.
8. Buchman's method is carefully explained in H.A. Walter, *Soul-Surgery: Some Thoughts on Incisive Personal Work* (New York: Association Press, 1919).
9. "Profiles: Soul Surgeon," *The New Yorker* (April 23, 1932): 22–23.
10. "Students Roused by Buchmanism," *The New York Times* (November 7, 1926): 6. In SCARP, Box 14, Folder 4. See also "Calls Upon Oxford to End Buchmanism," *The New York Times* (May 17, 1928), 7. In SCARP, Box 14, Folder 4; "Quits School Post in Buchmanism Row," *The New York Times* (February 20, 1927), 16. In SCARP, Box 14, Folder 4.
11. "Profiles: Soul Surgeon," 25. See also Richard Henry Edwards, *Undergraduates: A Study of Morale in Twenty-Three Colleges and Universities* (Garden City, NY: Doubleday, Doran and Company, 1928), 283.
12. Kemeny, *Princeton in the Nation's Service*, 211. See also "Report to President Hibben of Special Committee Appointed to Study Activities and Scope of the Philadelphian Society, December 31, 1926," 8. SCARP, Box 14, Folder 6.
13. "Buchman 'House Party,' " *Time* (July 18, 1927): 22.
14. "Students Roused by Buchmanism," 6.
15. Princeton University, Committee Appointed by President Hibben to Investigate the Activities of the Philadelphian Society, Second Meeting of Committee at Prospect on Tuesday, November 23, 1926, 9.
16. Ibid., "Charles Howard," 5. "Report to President Hibben of Special Committee Appointed to Study Activities and Scope of the Philadelphian Society, December 31, 1926," 8, 10–11. SCARP, Box 14, Folder 6. Just one year later, the Oxford University magazine *Isis* demanded that student leaders of the "semi-religious cult known as Buchmanism" be suspended from the university. "Calls Upon Oxford to End Buchmanism," 1.

17. Report of the President, to the Board of Directors of the Philadelphian Society for the year 1926–1927, 6. SCARP, Box 3, Folder 9.
18. Leitch, *A Princeton Companion.*
19. Roy B. Chamberlain, "Can Religion Recapture the Campus?" *The Christian Century* 47, no. 44 (October 29, 1930): 1312.
20. Robert Cooley Angell, *A Study in Undergraduate Adjustment* (Chicago: The University of Chicago Press, 1930), 85; James Bissett Pratt, "Religion and the Younger Generation," *The Yale Review* 12 (1923): 595; Edwards, *Undergraduates*, 281.
21. "Students and Religion," *Michigan Daily* (Tuesday, January 19, 1932). Office of Ethics and Religion SCA Records, Box 6, Folder "Clippings from Michigan Daily," Bentley Historical Library, University of Michigan; Hugh Hartshorne, *From School to College* (New Haven: Yale University Press, 1929), 254. See also Dean Hoge, *Commitment on Campus: Changes in Religion and Values over Five Decades* (Philadelphia: The Westminster Press, 1974), 47–51, 59.
22. Angell, *The Campus*, 184; Hoge, *Commitment on Campus*, 65; Edwards, *Undergraduates*, 259, 281; Pratt, "Religion and the Younger Generation," 594.
23. However, in a survey of twenty-three institutions that included large universities and smaller colleges, the Institute of Social and Religious Research found that 44% of men and 55% of women claimed to attend regularly. See "Changes in the Religious Attitudes of College Students," *Religious Education* 24 (February 1929): 160.
24. Edwards, *Undergraduates*, 244.
25. Pratt, *Religion and the Younger Generation*, 602.
26. Robert T. Handy, "The American Religious Depression, 1925–1935," *Church History* 29 (1960): 3–16.
27. Student Division, *After Fifty Years: The Story of Significant Trends in a Transition Year in the Student Christian Movement of the United States, 1926–1927* (New York: The Student Division, 1927), 4.
28. Ibid.
29. Bruce Curry, *Students and the Religion of To-Day* (New York: Council of Christian Associations, 1926), 16.
30. Milton Robert Allen, "A History of the Young Men's Christian Association at the University of Virginia" (Ph.D. diss., University of Virginia, 1946), 244–245.
31. Fass, *The Damned and the Beautiful*, 140.
32. Veysey, *The Emergence of the American University*, 263–276; Curry, *Students and the Religion of To-Day*, 5.
33. Angell, *The Campus*, 200–201.
34. Ibid., 189, 207.
35. "Report of Sub-Committee #4 to the Meeting of the Advance Program Commission, September 11, 1924," 2. AYSD, Box 40, Folder 577.
36. Alfred J. Henderson, "An Evaluation," 1930. University of Rochester Archives; Allen, "A History of the Young Men's Christian Association at the University of Virginia," 248.
37. Fass, *The Damned and the Beautiful*, 147.
38. Allen, "A History of the Young Men's Christian Association at the University of Virginia," 258. A faculty member at the University of Denver reported that fraternities considered it a disgrace to have "Y" men in their houses. "A Study of the

Relation of the Student YMCA to the College and University Situation: University of Denver." AYSD, Box 55, Folder 791.

39. Student Division, *After Fifty Years*, 5–6.
40. Ibid.
41. Honnold, "The History of the Y.M.C.A. of the University of Wisconsin, 139.
42. "Findings of the Student Section," *Int* 39, no. 3 (December 1919): 9.
43. "Reports of Student Movements, 1918–1919," 18. Student Work, Kautz Archives.
44. See "Minutes of the Midwest Field Council," Lake Geneva, Wisconsin (1925). Student Work, Kautz Archives.
45. Thomas S. McWilliams, "The Attractions of the Ministry to the College Man of To-Day," *Religious Education* 14, no. 4 (August 1919): 261.
46. Ibid.
47. Francis P. Miller, ed. *Religion on the Campus: The Report of the National Student Conference*, Milwaukee, December 28, 1926–January 1, 1927, 3.
48. Szasz, "The Stress on Character and Service in Progressive America," 145–156.
49. William Hutchison, *The Modernist Impulse in American Protestantism* (New York: Oxford University Press, 1992), 2. See also Sydney Ahlstrom, *A Religious History of the American People* (New Haven: Yale University Press, 1972), 787–788.
50. The Committee on Voluntary Study of the Council of North American Student Movements, "The Christian Social Order: The World-Wide Kingdom of God: 1913–1914." AWSCF, Box 67, Folder 544.
51. Harry F. Ward, "Social Unrest in the United States," *Int* 38, no. 9 (June 1919): 4; R.H. Edwards, "A Prophet to College Men," *Int* 38, no. 9 (June 1919): 4. Born in Rochester, New York, Rauschenbusch served as a pastor to a German Baptist congregation in the Hell's Kitchen region of New York City and also taught at Rochester Theological Seminary.
52. Walter Rauschenbusch, "The New Evangelism," in Robert T. Handy, ed. *The Social Gospel in America* (New York: Oxford University Press, 1966), 323–330. See also Edwards, "A Prophet to College Men," 4.
53. Rauschenbusch, "The New Evangelism," 102.
54. Ibid., 73.
55. Ibid., 67–70.
56. Henry P. Van Dusen, *In Quest of Life's Meaning: Hints toward a Christian Philosophy of Life for Students* (New York: Association Press, 1926), 81–82.
57. Ibid., 11.
58. Young Men's Christian Association, *The Student Christian Movement and a New Christian World* (New York: Association Press, 1917), 4, 10–11.
59. Ibid., 11.
60. Goodwin Watson, "Character and Religious Education in Y.M.C.A. Schools," *The Educational Council Bulletin* 1, no. 1 (1929): 14–15.
61. "The Need for a New Puritanism," *Int* 41, no. 5 (February 1924): 1.
62. See Hopkins, *A History of the YMCA in North America*, 516; *The Student Association in War and Reconstruction Years* (New York: Association Press, 1919).

63. Sherwood Eddy, *A Pilgrimage of Ideas* (New York: Farrar & Rinehart Publishers, 1934), 42. See also The Council of Christian Associations, *Education Adequate for Modern Times* (New York: Association Press, 1931), 139; Richard Henry Edwards, *Undergraduates: A Study of Morale in Twenty-Three Colleges and Universities* (Garden City, NY: Doubleday, Doran and Company, 1928), 283; H.H. Bushnell to Dr. John Anderson, (April 2, 1920). University of Florida Archives.
64. Allan A. Hunter, "A Christian Social Order: What Would It Be Like?" *Int* 52, no. 2 (November 1934): 36–38.
65. See *APJRM*, 95. See also Hopkins, *John R. Mott*, 116; Allan T. Burns, "Social Service by College Men," *Int* 29, no. 6 (March 1907): 126–127.
66. Theodore A. Greene, "The Place of the Christian Association in the College Community," *Int* 37, no. 6 (March 1920): 6. In this way, the YMCA was an excellent example of a shift from what educational historian Herbert Kliebard has called "social efficiency" educational philosophy to a "social reconstructionist" philosophy. See Herbert Kliebard, *The Struggle for the American Curriculum* (New York: Routledge & Kegan Paul, 1986).
67. *The Student Christian Movement and a New Christian World*, 15.
68. Student Department of the International Committee of Young Men's Christian Associations, *Forces Affecting Student Faith* (New York: Association Press, 1924), 12. See also Kenneth S. Latourette, "Morals and the Classroom," *Int* 39, no. 2 (November 1919): 5.
69. Miller, ed. *Religion on the Campus*, 150–155.
70. Ibid.
71. Ibid., 154–155.
72. Ibid., 3.
73. A. Bruce Curry, "How Radical Shall We Be?" *Int* 42, no. 9 (June 1925): 273.
74. Curry, *Students and the Religion of To-Day*, 15.
75. See Barton, "Out of the 'Y' and in again," 258.
76. A. Bruce Curry, "Campus Angles on Bible Study," *Int* 41, no. 6 (March 1924): 12.
77. Dan Gilbert, *Crucifying Christ in our Colleges* (San Diego: The Danielle Publishers, 1933).
78. *The Message of the Student Christian Association Movement* (New York: Association Press, 1928), 4, 6, 11–12.

Chapter Nine Building a Christian Youth Movement

1. Cited in Shedd, *Two Centuries*, 374.
2. Howard Kirschenbaum, *On Becoming Carl Rogers* (New York: Delacorte Press, 1979), 19.
3. Ibid., 23.
4. Carl R. Rogers, "An Experiment in Christian Internationalism," *Int* 39, no. 9 (June 1922): 1.
5. Kirschenbaum, *On Becoming Carl Rogers*, 28. Rogers' parents were again deeply opposed to his decision to attend Union Theological Seminary. Rogers noted that they believed that Union was "the devil in disguise." Rogers' father attempted to bribe him by noting that he would pay Carl's way if he chose to attend Princeton instead. Kirschenbaum, *On Becoming Carl Rogers*, 44.
6. Ibid., 39.
7. Ibid., 25.

8. Fass, *The Damned and the Beautiful*, 15.
9. Many leaders felt that youth were only acting in ways consonant with the larger culture. See Curry, *Students and the Religion of To-Day*, 9; Coe, *What Ails Our Youth?* (New York: Charles Scribner's Sons, 1925), 46.
10. Annual Report of the General Secretary of the Philadelphian Society to the Board of Directors, September 1, 1925. SCARP, Box 3, Folder 9. On this shift, see also Marty, *Righteous Empire*, 184.
11. Walter Rauschenbusch, *Christianity and the Social Crisis* (New York: Macmillan, 1907), 151.
12. "My Forty Years Have Taught Me to Trust Youth: Special Interview with Dr. John R. Mott," *The Methodist Times* 47, no. 2,360 (April 3, 1930): 1.
13. "The College Situation and Student Responsibility: A Series of Discussions Concerning the Advance Program of the Student Young Men's Christian Association," 17–18. AYSD, Box 74, Folder 599.
14. Curry, *Students and the Religion of To-Day*, 20.
15. Angell, *The Campus*, 33.
16. See "Supplementary Report of Sub-committee IX, Advance Program Commission," 1. AYSD, Box 40, Folder 583; "Report of the Advance Program Commission to the National Student Council, September 13, 1924," 2. AYSD, Box 39, Folder 564.
17. Student Department of the International Committee of Young Men's Christian Associations, *Forces Affecting Student Faith*, 7–8.
18. A.J. Elliott, "Social Forces in Collegiate Life that Must Be Made Constructive," *Int* 37, no. 8 (May 1919): 2. See also Lyman Hoover, "Fraternities and a Fraternal World," *Int* 42, no. 5 (January 1925): 115–116.
19. Edwin E. Aubrey, "Does the Fraternity Crush Individual Thinking?" *Int* 41, no. 7 (April 1924): 10–11.
20. Curry, *Students and the Religion of To-Day*, 15. See also Elliott, "Social Forces in Collegiate Life that Must Be Made Constructive," 1–2.
21. "Tentative Report of the Subcommittee on Education," 1924, 13–15. AYSD, Box 39, Folder 563. See also Tucker Smith, "The Effect of Athletics Upon Scholarship," *Int* 42, no. 2 (November 1924): 5–7.
22. Gordon Chalmers, "Shall We Have Representative Students on the Cabinet?" *Int* 42, no. 6 (March 1925): 194. See also "The Perils of Popularity," *Int* 36, no. 4 (January 1919): 2.
23. Harry Bone, "The New Association President," *Int* 42, no. 7 (April 1925): 215.
24. J.T. Hardwick, "Shall We Have Representative Students on the Cabinet?" *Int* 42, no. 6 (March 1925): 193.
25. The Student Division National, Council of the YMCA, *After Fifty Years*, 11.
26. Paul Super, *What Is the YMCA?* (New York: Association Press, 1922), especially 1–25.
27. "Reports of Student Movements, 1916–1917," 18–19. Student Work, Kautz Archives. See also Edwards, *Undergraduates*, 277.
28. "That Boy of Ours" (1927). University of Illinois Archives, Record Series Number 41/69/322, Box 5.
29. "False Gods," *Y's Indian* 2, no. 3 (December 1921): 2.

30. Van Dusen, *In Quest of Life's Meaning*, 72–73, 81–82.
31. Galen M. Fisher, *Religion in the Colleges* (New York: Association Press, 1928), 13–14.
32. The Student Division, "National Council of the YMCA" *After Fifty Years*, 6.
33. The Council of Christian Associations, *Education Adequate for Modern Times*, 10.
34. Zald, *Organizational Change*, 171.
35. Ibid., 60.
36. "An Outsider's View," *Int* 31, no. 6 (March 1910): 78.
37. Philpott, "A History of the Student YMCA.
38. These field councils were indeed created. In 1922, there was a national conference of field councils to determine policies on membership, organization, missionary and Bible study procedures, and life work guidance. See William P. Tolley, "National Conference of Field Councils," *Int* 39, no. 6 (March 1922): 3.
39. Zald, *Organizational Change*, 60–85.
40. Local associations were to support the National Council's work financially on a proportional basis. See Pence, *The YMCA and Social Need*, 147.
41. Ibid., 146–148.
42. Ibid. In addition to the National Student Council, there was also to be a National Student Committee, representing the student associations in relation to church bodies and other national or regional organizations.
43. Ibid.
44. "Proposals for the Supervision of Student Associations, Submitted by Action of the Third Annual Meeting of the National Council of the Young Men's Christian Associations of the United States of America, Chicago, Illinois, October 26–29, 1926." AYSD, Box 60, Folder 842.
45. "Proposals for the Supervision of Student Associations," 12.
46. Porter to Mott, June 16, 1927, 1–2. JRMP, Box 141, Folder 2234.
47. Philpott, "A History of the Student YMCA," 477.
48. "Recent Developments in the Relationships of the Student Young Men's Christian Association Movement to the Association Brotherhood, June 1927." AYSD, Box 60, Folder 842.
49. Philpott, "A History of the Student YMCA," 479–482.
50. Paul Super, *Formative Ideas in the YMCA* (New York: Association Press, 1929), especially 7–11, 98–102.
51. Ibid., 102. See also Clifford Putney, "Character Building in the YMCA, 1880–1930," *Mid-America: An Historical Review* 73, no. 1 (January 1991): 64–69.
52. A.J. Gregg, "Two Great Resources for Religious Education," *Association Boys' Work Journal* 1, no. 1 (November 1, 1927): 10–11. See also H. Parker Lansdale, "A Historical Study of YMCA Boys' Work in the United States, 1900–1925" (Ph.D. diss., Yale University, 1956), 181; Super, *Formative Ideas*, 34.
53. John Dewey, *How We Think* (Boston: D.C. Heath & Co., 1910).
54. Ibid.
55. Committee of Student Secretaries, Central Region, *A Study of the Present Position*, 19.
56. See "Report of Sub-Committee #4 to the Meeting of the Advance Program Commission, September 11, 1924," 3–4. AYSD, Box 40, Folder 577.

57. Annual Report, Young Men's Christian Association of the University of Virginia, 1932–1933, 1. University of Virginia Archives.
58. As religious historian William Hutchison pronounced, the combination of progressive education and liberal religion indeed represented a "match made in heaven." Hutchison, *The Modernist Impulse in American Protestantism* (New York: Oxford University Press, 1992), 219. See also Sydney Ahlstrom, *A Religious History of the American People* (New Haven: Yale University Press, 1972), 781.
59. Elliott, *Training an Adequate Leadership*, 2–3; Richard H. Edwards, "Some Glimpses and Principles of Service," *NAS* 2, no. 6 (March 1914): 270–272.
60. Elliott, *How Jesus Met Life Questions* (New York: Association Press, 1920), 7.
61. Elliott, *Building a New World* (New York: Association Press, 1918).
62. See also A. Bruce Curry, "Better Bible Discussion Groups," *Int* 41, no. 5 (February 1924): 26–27.
63. In 1926, an estimated 1,100 fraternity students were studying Curry's curriculum. "Report of the Retiring President to the Members of the Young Men's Christian Association at the University of Illinois" (July 1, 1926), 1. University of Illinois Archives, Record Series Number 41/69/322, Box 11.
64. Curry, *Facing Student Problems*, xiv-xv. See also "Reports on Curry Meetings," 3. AYSD, Box 40, Folder 577.
65. The war, in fact, stimulated discussions along these lines in YMCA groups around the country. In Florida, the campus YMCA proposed a more vigorous campaign of sex education because of the high incidence of venereal disease among servicemen, especially those from the state of Florida. Winfred Scott Hall was asked to come to campus to address students on "The Sex Life of Man," and "The Social Evil and Its Cure," among others. W.A. Lloyd to A.A. Murphree, 6 September 1918. University of Florida Archives. A book length study of such issues, Arthur Herbert Gray's *Men, Women, and God* (New York: George H. Doran Co., 1922) was a wildly popular text in the 1920s. Exner did not concur with those who spoke of the colleges as morally depraved institutions. See M.J. Exner, *Friend or Enemy?* (New York: Association Press, 1917); M.J. Exner, *Problems and Principles of Sex Education: A Study of 948 College Men* (New York: International Committee of Young Men's Christian Associations, 1915).
66. "Young Men Talk Religion," 1927. University of Illinois Archives, Record Series Number 41/69/322, Box 11.
67. "Firesides Popular," *Y's Cyclone* (September 22, 1932): 3.
68. "Report of the Retiring President to the Members of the Young Men's Christian Association at the University of Illinois" (June 16, 1927), 5–6. University of Illinois Archives, Record Series Number 41/69/322, Box 11.
69. "The College Situation and Student Responsibility," 50–51; F.E. Morgan, "Helping Freshmen to Think," *Int* 41, no. 4 (January 1924): 16.
70. Elliott, *Training an Adequate Leadership*, 2.
71. Committee of Student Secretaries, Central Region, *A Study of the Present Position*, 33.
72. Ibid., 23–27, 34.
73. David R. Porter, *The Necessity of the Student Christian Movement* (New York: The Student Division, National Council of Y.M.C.A., 1928), 8.

74. Porter, "Can You Kill a Student Association?" *Int* 38, no. 6 (March 1921): 7.
75. Committee of Student Secretaries, Central Region, *A Study of the Present Position*, 11.
76. At the Conference on Religion in Universities, Colleges, and Preparatory Schools, held at Princeton in 1928, a ringing endorsement of a continued YMCA presence in higher education was provided along these lines. See *Religion in the Colleges*, especially 20–21.
77. Porter, *The Necessity of the Student Christian Movement*, 7. AYSD, Box 54, Folder 773
78. Such conceptions were formalized by the YMCA National Council in 1931. "Report of the General Secretary to the Board of Directors of the Young Men's Christian Association of the University of Illinois, Friday, October 16, 1931," 11. University of Illinois Archives, Record Series Number 41/69/322, Box 5.
79. Curry, "How Radical Shall We Be?" 274.

Chapter Ten The Curriculum of the Kingdom

1. "Charter of the Advanced Program Commission," 1. AYSD, Box 39, Folder 563.
2. "Report of Sub-Committee #4 to the Meeting of the APC, September 11, 1924," 3. AYSD, Box 40, Folder 577.
3. "A Message from the National Council of Student Associations to the Officers, Members and Friends of Student Associations, October 14, 1924," 1–2. Student Work, Kautz Archives.
4. Morgan, *Student Religion during Fifty Years*, 151–152, 172.
5. *The Student Association in War and Reconstruction Years* (New York: Association Press, 1919), 17–19. On this theme, see especially Putney, "Character Building in the YMCA, 1880–1930," 63; "What Does the War Mean to Us?" *NAS* 3, no. 5 (February 1915): 1–2; Fisher, *Public Affairs and the YMCA*, 82–83.
6. J. Lovell Murray, *The Call of A World Task in War Time* (New York: Student Volunteer Movement, 1918). See also "Teacher's Oaths," *Int* 54, no. 1 (October 1936): 21.
7. "The Student Christian Movement in a New Christian World," 32–33. See also David R. Porter, "American Students' Friendship Fund for Prisoners, 1917." AYSD, Box 54, Folder 772; John R. Mott, *Students of America for the Students in European War Prisons* (New York: Student Department, Young Men's Christian Association, 1916).
8. See Helen Ogden, "The Friendship Fund—again!" *Int* 41, no. 2 (November 1923): 18–19. See also William Howard Taft, ed. *Service with Fighting Men: An Account of the Work of the American Young Men's Christian Associations in the World War*, vol. 1 (New York: Association Press, 1922), 235.
9. Fisher, *Public Affairs*, 86; H.L. Seamans, ed. *The Work of the Student Young Men's Christian Association* (New York: The General Board of the Young Men's Christian Association, 1927), 30–31.
10. S. Ralph Hartlow, "The Indianapolis Convention," *Int* 39, no. 4 (January 1922): 5. See also Henry P. Van Dusen, "Student Opinion and the World Court: The Report of the National Director of the World Court Committee of the Council of Christian Associations, January 2, 1926." AWSCF, Box 64, Folder 523.

11. Van Dusen, "Student Opinion and the World Court," 5–6.
12. Rogers, "An Experiment in Christian Internationalism," 1.
13. "Internationalism and the Christian Student Movement," *Int* 40, no. 6 (March 1923): 1.
14. Council of Christian Associations, "Christian World Education Scrapbook, 1929." AYSD, Box 40, Folder 586.
15. David R. Porter, "Maintaining Christian Ideals in War Time" AYSD, Box 54, Folder 772; Lawrence J. Dennis, *From Prayer to Pragmatism: A Biography of John L. Childs* (Carbondale, IL: Southern Illinois Press, 1992), 37.
16. See Edward Hachtel, John Nevin, and Luther Tucker, "Report on Reserve Officers Training Corps and Citizens Military Training Camps, 1925." AYSD, Box 55, Folder 794. Frank Olmstead, "The Christian Student and the R.O.T.C.," *Int* 42, no. 1 (October 1924): 13–14; Kenneth S. Latourette, "Are Our Campuses a Menace to Peace?" *Int* 42, no. 1 (October 1924): 4–5.
17. Miller, *Religion on the Campus*, 93. See also Morgan, *Student Religion during Fifty Years*, 44.
18. Many chapters utilized a discussion series entitled "To End War." See "The Disarmament Conference," *Int* 39, no. 2 (November 1921): 1; Frank Olmstead, "Students and Disarmament," *Int* 39, no. 3 (December 1921): 8.
19. James Green, "Address to the World's Disarmament Conference, February 6, 1932," 2–3. AYSD, Box 52, Folder 752.
20. Shedd, *Two Centuries*, 402. See also "Students and Disarmament." AYSD, Box 52, Folder 750.
21. By 1925, the number of foreign students in American colleges and universities stood at 14,000, up from 5,000 in 1910. *Report of the Lake Forest Summer School 1911*, 162. AYSD, Box 68, Folder 925.
22. Committee on Friendly Relations among Foreign Students, *Envoys Extraordinary* (New York: Committee on Friendly Relations Among Foreign Students, 1930), 5.
23. Council of Christian Associations, "Christian World Education Scrapbook," 1929. AYSD, Box 40, Folder 586.
24. Eighth Annual Report of the Work of the Committee on Friendly Relations among Foreign Students of the University of Illinois, 1925–1926, 2. University of Illinois Archives, Record Series Number 41/69/322, Box 12.
25. "Our Guest Students," *Intercollegian Program Service* 2, no. 2 (1930–1931): 1–3. "Foreign Students in the United States," *Int* 40, no. 6 (March 1923): 13.
26. Committee on Friendly Relations among Foreign Students, *Bridges of Understanding* (New York: Committee on Friendly Relations among Foreign Students, 1930), 7.
27. Morgan, *Student Religion during Fifty Years*, 184.
28. Intercollegiate Disarmament Council Model World Economic Conference. AYSD, Box 52, Folder 757.
29. Council of Christian Associations, *Toward a New Economic Society* (New York: Council of Christian Associations, 1931), 8.
30. Ibid., 10–11, 45–46.
31. Ibid., 69–70.
32. Ibid., 11–12.

33. "The Industrial Groups," *Int* 41, no. 1 (October 1923): 25. Cherrington was later acting chancellor of the University of Denver. See Fisher, *Public Affairs and the YMCA*, 84. For a similar program in New York, see Richard H. Edwards, "Summer Service By College Men," *Int* 37, no. 1 (October 1919): 9–10.
34. R.M. Cherrington, "The Denver Group," *Int* 38, no. 2 (November 1920): 8–9; F. Ernest Johnson, "Students in Overalls," *Int* 39, no. 1 (October 1921): 5; James Myers, *Religion Lends a Hand: Studies of Churches in Social Action* (New York: Harper and Brothers, 1929), 83–85.
35. Shedd, *Two Centuries*, 189–192.
36. The program of African American chapters remained somewhat more traditional than that in white colleges, Bible study remaining a stronger centerpiece of religious work. Black chapters developed a keen interest in missionary work, supporting Max Yergan in South Africa. As the first black missionary supported by the black colleges of the United States, Yergan's work became an impetus for greater international awareness among these chapters. "Report: Student Division of the National Council of YMCA's, 1928–1929," 24. AYSD, Box 42, Folder 619.
37. "Progress in Southern Student Christian Life, 1915–1916," 1. AYSD, Box 64, Folder 896; Homer Grafton, "The Race and Neighbor Question Again," *Int* 42, no. 6 (March 1925): 197.
38. "The Race Question and Indianapolis," *Int* 41, no. 2 (November 1923): 2.
39. A.M. Trawick, "The Trend of the Races," *Int* 40, no. 8 (May 1923): 9.
40. J.H. Oldham, *Christianity and the Race Problem* (New York: Association Press, 1924), 215.
41. Ibid., 223.
42. "Negro Association Work," *Int* 42, no. 7 (April 1925): 211.
43. "Negroes in Northern Colleges," *Int* 41, no. 6 (March 1924): 3.
44. See "Democracy Triumphant," *Int* 36, no. 4 (January 1919): 1. Several state student conferences, including those in Arkansas and Virginia, were held jointly beginning in the mid-1920s.
45. "Southern Regional Field Council Meeting, January 9–12, 1930," 5. Student work, Kautz Archives.
46. For their part, many white students within the movement were also critical of what they saw as an unjustifiable compromise of principle. See Will Alexander, "An Effort to Create Racial Good Will," *Int* 39, no. 9 (June 1922): 4.
47. Weatherford, "Colored YMCA, the Interracial Committee and Related Subjects," 15. Colored Department Records, Kautz Archives.
48. Ibid., 22.
49. In 1928, there was half a million dollar cut in the budget of the National Council. John Dillingham, responsible for black student work in the south, was cut. R.P. Hamlin, in charge of black city associations, was also cut. Another was expected to be cut when Tobias was arguing for the merger. See Weatherford, "Colored YMCA," 19–22.
50. Ibid., 13–14.
51. Ibid., 37.
52. Fisher, *Public Affairs and the YMCA*, 88; Weatherford, "Colored YMCA," 47–49.
53. Fisher, *Public Affairs and the YMCA*, 88.
54. Regional organizations were not integrated in the Southeast until 1937, the same year that the King's Mountain conferences were ended in order to forge an

interracial southern conference at Talladega College. See Richard I. McKinney, *Religion in Higher Education among Negroes* (New York: Arno Press, 1972), 88. In 1944, all conferences except Blue Ridge were interracial. See Sherwood Eddy, *A Century with Youth: A History of the YMCA from 1844 to 1944* (New York: Association Press, 1944), 88.

55. At Vanderbilt, former YMCA student President Morton King recalled that attempts to join the YMCA and YWCA on that campus were halted by racial issues. Because the YWCA had achieved significant integration at the national and regional levels, King suggested, "The white YWs did not want to jeopardize that by getting too close to the white YM." In addition, issues of cooperation were far more complicated when both race and gender were involved. While same-sex religious meetings could often be held on an interracial basis, he claimed, "Mixing race AND gender was a 'no-no.' " King, "The Work at Vanderbilt," Kautz Archives.
56. Fass, *The Damned and the Beautiful*, 45.
57. Eddy, *A Century with Youth*, 89–90; Owen Pence, "Why the Student Young Men's Christian Association Is Interested in Life Work Guidance," *Int* 42, no. 6 (March 1925): 175.
58. "The Organization of Student Christian Work: A Report of the Commission on Consultations about Student Christian Work, March 1934–March 1935," 67.
59. See "What's the Idea, Mr. Eddy?" *Detroit Free Press* (January 18, 1927). Office of Ethics and Religion SCA Records, Box 6, Folder "SCA Miscellaneous (3)," Bentley Historical Library, University of Michigan; "Hobbs and Reed Oppose Eddy in Debate on Maintaining Present National Preparedness System," *Ann Arbor Times* (February 22, 1927). Office of Ethics and Religion SCA Records, Box 6, Folder "SCA Miscellaneous (3)," Bentley Historical Library, University of Michigan.
60. Student Division, *After Fifty Years*, 18–19.
61. "The Young Democracy," *Int* 37, no. 7 (April 1920): 12.
62. "The College Situation and Student Responsibility," 75. AYSD, Box 74, Folder 599.
63. "Statements Regarding the Use of Stiles Hall by Student Groups of the University of California, 1934." University of California Archives. More than ten years later in a less hospitable political climate, the campus chapter was accused by Senator Jack Tenney of harboring communist sensitivities. "Senator Jack Tenney's Criticism of the YMCA at U.C., March 27, 1947." University of California Archives.
64. Weatherford, *Student Secretaries in Training*, 5–6; "The Findings of the Commission on the Training of Student Secretaries, 1924." AYSD, Box 56, Folder 799. Out of 173 secretaries in 1925, two possessed a Ph.D., 37 possessed an M.A., 18 had earned a B.D., and 5 held both an M.A. and a B.D. "Student Secretaries Now in the Work: A Summary, April 6, 1925." AYSD, Box 56, Folder 803; "Are Student Secretaries Students?" AYSD, Box 56, Folder 800. By 1930, the average age of the college secretary was 36. Of the106 secretaries surveyed in 1933, 92 were married, and 67 had children. By 1931, a full 25% of college secretaries remained in the profession for fifteen years or more. See "Personnel Study of the Student YMCA Secretaryship." CPSP, Box 88, Folder 836; George H. Menke, "The Secretary of the Student YMCA in the American System of Higher Education." Unpublished Paper, Yale University,

1934. CPSP, Box 88, Folder 836. In 1918, the department of religious education at the University of Southern California's Maclay College of Theology started offering a two-year secretarial course. Manuel P. Servin and Iris H. Wilson, *Southern California and Its University* (Los Angeles: The Ward Ritchie Press, 1969), 80. Training for the secretaryship could also be achieved through the fellowship plan by which graduated college students could engage in an internship with an existing YMCA secretary. "The Fellowship Plan: A Way into the Secretaryship." AYSD, Box 69, Folder 934.

65. "Personnel Study of Student YMCA Secretaries in Universities of the USA." CPSP, Box 88, Folder 837.
66. Ibid. When asked about vital social interests in that year, forty-four mentioned abolition of war, forty-four the improvement of labor, but only fourteen discussed the improvement of race relations as a fundamental task.
67. University of Nebraska YMCA to Clarence Shedd, March 21, 1928, 1–2. CPSP, Box 88, Folder 839.
68. Cyrus T. Barnum to Clarence Shedd, March 30, 1928. CPSP, Box 88, Folder 839.
69. Jack E. Boyd to Clarence P. Shedd, March 26, 1928. CPSP, Box 88, Folder 839.
70. Zald, *Organizational Change*, 85.
71. Ibid., 84.
72. Morgan, *Student Religion during Fifty Years*, 170.
73. See George M. Marsden, *Fundamentalism and American Culture: The Shaping of Twentieth-Century Evangelicalism* (New York: Oxford University Press, 1980).
74. Shedd, *The Church Follows Its Students*, 89.
75. See Keith Hunt and Gladys Hunt, *For Christ and the University: The Story of Intervarsity Christian Fellowship of the U.S.A., 1940–1990* (Downers Grove, IL: InterVarsity Press, 1991), 56–73.
76. See H. Richard Niebuhr, *The Kingdom of God in America* (New York: Harper and Row, 1937), 193.
77. "What of the Future of Student YMCA's?" AYSD, Box 43, Folder 631.
78. National Student Council, "What Kind of Religion We Believe in," January 1935. AYSD, Box 72, Folder 958; Lynn Harold Hough, "Christianity on the Campus," *The Intercollegian* 53, no. 5 (March 1936): 131–132.
79. Harrison Elliott, *Can Religious Education Be Christian?* (New York: Macmillan, 1940): 46–48.
80. On this theme, see John Patrick Diggins, *The Promise of Pragmatism: Modernism and the Crisis of Knowledge and Authority* (Chicago: The University of Chicago Press, 1994), 237.
81. Graham, "The Present Situation and Outlook among Students," 12.

Index

CPSIA information can be obtained
at www.ICGtesting.com
Printed in the USA
LVOW05s1030080817
544240LV00015B/402/P